ADVANCES IN
X-RAY ANALYSIS

VOLUME 3

Edited by

WILLIAM M. MUELLER

Proceedings of the Eighth Annual Conference on Applications of X-Ray Analysis Held August 12-14, 1959

Sponsored by
University of Denver
Denver Research Institute

Distributed by
Plenum Press, New York

ISBN-13: 978-1-4684-7403-9   e-ISBN-13: 978-1-4684-7401-5
DOI: 10.1007/978-1-4684-7401-5

FOREWORD

It is interesting to observe the ever increasing versatility of X-ray analysis as evidenced by the wide range of application to the myriads of problems confronting the technological community, a versatility limited only by the imagination and ingenuity of the scientist, the designer of X-ray equipment, and the novice or student. Tomorrow's engineering alloys will undoubtedly be influenced by today's extremely low- and very high-temperature X-ray research. New and continued insight into the basic architecture of crystalline materials is being achieved by studies of lattice imperfection, recrystallization habit, and phase transformation. Techniques for identification and analysis of minerals by X-ray diffraction and fluorescence are equally amenable to pathological and physiological diagnosis. The experimental setup of this month may well become an instrument for routine process control next month. And such developments occur so rapidly in so many different laboratories that it is difficult to keep abreast of this tidal wave of information.

The dictates of this nation's economy and its struggle for technological supremacy demand a full awareness of the accomplishments of one's associates. Such awareness is most effectively obtained through personal contact, where the beginner can benefit from the experiences of the expert, the basic researcher and the applied researcher can exchange views, and the creative research of each is nurtured by the sharing of mutual or associated problems. The stimulus of a conference dedicated solely to the informal and free interchange of information concerning the most recent progress in the field of X-ray analysis is a potent factor in the rapid development of new applications for X-ray diffraction, fluorescence, and microscopy.

This then is the purpose of the ANNUAL CONFERENCE ON APPLICATIONS OF X-RAY ANALYSIS. We continually strive to present a program which provides a comprehensive survey of the most important X-ray research in progress, to bring together for a few days the leaders in this ever so important field of endeavor, and to publish as quickly as possible a compilation of the discussions for reference by others. In so doing, it is our hope that valuable time of our associates will be saved, for time is the researcher's most priceless commodity.

JAMES P. BLACKLEDGE

iii

PREFACE

.This volume contains twenty-three of the papers presented at the Eighth Annual Conference on Applications of X-Ray Analysis, sponsored by the University of Denver at the Stanley Hotel, Estes Park, Colorado, on August 12, 13, and 14, 1959. Also included are two papers that were presented at the Seventh Conference, held in 1958, but approved for publication too late to be included in the Proceedings of that conference.

The annual conferences on applications of X-ray analysis have as their purpose the presentation of up-to-date information in the fields of X-ray absorption fluorescence, diffraction, and instrumentation in an atmosphere that stimulates discussion and cross-fertilization of new ideas. The Proceedings provide a permanent record of the developments reported.

Financial aid provided by the Office of Naval Research permitted the University of Denver to invite two prominent European scientists, Dr. Albert Franks of the National Physical Laboratory, England, and Dr. Volkmar Gerold of the Max Planck Institut für Metallforschung, Germany, to participate in the conference. The contributions of these colleagues added considerably to the value of the meeting. The assistance of the Office of Naval Research is greatly appreciated.

Much of the credit for the success of the conference is due to the following individuals who chaired the various technical sessions:

    Mr. Arthur A. Chodos, California Institute of Technology, Pasadena, California

    Dr. G. M. Rassweiler, General Motors Research Laboratories, Detroit, Michigan

    Dr. William L. Fink, Aluminum Company of America, New Kensington, Pennsylvania

    Dr. B. D. Cullity, University of Notre Dame, Notre Dame, Indiana

    Dr. Irwin I. Bessen, Jones and Laughlin Steel Corporation, Research Laboratories, Pittsburgh, Pennsylvania

The continued encouragement of Mr. James P. Blackledge, the able assistance of Mrs. Esther Marie Capps in making conference preparations, the devoted efforts of Mrs. Marie Fay and Mrs. Mildred Cain in the preparation of manuscripts, and the skillful art work of Mr. John Silver are gratefully and sincerely acknowledged.

WILLIAM M. MUELLER

v

TABLE OF CONTENTS

ANALYSIS OF ALUMINUM—NICKEL DIFFUSION COUPLES BY X-RAY ABSORPTION

Paul Lublin
Sylvania Research Laboratories, Bayside, New York

ABSTRACT

The X-ray technique for analyzing diffusion couples employs a monochromatic X-ray beam which impinges on a very fine slit, a holder for aligning the diffusion zone with the slit, and a scintillation counter for the high counting rates which have to be measured. Sample preparation, experimental instrumentation, and the application of the technique to the aluminum—nickel system will be discussed.

INTRODUCTION

An investigation of the aluminum—nickel diffusion system was started at Sylvania Research Laboratories by L. S. Castleman and L. L. Seigle.[1] It was their desire to study the formation and growth of the intermetallic layers in the aluminum—nickel system as an aid to the understanding of the diffusion-bonding process between these metals. They measured increases in thicknesses of these intermetallic layers as a function of interdiffusion time, interdiffusion temperature, and applied pressure. They found that an aluminum—nickel diffusion couple annealed at 600°C for 340 hours contained all the phases that are thermodynamically possible at the diffusion temperature. The phases present were $NiAl_3$ (β), Ni_2Al_3 (γ), $NiAl$ (δ), and Ni_3Al (ϵ). It was felt that X-ray absorption techniques would be useful in establishing the compositional limits of the various phases as described in the published phase diagram[2] which is shown in Figure 1. This technique has the advantage that systems with steep concentration gradients can be analyzed where chemical analysis is very difficult. Moreover, it has the advantage that the high-temperature phase need not be retained at room temperature for the determination of phase boundaries.

THEORY

The use of X-ray absorption techniques for the analysis of metallic diffusion couples was investigated originally by R. E. Ogilvie[3] of the Massachusetts Institute of Technology.

---

[1] Superscripts pertain to references at the end of the paper.

1

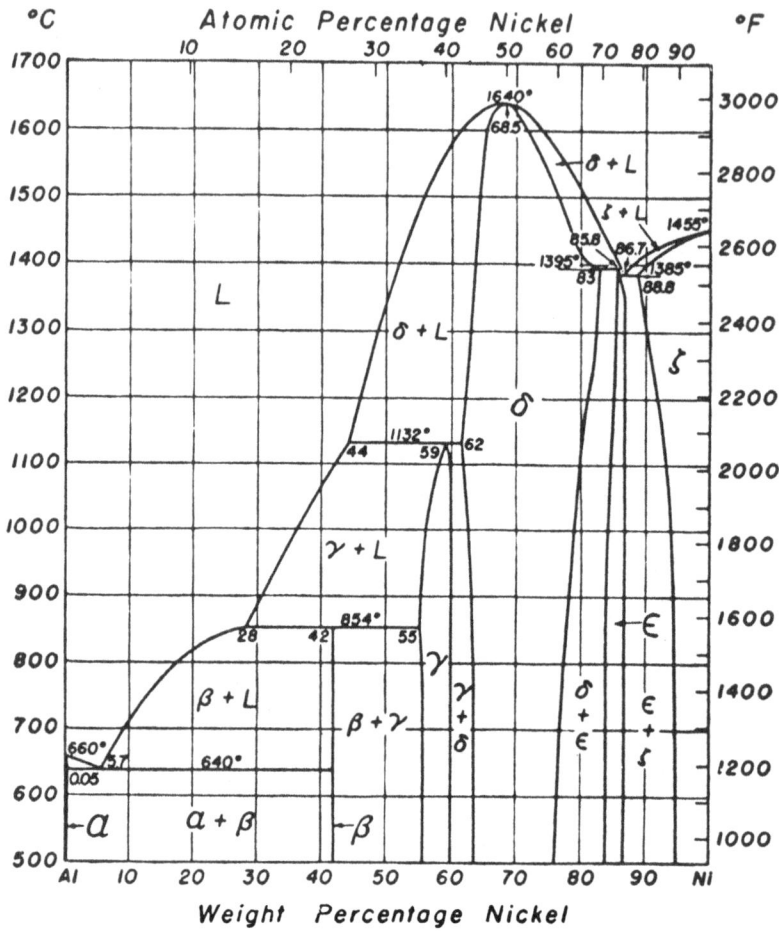

Figure 1. Phase diagram of the aluminum–nickel system.

If a monochromatic X-ray beam is passed through a thin metal specimen, the intensity of the transmitted beam is given by the equation

$$I = I_0 \exp\left(-\frac{\mu}{\rho}\,\rho t\right) , \qquad (1)$$

where I_0 is the initial intensity, μ/ρ is the mass absorption coefficient of the metal for the wavelength used, ρ is the density of the metal, and t is the thickness of the metal in centimeters.

The mass absorption coefficient can be obtained from the literature or measured experimentally. The advantage of using the mass instead of the linear absorption coefficient is that the mass absorption coefficient is independent of the physical condition or state of the absorber.

If a homogeneous binary alloy of elements A and B is placed in the path of a monochromatic X-ray beam, the transmitted intensity is given by

$$I = I_0 \exp\left\{-\left[\left(\frac{\mu}{\rho}\right)_A W_A + \left(\frac{\mu}{\rho}\right)_B W_B\right]\rho_{A+B}\, t\right\}, \qquad (2)$$

where W_A and W_B are the weight fractions of elements A and B respectively, and ρ_{A+B} is the density of the alloy.

If a sample of pure A of the same thickness as the alloy is placed in the X-ray beam, the relation obtained is

$$\frac{\ln(I/I_0)_{A+B}}{\ln(I/I_0)_A} = \left[W_A + \frac{(\mu/\rho)_B}{(\mu/\rho)_A} W_B\right]\frac{\rho_{A+B}}{\rho_A}. \qquad (3)$$

When using this expression, the density as a function of composition must be known. Then the thickness is eliminated provided the thickness of A is the same as that of the diffused zone. The ratio of the mass absorption coefficients can be obtained from the literature or measured experimentally from the ends of the diffusion couple, using the following expression:

$$\frac{\ln(I/I_0)_B}{\ln(I/I_0)_A} = \frac{(\mu/\rho)_B\rho_B}{(\mu/\rho)_A\rho_A}. \qquad (4)$$

However if the density as a function of composition is not known, and it is desirable to eliminate the dependency on uniform thickness, it is possible to use an experimental technique employing two X-radiations. The following expression, using copper K_{α_1} and molybdenum K_{α_1} is an example:

$$\frac{\ln(I/I_0)_{Cu}}{\ln(I/I_0)_{Mo}} = \frac{\left[W_A(\mu/\rho)_A + W_B(\mu/\rho)_B\right]_{Cu}}{\left[W_A(\mu/\rho)_A + W_B(\mu/\rho)_B\right]_{Mo}}. \qquad (5)$$

This equation is then independent of the density of the alloy as a function of composition and uniform specimen thickness is not required.

EXPERIMENTAL

Sample Preparation. The aluminum−nickel couples were made from high-purity material. Half-inch-diameter disks were punched from strip material, were given a grain-coarsening

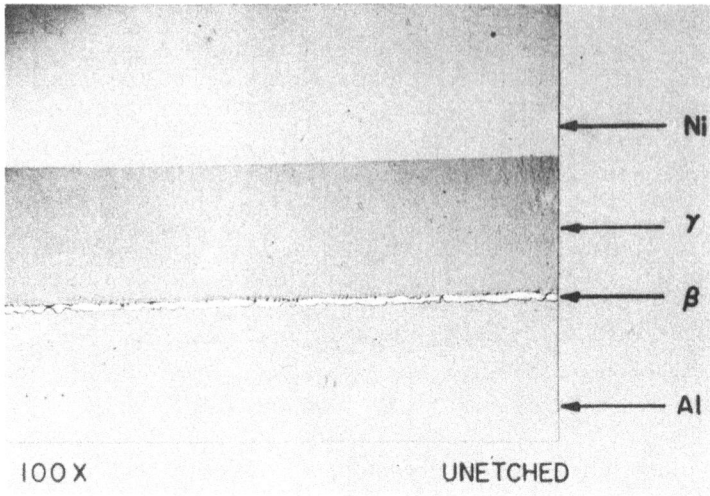

Figure 2. Al—Ni couple diffused at 600°C for 96 hours under a pressure of 5 tons / in.[2].

anneal, and then were chemically cleaned. The furnace used and conditions for diffusion of the couples are described elsewhere.[1] After diffusion, cross sections were prepared and examined metallographically. The X-ray samples were obtained from these cross sections. An example is shown in Figure 2. To prepare samples for the X-ray examination, a cross section was mounted on a glass disk, polished, and then etched to produce a thickness of 5 to 6 mils.

Selection of Radiation. The selection of the radiation for the analysis was important. A strong characteristic line was used such that its wavelength was between the absorption edges of the two elements. In order to use equation (5) a second radiation was chosen such that its wavelength was much shorter than the absorption edge of the heavier element.

Procedure. The standard Norelco X-ray diffractometer with associated counters and electronics was used for the analysis. The monochromatic X-ray beam was obtained by using a silicon single crystal oriented such that its face was parallel to the (111) plane. The crystal was mounted on the axis of the goniometer. Because the second-order reflection of the (111) plane is not permitted, the $\lambda/2$ wavelength is not diffracted. The X-ray tube is operated below the excitation voltage of $\lambda/3$.

The apparatus for holding the sample, slit, and traversing mechanism was similar to the one which was constructed by J. Hilliard[4] at the Massachusetts Institute of Technology. The device was modified so that it could be mounted on the Norelco goniometer in front of the counter, as shown in Figure 3. The analyzer consisted of an alignment device, a slit, and a specimen carriage supported by free-running ball bearings. The carriage was driven easily by a micrometer which could be read to 0.0001 in. The slit used was approximately 0.0006 in. wide and 0.125 in. long.

The alignment device was used to prealign the diffusion zone with the slit. An image of the slit was obtained on a fine-grain photographic plate which was mounted on a specimen holder in place of a specimen. The plate and holder were then positioned in the alignment jig under a microscope and the slit image rotated so as to line up with the cross hairs of the microscope. A new holder was positioned under the microscope and the sample cemented to this so that the diffusion zone lined up with the cross hairs. Figure 4 shows the jig and sample carriage. Then the sample was fastened to the carriage and the analysis begun.

In order to locate the diffusion zone, a slow-speed synchronous motor was fastened to the micrometer and the sample driven across the X-ray beam while a chart recording was made. This is illustrated in Figure 5. Starting at one end of the diffusion zone, the specimen was moved in increments of 0.0005 in. across the X-ray beam using fixed count techniques to record the intensity of the transmitted beam.

In order to use equation (5) it was necessary to have the same starting point for both copper and molybdenum radiations. This was accomplished by cementing a 1-mil tungsten wire to the pure aluminum section parallel to the diffusion zone. This proved to be very satisfactory.

The measurement of I_0 can reach 100,000 counts/sec; therefore, the Geiger tube cannot be used without a series of absorbing foils. A scintillation counter with high speed resolution and good spectral efficiency is more satisfactory. However, at counting rates above 40,000 counts/sec it again becomes necessary to prepare a linearity curve for the scintillation counter.

Before starting the analysis, a working curve was obtained by arbitrarily placing various weight fractions in the right-hand side of equation (5) and computing the left-hand value. This is shown in Figure 6. From the experimentally measured values of the intensities, the left-hand side of equation (5) was calcu-

Figure 3. Traversing mechanism mounted on goniometer.

Figure 4. Jig and sample carriage.

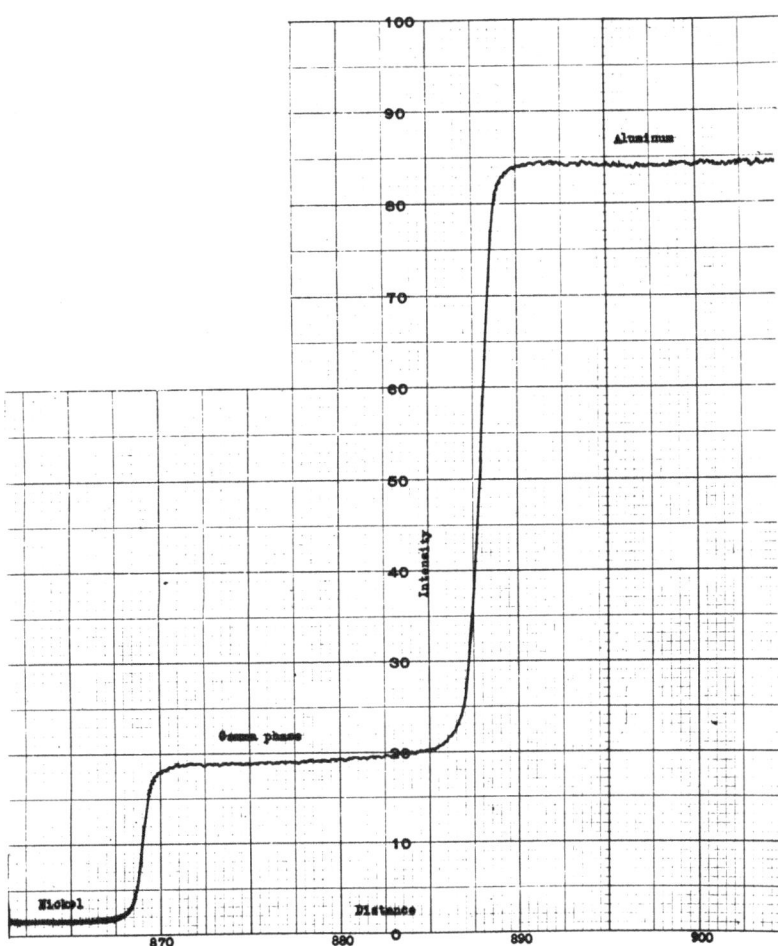

Figure 5. Chart recording of X-ray transmission of copper $K\alpha_1$ through an aluminum−nickel diffusion couple. (Approx. 5 mil thickness). Counting rate 51,200 counts/sec (full scale).

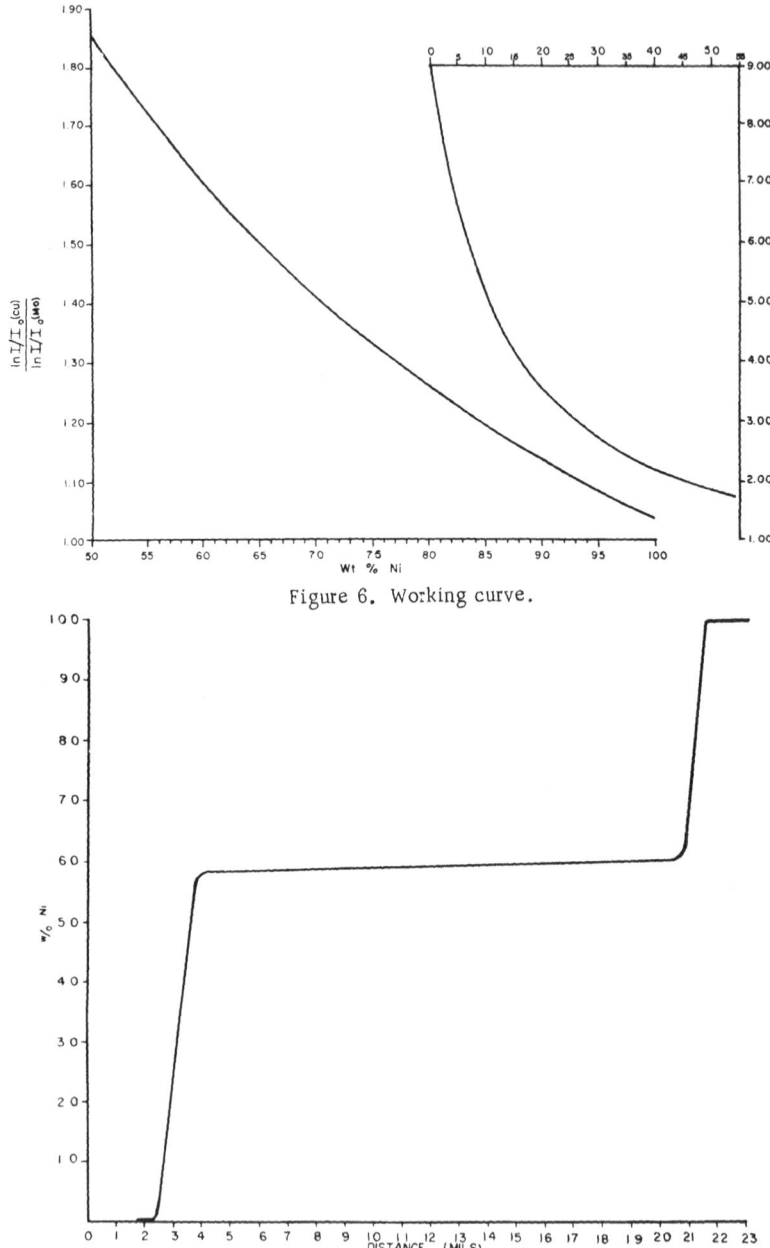

Figure 6. Working curve.

Figure 7. Plot of nickel content across diffusion couple.

lated and the weight percent for each measurement was obtained from the working curve.

The final form of the data consisted of a plot of weight percent nickel vs distance along the couple. This is illustrated in Figure 7.

In order to reduce the analysis time, the calculations were programed on a Librascope LGP-30 computer.

RESULTS AND DISCUSSION

The only phase which appeared in the X-ray analysis was the gamma phase. The compositional limits of this phase obtained by X-rays agreed fairly well with the phase diagram. The other phases did not appear because their rate of growth was extremely slow under the conditions employed. The beta phase was detected microscopically but was missed by the X-rays as it was less than 1 mil wide. Gamma nickel diffusion couples heated and diffused above the eutectic temperature revealed the existence of the delta and epsilon phases in widths not exceeding 2μ.

There are a number of experimental difficulties which can limit the accuracy of the X-ray analysis. The most difficult part of the analysis is the preparation of a uniformly thick sample. This is especially troublesome where the difference in hardness between the two elements is large. There are errors involved in alignment of the diffusion couple with the slit, errors due to impurities, mass absorption coefficients, counting errors, and porosity. Many of these can be minimized by using a slit length that is as short as possible. Counting errors are usually never greater than 1%. Despite the limitations, the technique has many advantages over other methods.

CONCLUSIONS

The X-ray absorption analysis technique has been applied to aluminum—nickel diffusion couples. The results are in agreement with the phase diagram as published in the literature. This technique is well suited to the study of diffusion couples. It can be used to determine concentration gradients, porosity in the diffusion zone, and phase boundary limits. The technique has the advantage that the high-temperature phase need not be retained, because the chemical composition and not the phases present determines the attenuation of the X-ray beam.

REFERENCES

[1] L. S. Castleman and L. L. Seigle. "Layer Growth during Interdiffusion with Al—Ni Alloy Systems," Transaction A.I.M.E., Vol. 212, October 1958, p. 589.

[2] W. L. Fink and L. A. Willey, "Constitution of Binary Alloys (Al—Ni)," Metals Handbook A.S.M. 1948, p. 1164.

[3] R. E. Ogilvie, "Quantitative Analysis by X-ray Absorption," Ph.D. Thesis, Department of Metallurgy, Massachusetts Institute of Technology, May 1955.

[4] Personal communication to L. S. Castleman of Sylvania Research Laboratories.

APPLICATIONS OF OPTICAL AND ELECTRONIC DISPERSION TO X-RAY ABSORPTION-EDGE SPECTROMETRY

Charles G. Dodd
University of Oklahoma, Norman, Oklahoma

ABSTRACT

The applications of X-ray emission or fluorescent spectrography to chemical analysis have increased spectacularly in recent years, but little attention has been paid to the potentialities of X-ray absorption techniques. Monochromatic X-ray absorption-edge spectrometry, in particular, is most promising. The ultimate sensitivity of absorption-edge spectrometry probably will be less than that of fluorescent analysis, but this disadvantage may be outweighed by the convenience, economy, and absence of matrix effects with the former method. Both methods appear limited in application only to certain elements.

A pulse height analyzer coupled with scintillation and proportional counter detection has been found to permit an increase in sensitivity of absorption-edge spectrometry, primarily because controlled window widths may be utilized in determining transmitted X-ray intensities with a scaler. Further work has led to the development of a new rapid, convenient technique known as "differential pulse amplitude distribution (PAD) peak height analysis." Work carried out during the development of the new method is described.

INTRODUCTION

The concept of absorption-edge spectrometry was developed in 1925 by Glocker and Frohnmayer[1] but has not enjoyed wide application in competition with other physical analytical tools because suitable instrumentation was not available. Engstrom[2] more recently has thoroughly analyzed the procedure and applied it to biological samples. The development of commercially available stable X-ray sources, precise goniometers for X-ray diffraction, and pulse-height discrimination equipment have suggested a further reconsideration of this old technique. X-ray absorption-edge spectrometry possesses the great advantage of little or no matrix effect, although its sensitivity is less than

---

[1]Superscripts pertain to references at the end of the paper.

that of X-ray fluorescence methods. It has been found that it also is cheaper and simpler than the latter. An exploratory investigation on the determination of strontium in brines by the writer in 1955 was most promising,[3] and Barieau[4] has applied it to the determination of molybdenum and zinc.

THEORETICAL BACKGROUND

Absorption Coefficients. When a beam of X-rays passes through a thin layer of matter, the fraction dI/I absorbed of the incident intensity I of the beam is proportional to the thickness of the layer, dx. This is expressed by the equation

$$\frac{dI}{I} = -\mu dx \ , \tag{1}$$

in which the proportionality constant, μ, is known as the "linear absorption coefficient." The negative sign indicates a decrease in the intensity of the transmitted beam. Integration of equation (1) may be accomplished if μ is truly constant to give

$$\ln \frac{i}{I} = -\mu x \tag{2}$$

or

$$i = I \exp(-\mu x) \ ,$$

in which I is the intensity when $x = 0$ and i the intensity of the transmitted beam. Examination of equation (2) indicates the dimensions of μ are reciprocal length; indeed, it is the fractional decrease in intensity per unit length of path through the absorbing medium. A more useful quantity for chemical analysis by X-rays is the "mass absorption coefficient" written as μ/ρ or μ_m. (ρ is the density of the matter traversed by the beam.) Equation (2) then becomes:

$$i = I \exp\left(-\frac{\mu}{\rho}x\rho\right) \ . \tag{3}$$

The mass absorption coefficient is a measure of the fraction of energy in the incident beam absorbed when a beam of unit cross section traverses unit mass of material. It is useful because it is characteristic of the chemical elements in the absorbing material essentially independent of their chemical or physical state.

Equations (1) through (3) are valid only if the absorption coefficient is constant, but this is true only if the incident X-rays are monochromatic or homogeneous, i.e., of only one wavelength. If the incident beam is polychromatic most of the "softer" rays of longer wavelength will be absorbed in the first layers of matter penetrated by the beam and the spectral distribution of energy in the transmitted beam will be shifted to shorter wavelengths. For a given element the mass absorption coefficient may be expressed empirically as a function of wavelength by an equation having the form of the following:

$$\mu/\rho = C\lambda^n Z^m + b, \qquad (4)$$

in which C is a constant for a given element in a given wavelength range, λ is the wavelength, Z the atomic number of the element, and b is an expression for the scattering (discussed below) and also a function of λ and Z; the exponent n varies between 2.5 and 3.0 and m is approximately 4.

Absorption Mechanisms. When X-rays traverse matter the predominant absorption mechanism is photoelectric in nature, with scattering making a minor contribution to the total absorption (except for elements of low atomic number and for wavelengths shorter than 0.5 A). Secondary rays scattered from material traversed by the primary X-ray beam have essentially the same wavelength as the incident beam. They are generated by the acceleration of electrons in the irradiated material to the same frequency as the incident electromagnetic wave. According to electrodynamic theory, such an oscillating electric charge must radiate rays of essentially the same frequency. Simultaneously fluorescent rays are produced by a photoelectric mechanism. The primary X-ray beam ejects electrons from some of the atoms. This leaves these atoms in an ionized condition, and in returning to the normal state, energy is liberated when other electrons fall down to the energy level of the electrons that were ejected. The total mass absorption coefficient μ/ρ thus can be represented as the sum of two terms, the mass transformation coefficient (or fluorescent or photoelectric absorption coefficient) and the mass scattering coefficient:

$$\frac{\mu}{\rho} = \frac{\tau}{\rho} + \frac{\sigma}{\rho} . \qquad (5)$$

The mass transformation coefficient τ/ρ increases with both the wavelength of the radiation and the atomic number of the absorber. The value of the mass scattering coefficient σ/ρ (which includes both coherent and incoherent or Compton scattering) is small in comparison with τ/ρ and may be neglected in most applications concerned with chemical analysis by X-ray absorption, and the term b in equation (4) may, likewise, be dropped (unless wavelengths appreciably less than 0.5 A are employed).

X-Ray Photoelectric Absorption Edges. If the mass absorption coefficient of an element is determined as a function of wavelength, sharp discontinuities are observed at X-ray wavelengths characteristic of the element. These discontinuities are known as "absorption edges" or "absorption limits," and they are found to occur at wavelengths slightly shorter than the characteristic emission line, the K. At longer wavelengths are found the three L absorption edges with a similar relationship to the characteristic edges. At an absorption edge the magnitude of the sudden drop in mass absorption coefficient is referred to as the "absorption edge jump."

Although X-rays obey an absorption law [equation (1)] similar to that obeyed by light rays, the photoelectric mechanism accounting for most of the absorption of X-rays in matter results in the unique spectral properties described above. Characteristic K emission lines of an element in an irradiated absorber are not excited unless the incident X-ray beam contains photons of high enough energy to eject electrons from the K shell completely outside the atom. The critical photon energy below which K absorption does not occur is given by

$$W_K = h\nu_K = \frac{hc}{\lambda_K} , \qquad (6)$$

in which ν_K and λ_K are the frequency and wavelength respectively corresponding to the K absorption edge. Similar relationships hold for the L and M edges. Furthermore, the frequencies (and wavelengths) of the characteristic emission lines are simply related to the frequencies of the absorption limits. For example, the K_{α_1} line is radiated when an electron at the L_{111} level in the L shell falls into the K shell, thus:

$$\nu_{K_{\alpha_1}} = \nu_K - \nu_{L_{111}} . \qquad (7)$$

The absorption edge generally is considered to be a sharp discontinuity. This is not strictly true; the position and structure of the edge do depend on the chemical state of combination of an element and possibly its physical state. These fine structure effects are of a different order of magnitude from the coarser phenomena discussed above. When performing chemical analysis by X-ray absorption it is not usually possible to resolve the structure of an absorption edge with the instruments employed.

The Absorption-Edge Method of Chemical Analysis.
Equation (4) above describes the absorption of monochromatic X-radiation when the absorbing sample consists of one element. If several elements are present and monochromatic radiation is employed, the appropriate equation is

$$\iota = I \exp\left[-\sum_j \left(\tfrac{\mu}{\rho}\right)_j q_j\right] \qquad (8)$$

in which q is the quantity of the j th element expressed in grams per square centimeter.

The peculiar virtue of the method of X-ray absorption-edge chemical analysis is that it is possible, under the proper conditions, to determine the mass of one element in an absorbing sample in the presence of large amounts of other elements. Furthermore, the method is unique with respect to its high degree of specificity (rivaled only by X-ray fluorescence analysis). In order to attain this specificity, together with sensitivity approaching microanalysis, it is necessary to measure absorption of the monochromatized X-ray beams at selected wavelengths on each side of an absorption edge.

Let the mass of the sample be Q g/cm^2 and let X be the mass of the desired element in grams per square centimeter of sample cross section. [$X = x\rho$ of equation (3).] If the absorption edge of the element lies at a wavelength λ_K, select suitable wavelengths λ_1 and λ_2, where subscript 1 refers to the short-wavelength side of the edge and subscript 2 to the long-wavelength side ($\lambda_1 < \lambda_K < \lambda_2$). For each selected wavelength we may write equation (8) as

$$i_1 = I_1 \exp -\left[\frac{\mu_1}{\rho}X - \frac{\mu_1'}{\rho}(Q - X)\right]$$

$$i_2 = I_2 \exp -\left[\frac{\mu_2}{\rho}X - \frac{\mu_2'}{\rho}(Q - X)\right], \tag{9}$$

where the primes are used for a hypothetical mass absorption coefficient for $(Q - X)$, i.e., everything in the sample except the desired element.

If the quantity $(Q - X)$ is eliminated from equations (9) one obtains:

$$X = \frac{\ln\frac{i_2}{I_2}\left(\frac{\mu_1'}{\mu_2'}\right) - \ln\frac{i_1}{I_1}}{\frac{\mu_1}{\rho} - \frac{\mu_2}{\rho}\left(\frac{\mu_1'}{\mu_2'}\right)}. \tag{10}$$

Of the quantities in equation (10), i_1/I_1 and i_2/I_2 can be determined experimentally and μ_1, μ_2, and ρ can be measured, found in available tables, or calculated, but μ_1'/μ_2' remains to be determined. Referring to equation (4) above and neglecting the term b, we have:

$$\left(\frac{\mu_1'}{\mu_2'}\right) = \frac{\lambda_1^n \sum_j C_j Z_j^m}{\lambda_2^n \sum_j C_j Z_j^m} = \left(\frac{\lambda_1}{\lambda_2}\right)^n. \tag{11}$$

The exponent n varies continuously from 3.0 to 2.3 when $Z\lambda$ varies from 8 to 790. Thus, equation (10) can be written as:

$$X = \frac{\ln \frac{i_2}{I_2}\left(\frac{\lambda_1}{\lambda_2}\right)^n - \ln \frac{i_1}{I_1}}{\left(\frac{\mu_1}{\rho} - \frac{\mu_2}{\rho}\right)\left(\frac{\lambda_1}{\lambda_2}\right)^n} . \tag{12}$$

If the wavelengths λ_1 and λ_2 are close enough to the edge so that the term $(\lambda_1/\lambda_2)^n$ may be neglected, and if the intensity of the incident radiation at λ_1 and λ_2 is equal, or if corrections are made so that $I_1 = I_2$, equation (12) may be simplified to the more convenient working form:

$$i_1 = i_2 \exp -\left[\left(\frac{\mu_1}{\rho} - \frac{\mu_2}{\rho}\right)X\right] ,$$

or

$$X = \frac{2.303 \log(i_2/i_1)}{\mu_1/\rho - \mu_2/\rho} . \tag{13}$$

Equation (13) also may be shown to follow directly from equation (3) if the same assumptions are made.

If measurements cannot be made close enough to the edge to permit $(\lambda_1/\lambda_2)^n$ to be neglected, two alternative procedures may be employed. First, the exponent n may be determined on substances of known composition and a working curve drawn up for a given type of analysis. Equation (12) is employed to calculate X. This procedure permits adequate corrections to be made for the absorption of the solvent, but if other elements are present in solution in large and varying amounts it may be difficult or impossible to determine n precisely enough. The second alternative procedure is then preferred.

Again let us utilize equation (4). Take logarithms of both sides of the equation (ignoring the scattering term b) to obtain for a given element:

$$\log \frac{\mu}{\rho} = n \log \lambda + \log C Z^m . \tag{14}$$

Equation (3) can be transformed to

$$\log \log \frac{I}{i} = \log \frac{\mu}{\rho} + \log X - \log 2.303 , \tag{15}$$

which upon combination with equation (14) gives

$$\log \log \frac{I}{i} = n \log \lambda + \log C\ Z^m + \log X - \log 2.303 , \quad (16)$$

or

$$\log \log \frac{I}{i} = A \log \lambda + B .$$

Evidently $\log \log I/i$ may be plotted versus $\log \lambda$ for two or three wavelengths on each side of the edge. The two straight lines obtained on such a plot may be extrapolated to their inter- section with a vertical line drawn at λ_K . The points of inter- section permit I_1/i_1 , I_2/i_2, and i_2/i_1 to be calculated for use in equation (13) for the calculation of X. In setting up a procedure for determination of a particular element it would be most ex- pedient to make plots of $\log \log I/i$ vs $\log \lambda$ for a series of known concentrations of the desired element. The heights of the discontinuities on the plots (vertical distances between inter- sections on a line at $\log \lambda_K$ would be direct functions of known concentrations and might be plotted as calibration curves for determination of the element. The corrections for variation in intensity of the incident X-ray beam at different wavelengths [applied to obtain equation (13) from (12)] might be eliminated by making measurements always at the same wavelengths (same 2θ setting of the spectrogoniometer). All other experimental variables should be set up and operated under standard conditions and the calibration checked from time to time. (In the above λ_L would replace λ_K at L edges.)

Sensitivity and Range of Applicability to Various Chemical Elements. If equation (13) is employed in the anal- ysis. the sensitivity or the precision to be expected with various concentrations of the desired element may be calculated. Equa- tion (13) may be differentiated to obtain

$$\frac{dX}{X} = \frac{1}{(\mu_1/\rho - \mu_2\rho)X} \cdot \frac{d(i_2/i_1)}{i_2/i_1} . \quad (17)$$

The derivative dX/X represents the uncertainty in the deter- mination of the concentration of a desired element corresponding to the experimental uncertainty in the measurement of the in- tensity ratio i_2/i_1. It would be recalled that X represents the concentration of the element in grams per square centimeter. The use of a sample cell 10 cm long, in place of a 1-cm cell, would increase X or reduce dX/X by a factor of ten. The experi- mental precision in the measurement of (i_2/i_1) is not known with certainty but should be of the order of one or two percent if the method is applicable to the unknown element.

The actual range of chemical elements to which the method can be applied is not fully known. Using K absorption edges, the highest-atomic-number element that can be used corresponds to the voltage at which the X-ray tube is operated according to the equation

$$h\nu_K = \frac{hc}{\lambda_K} = \frac{12,407}{V} \, , \tag{18}$$

where h is Planck's constant, c the velocity of light, and V the voltage impressed upon the X-ray tube. With a tube rated at 50 kv, europium or gadolinium, element 63 or 64, would be the highest atomic number determinable at its K edge. With a tube rated at 60 kv, the K absorption edges could be used up to element 69, thulium. On the other hand, the L absorption edges can be used effectively for higher-atomic-number elements.

Considering elements of lower atomic number, the writer has applied K absorption-edge spectrometry successfully to strontium and bromine; however, when determining elements in the range from about titanium or vanadium, No. 22 or 23, to bromine, No. 35, the experimental problem of obtaining a measurable attenuation of the X-ray beam is a challenging one. The writer has experienced difficulty in working with manganese, element 25, for example. Barieau[4] has determined zinc, No. 20, however, and his reported sensitivity was about 200 ppm. The difficulties encountered become greater at X-ray wavelengths below about 1.2 to 1.3 A. The problem simply is to obtain a measurable signal without absorbing essentially all the incident X-ray beam, according to equations (9). Sample path lengths must be short to avoid excessive absorption, but the experimental problem can be solved. Engstrom[2], for example, has worked with carbon, element 6, having a K edge at 43.6 A, by using a vacuum spectrograph and microtome-sliced samples. The techniques developed by Henke[5,6,7] for generating ultrasoft X-rays are of help in such applications. For more routine analytical determinations, however, the lower limit of applicability of K-edge spectrometry remains to be determined.

EXPERIMENTAL INVESTIGATIONS

Previous Exploratory Studies. The advantage of simplicity and economy possessed by X-ray absorption-edge spectrometry relative to X-ray fluorescence spectrometry and the indicated absence of any appreciable matrix effect in the former

Figure 1. Experimental arrangement. Improvised glass cell and available calcite crystal in place on General Electric XRD-3 spectrogoniometer.

stimulated an exploratory study by the writer in 1955 applied to the determination of strontium in oil field brines. A General Electric XRD-3 spectrogoniometer was utilized with an available small single crystal of calcite placed in the sample holder in the position used for X-ray diffraction samples. The experimental arrangement is shown in Figure 1.

The absorption edge of strontium was determined experimentally by utilizing a filter of finely crushed strontium chloride grains sandwiched between two pieces of drafting tape. By scanning the wavelength region of the strontium K absorption edge it was found that readings on the short-wavelength side might be taken conveniently at a diffraction angle of 14.47° 2θ and the readings at the long-wavelength side at an angle of 14.75° 2θ. The glass absorption cell was made from 22-mm ID Pyrex glass tubing to which a side arm was first blown for use as a filling tube. Windows of thin aluminum foil were cemented on the ends of the 22-mm tubing at a separation corresponding to an X-ray path length of 23 mm. A support for the cell was fabricated from a block of aluminum so that it could be attached rigidly to the frame carrying the slits in front of the Geiger tube counter and so that the cell could readily be repositioned in the same place relative to the X-ray beam and the counter.

The procedure employed was to fill the cell with deionized water and measure the X-ray intensity at angles of 14.47 and 14.75° 2θ. Subsequently, transmitted intensities were measured with the cell filled with a solution containing 10 ppm, 100 ppm, and finally 1000 ppm of strontium chloride. All measurements were made with a molybdenum-target X-ray tube operated at 45 kv and 16 ma. Determinations of X-ray intensity were made with the counter scaling circuit, using the timer set to measure the time required to count 16,384 counts. Recorded data were the average of 5 measurements. Although monochromatic X-radiation diffracted from a crystal was employed in this work, the intensity of the beam after passage through the experimental cell was still of the order of 1000 counts/sec.

The experimental results are tabulated in Table I as measurements of transmitted X-ray intensity in counts/sec. Measurements were made with the experimental cell filled in turn with deionized water and with solutions of 10, 100, and 1000 ppm strontium. If the intensity of the incident X-ray beam had been equal at λ_1 and λ_2 (corresponding to 14.47 and 14.75° 2θ, respectively). the transmitted intensities measured with strontium solutions in the cell would have been larger at λ_2 than at λ_1. The opposite was found to be the case. as is seen by comparing measurements of i_1 and i_2 in columns 2 and 3 of Table I, except for the solution containing 1000 ppm of strontium. This was caused by unequal intensity of the X-ray beam incident at the two wavelengths. In order to correct for this it was necessary to multiply all the readings at i_2 by the ratio (i_1/i_2) measured with deionized water in the cell. Corrected intensities for i_2 are shown in column 4 of Table I.

The observed transmitted intensity ratios, i_2/i_1, calculated from the data in columns 2 and 4, are presented in column 5 and compared with predicted ratios, calculated with the help of equation (13) in column 6. The experimentally observed value of the ratio at 1000 ppm is seen to be 1.201, in reasonably good agreement with the predicted value of 1.282, considering the approximations involved in the use of equation (13) and the preliminary nature of the experimental measurements. On the other hand, the method employed was not sensitive enough to detect the presence of 10 ppm of strontium. The lower limit of sensitivity appeared to be greater than 10 ppm but less than 100 ppm of strontium.

Table I.   Transmitted X-Ray Intensities

Concentration of strontium (as $SrCl_2$) in absorption cell (ppm Sr)	i_1 at 14.47° 2θ or 0.763 A (cts/sec)*	i_2 at 14.75° 2θ or 0.778 A (cts/sec)*	Corrected i_2 (cts/sec)	i_2/i_1	
				observed	predicted
0	803.5	716.4	–	–	–
10	758.2	675.4	757.5	0.999	1.003
100	645.5	585.1	656.2	1.017	1.025
1000	520.0	556.9	624.6	1.201	1.282

* Note: Arithmetic average of 5 measurements.

An argon-filled Geiger tube counter was used and no effort was made to correct the measured counting rates for coincidence losses. Barieau[4] in 1957 described a similar application of X-ray absorption-edge spectroscopy applied to the determination of molybdenum and zinc. He devised a simple procedure for determining the effective dead time of the Geiger counter tube used in his measurements. By this means he was able to obtain good values for the mass absorption coefficient of pure molybdenum, using foils of 0.001 and 0.003 in. thicknesses. He also was able to obtain theoretical calibration curves corresponding to equation (13) above with a slope directly proportional to the jump at the k absorption edge.

Barieau further applied his procedure to the determination of briquetted solids using a sample rotating device in the X-ray beam. He stated he was planning to initiate measurements with a scintillation counter and a pulse-height analyzer, but published results of this work have not come to the writer's attention.

Probably the most extensive discussion of X-ray absorption-edge spectroscopy has been published by Engstrom.[2] Engstrom, however, was interested in applying his results to thin sections of biological specimens prepared with the aid of a microtome. With a vacuum spectrograph and the use of photographic film to detect the transmitted X-rays he was able to determine the carbon, nitrogen, and oxygen content of individual living cells.

Figure 2. Experimental arrangement. Teflon cell, LiF crystal, and a scintillation counter in place on the Norelco goniometer.

Application of Pulse-Height Dispersive Analysis. Recently, through the assistance of an Atomic Energy Commission research grant, the writer has been able to extend the earlier work described above by investigating the addition of modern pulse-height analysis equipment to optical monochromatization obtained by single lithium fluoride and topaz crystals on a Norelco goniometer. Norelco crystal analysis X-ray tubes having tungsten, copper, and chromium targets have been used in this work. The R.I.D.L.-type pulse-height analyzer used is the latest design supplied by Norelco, having a resolving time of $1 \mu sec$, sufficiently small to count all pulses from scintillation and proportional counters. The equipment is shown in Figure 2. With a cadmium salt sample in the beam, the transmitted X-ray spectrum is shown in Figure 3, compared with the same spectrum with pure water in the sample cell. To determine optimum goniometer angles for measurements on each side of an absorption edge it has been found satisfactory to make such scans at a slow rate of $\frac{1}{8}$ deg/min.

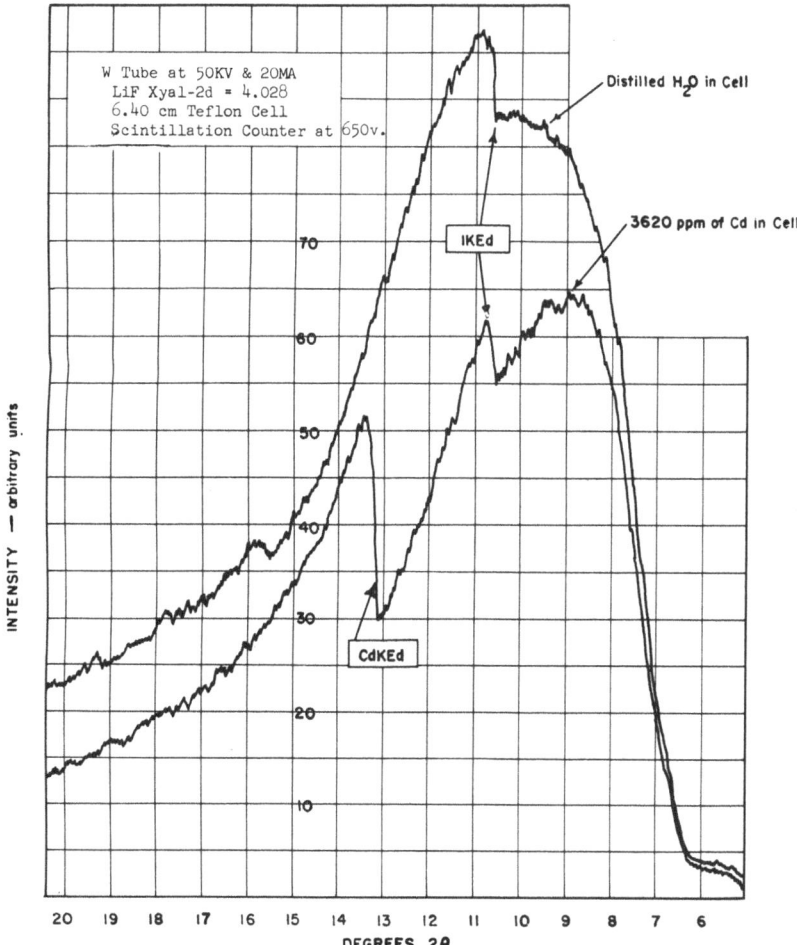

Figure 3. Transmitted X-ray spectra with water and with cadmium solution in cell.

In Figure 4 are shown traces of the integral and differential pulse amplitude distribution (PAD) curves with the goniometer set on each side of the K absorption edge of cadmium. Integral curves of this type are used to determine appropriate base line voltage settings and window settings for pulse-height discrimination. Following the procedures described above, intensities of transmitted X-ray beams were determined on each side of the absorption edge and calibration curves determined with the use of the scaler at a fixed count or a fixed time. The results were similar to those obtained with Geiger tube counters except

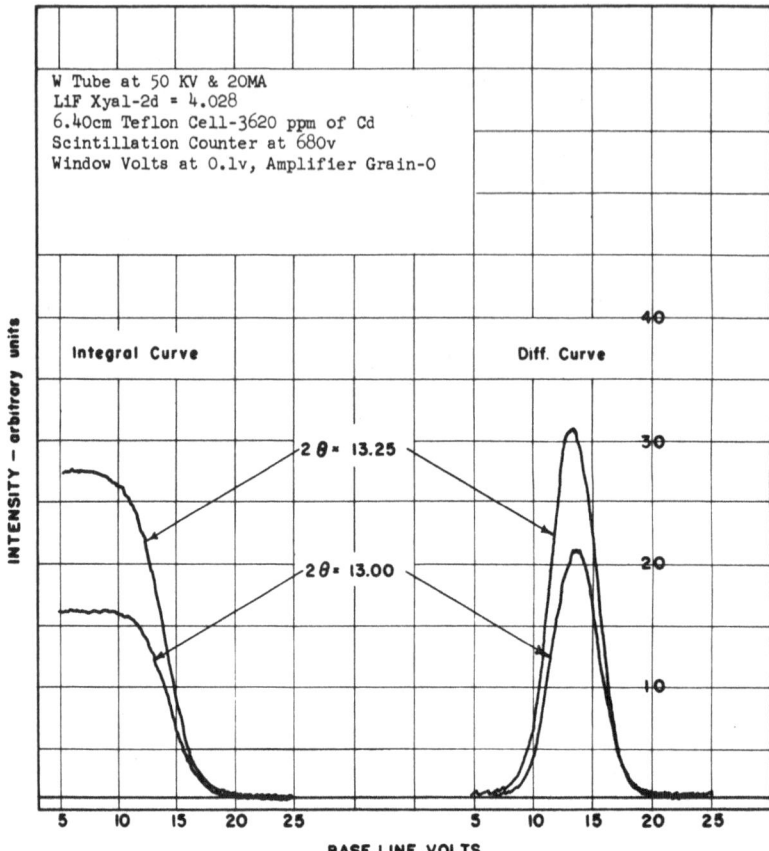

Figure 4. Integral and differential pulse amplitude distribution curves at cadmium K edge.

for increased sensitivity and less interference from noise in the counting circuitry and extraneous high-energy radiation.

At this stage of the work it was noted that a greater ratio of i_2/i_1 was obtained when the window of the pulse-height analyzer was set for a voltage appreciably less than that indicated to be desirable according to the integral or differential PAD curve. The reason for this was not clear until differential PAD curves were recorded on top of each other for goniometer settings on each side of the absorption edge, as shown on the right-hand side of Figure 4. The reason for the increased sensitivity with the smaller window width settings is simply that more of the differential PAD curve measured at the short-wavelength side of the edge was eliminated from the resulting count than was eliminated from the measurement at the long-wavelength side of the edge. It then became apparent that this was a technique that might be utilized to increase the sensitivity of X-ray absorption-edge spectrometry, but, if this were done, the resulting calibration curves would not follow equation (13) exactly.

Using a window voltage of 5.8 v, about half that indicated by the integral curve of Figure 4, calibration curves for cadmium, strontium, and barium were determined with the scaler at fixed time counts. These are shown in Figures 5, 6, and 7. Sensitivity to well below 100 ppm of the respective metal ions is indicated with good technique. The window voltage settings used for determining strontium and barium were approximately one-half the window indicated by the respective PAD integral curves.

To explore the possibility of attaining greater precision in the presence of other elements (the matrix effect) a determination of cadmium was made according to equation (16). This is shown in Figure 8.

Differential PAD Peak Height Analysis. In order to make measurements that would correspond to the theoretical development above, the use of a planimeter to measure areas of the two differential PAD curves was considered. This turned out to be a clumsy and time-consuming procedure. An alternative, that of measuring the height above the effective base line of an integral curve on each side of the edge, was also tried. It was found, however, that relatively good plateaus on the base lines were needed on each side of the edge and these conditions did not always prevail. The time required to make the integral curves and to measure the heights needed to obtain the ratio of transmitted intensity also was as long as that required for the scaler measurements. What was desired at this stage of the work was to develop a procedure whereby the measurements could be simplified and the determination time shortened.

The use of differential PAD curves then was reconsidered and it became apparent that suitable sharp peaks could be obtained on these curves if an unusually narrow width of approximately 0.1 v was used on the pulse-height analyzer. A narrow window voltage of this magnitude is of no advantage in X-ray diffraction or in X-ray fluorescent spectrography, but in the present application the narrow window made it possible to obtain better resolution, minimum W/A ratios (W is peak width at half height, and A the amplitude in terms of base line voltage), and hence PAD peaks that could be used by simply measuring their height. In order to determine the optimum settings to obtain the

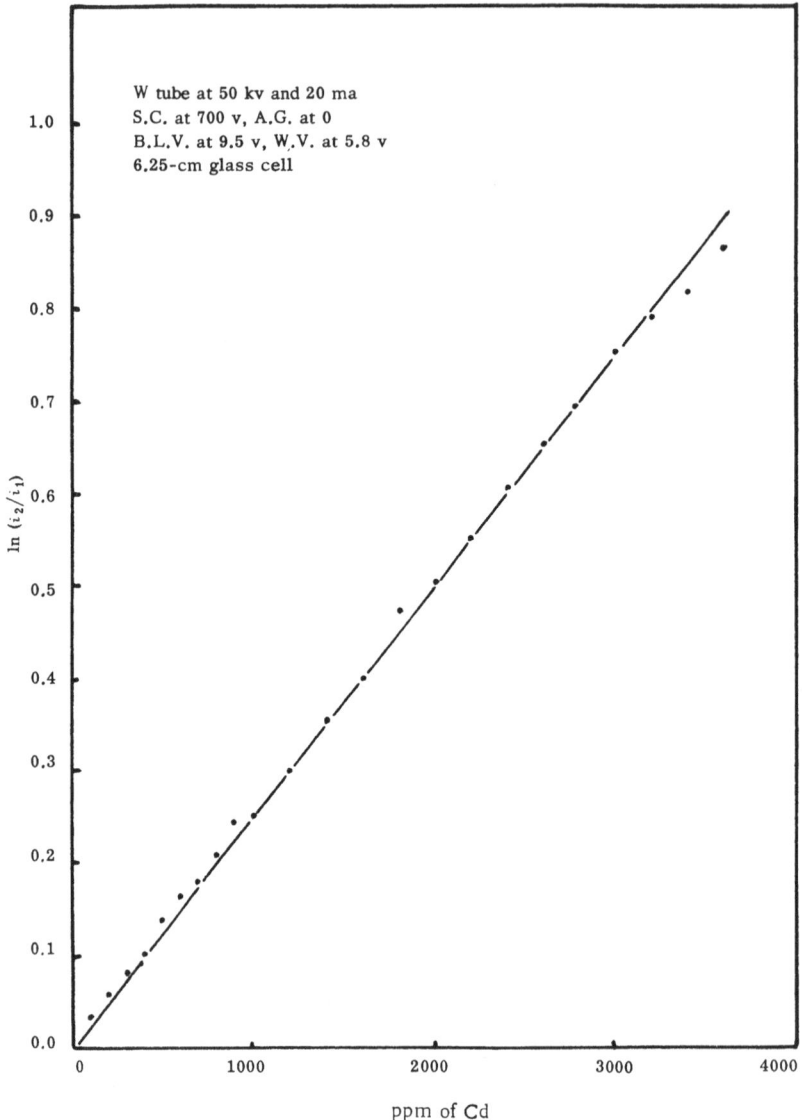

Figure 5. Calibration curve for cadmium solutions.

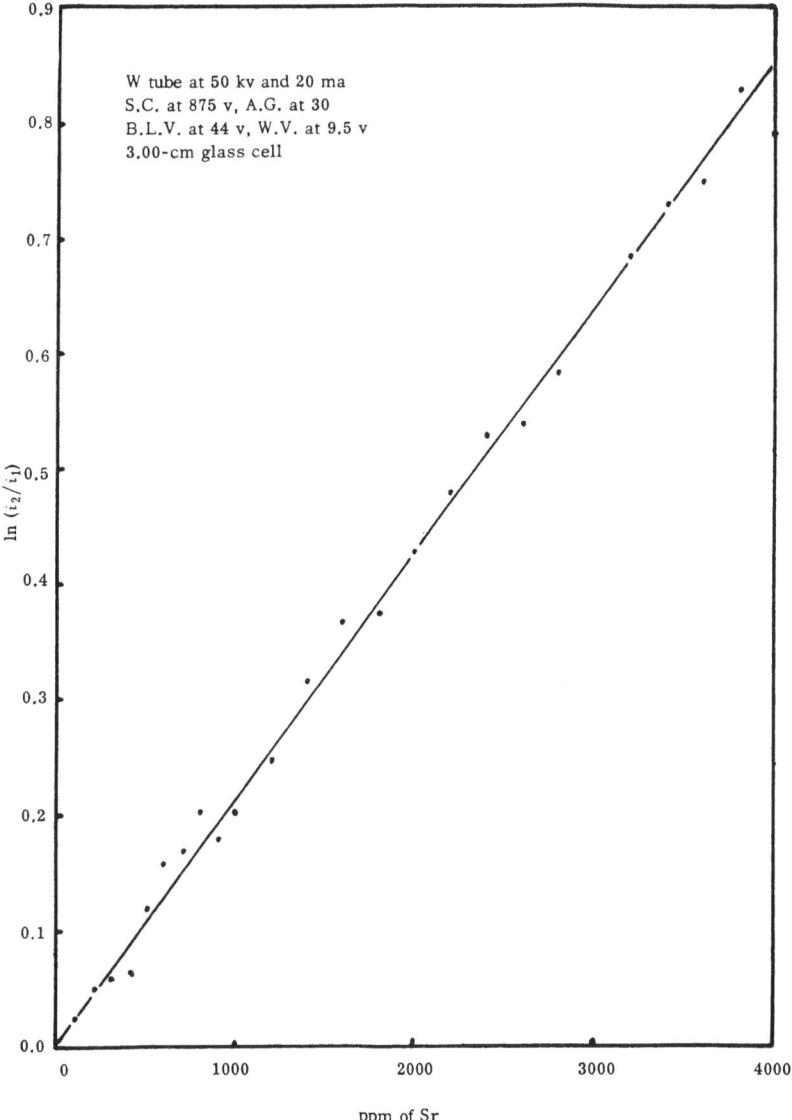

Figure 6. Calibration curve for strontium solutions.

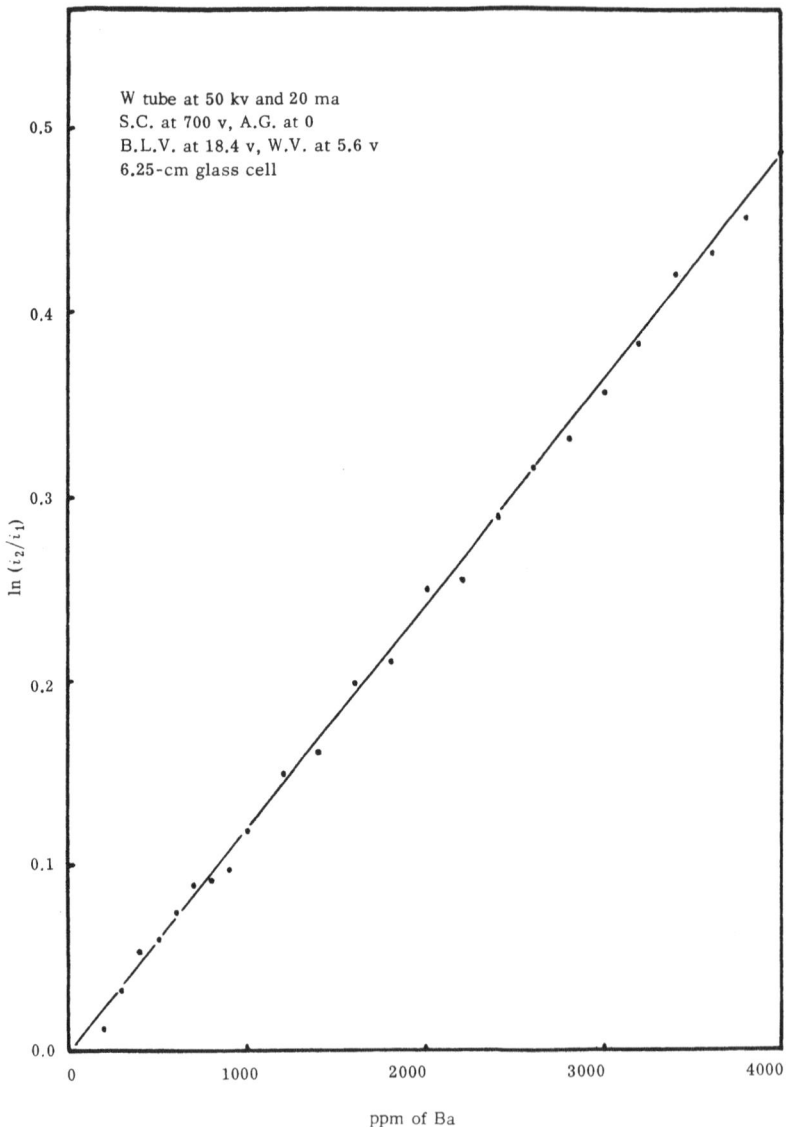

Figure 7. Calibration curve for barium solutions.

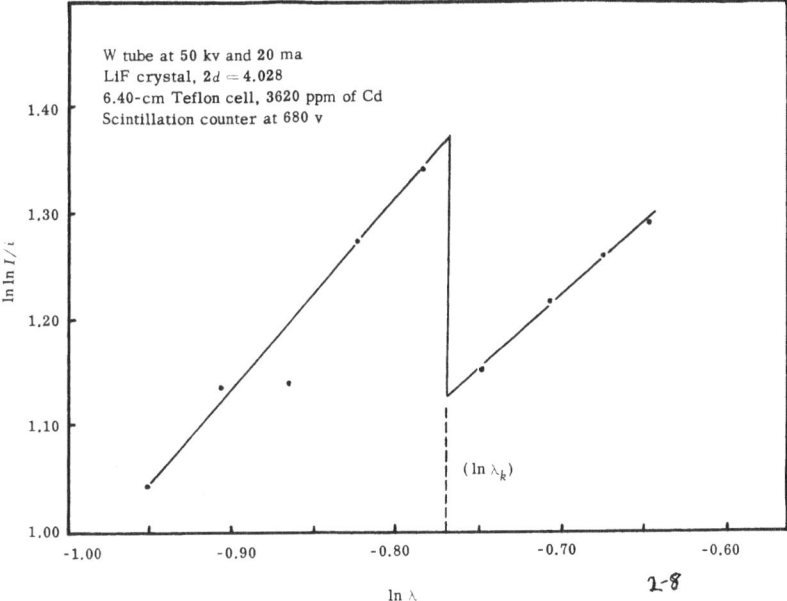

W tube at 50 kv and 20 ma
LiF crystal, $2d = 4.028$
6.40-cm Teflon cell, 3620 ppm of Cd
Scintillation counter at 680 v

Figure 8. Extrapolation plot to determine cadmium precisely in the presence of possible interfering elements, according to equation (16).

best PAD peaks of this type a study was made of the effects of scintillation counter high voltage, pulse-height analyzer amplifier gain, and pulse-height window voltage. The resulting curves are shown in Figures 9, 10, 11, 12, and 13 using a lithium fluoride crystal. Figure 14 displays data obtained with a topaz crystal and should be compared with Figure 12. The enhanced resolution obtained with the topaz crystal at the cadmium K edge is apparent, but the LiF gives at least five times as great intensity. The form of the recorded traces of differential PAD curves as affected by amplifier gain setting is illustrated in Figure 15a. Using optimum settings determined for the cadmium K edge, a series of three differential PAD peaks are shown in Figure 15b, measured on each side of the edge. With a series of reproducible peaks of this sort it is easy to draw a horizontal straight line and obtain a good mean value for the intensity of the peak on each side. The ratio of intensities then can be used to obtain a calibration curve for cadmium, or any other element of interest within the range of application of the method; and a chemical analysis may be made in a rapid and convenient manner. The resulting calibration curve for cadmium in aqueous solution is

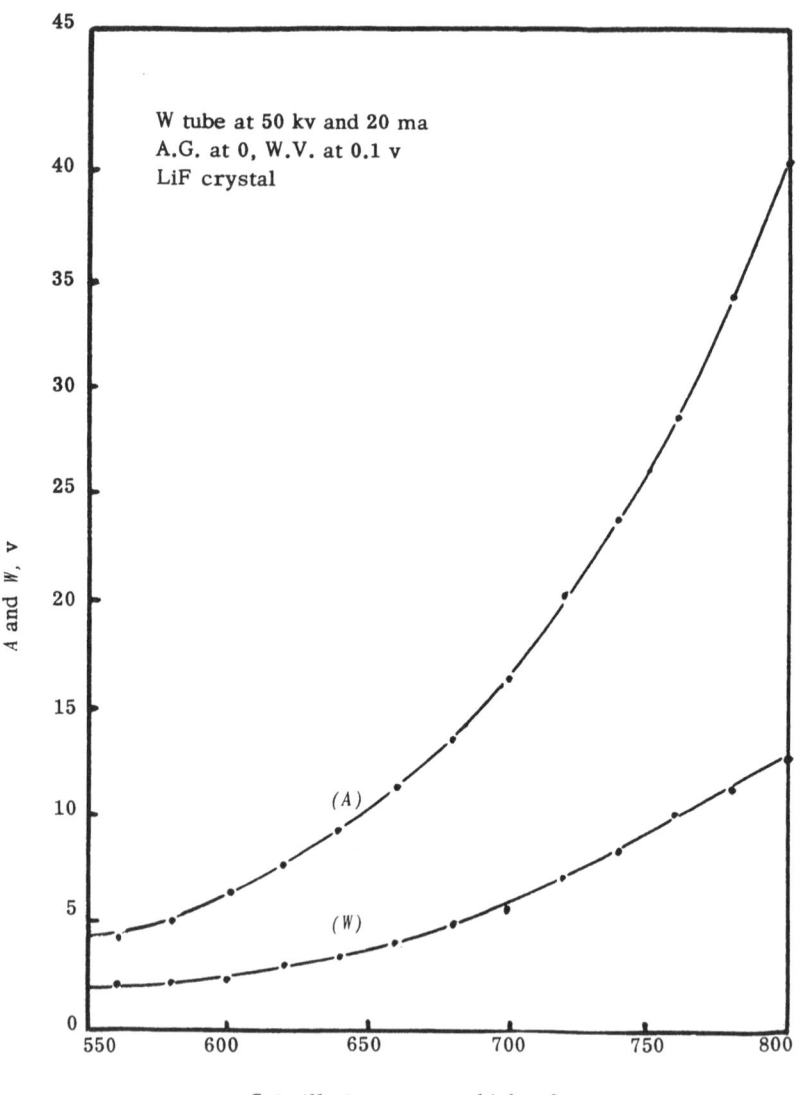

Figure 9. Effect of scintillation counter voltage on peak amplitude and width at half height at cadmium K edge (LiF crystal).

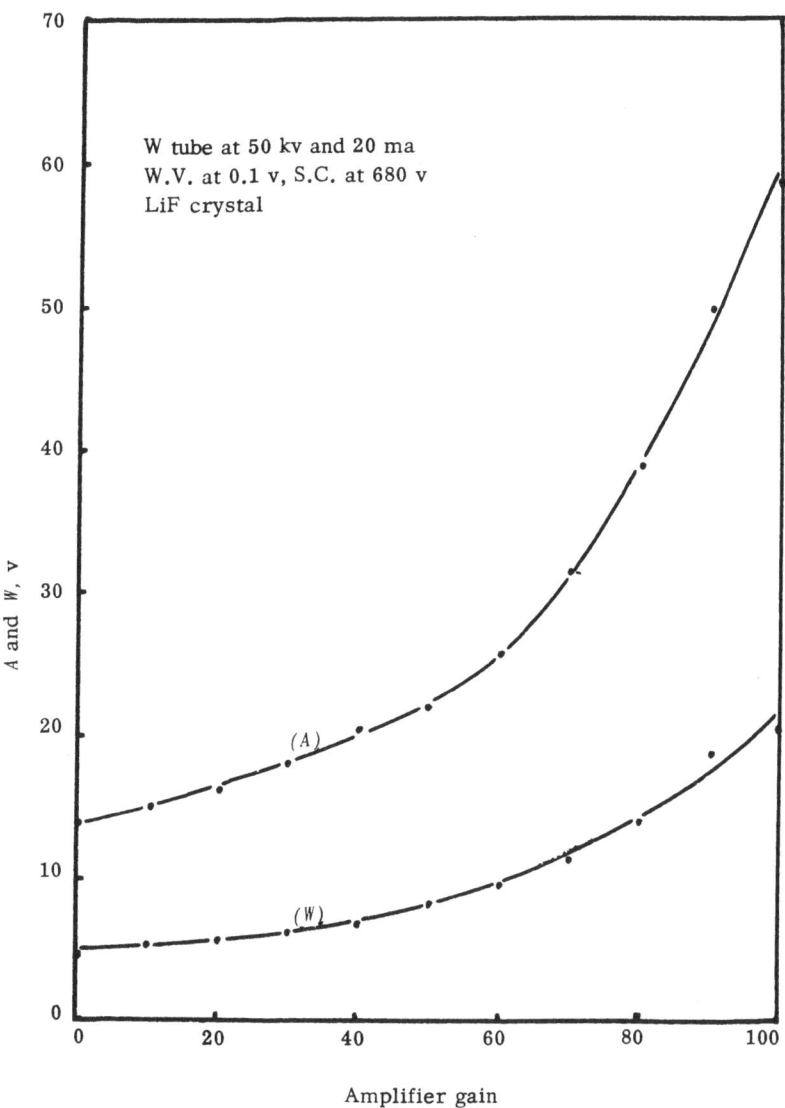

W tube at 50 kv and 20 ma
W.V. at 0.1 v, S.C. at 680 v
LiF crystal

A and *W*, v

Amplifier gain

Figure 10. Effect of pulse-height amplifier gain on peak amplitude and
width at half height at cadmium *K* edge (LiF crystal).

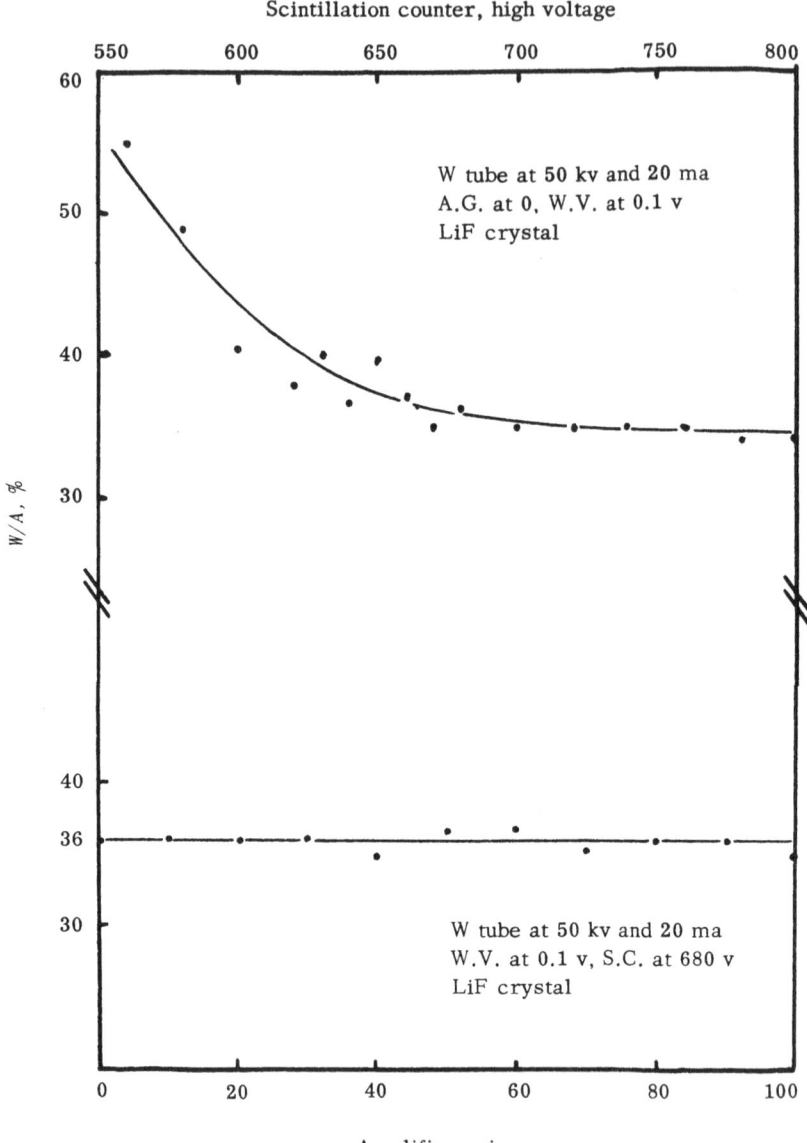

Figure 11. Half peak-width-to-amplitude ratios as functions of scintillation counter voltage and amplifier gain at cadmium K edge (LiF crystal).

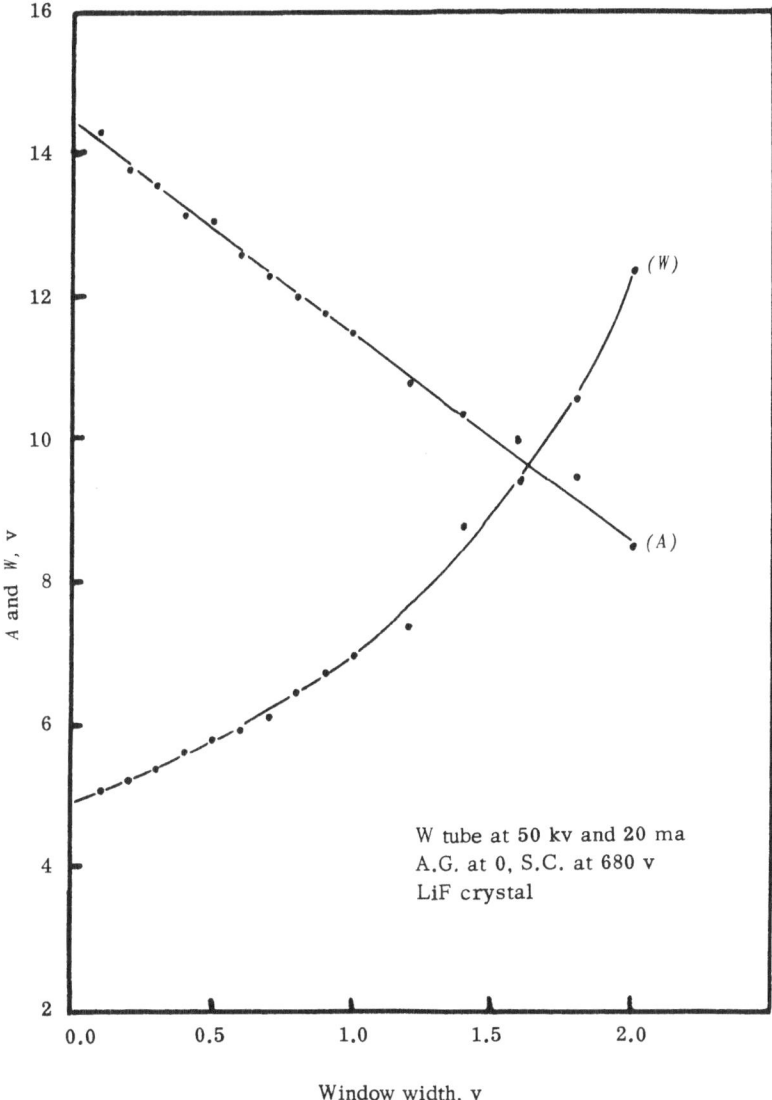

Figure 12. Effect of pulse-height window width on peak amplitude and width at half height at cadmium K edge (LiF crystal).

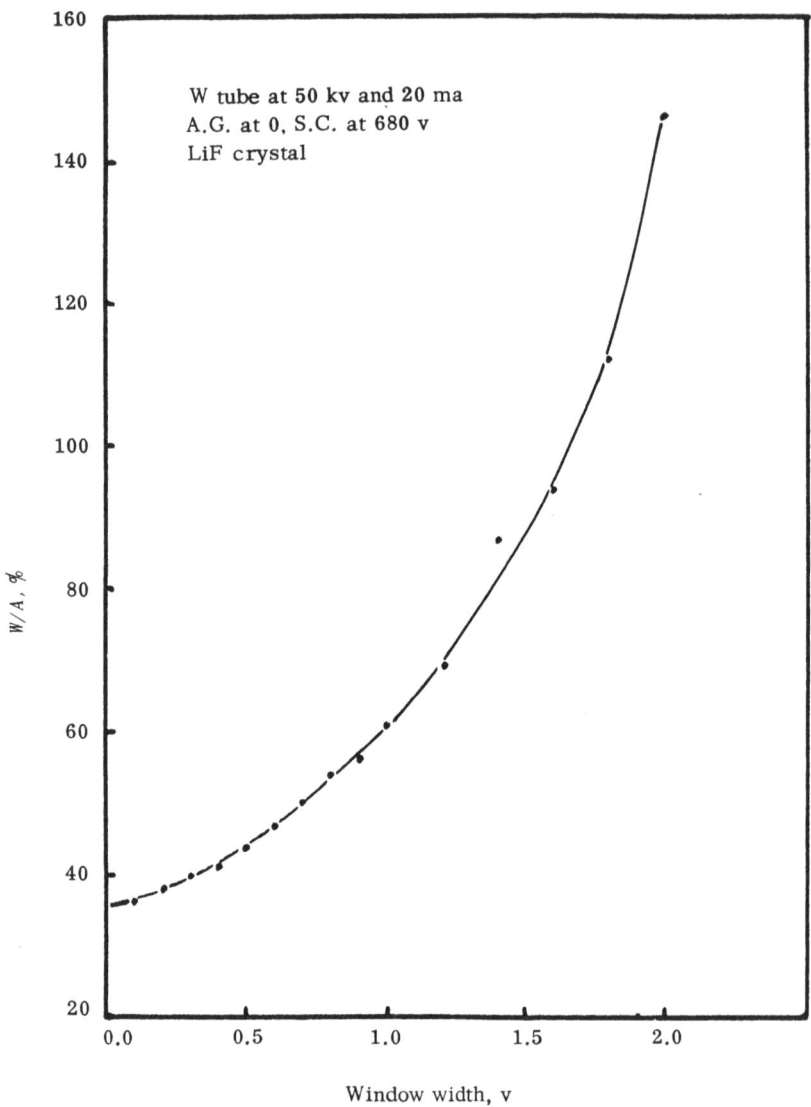

Figure 13. Half peak-width-to-amplitude ratios as a function of pulse-height window width at cadmium K edge (LiF crystal).

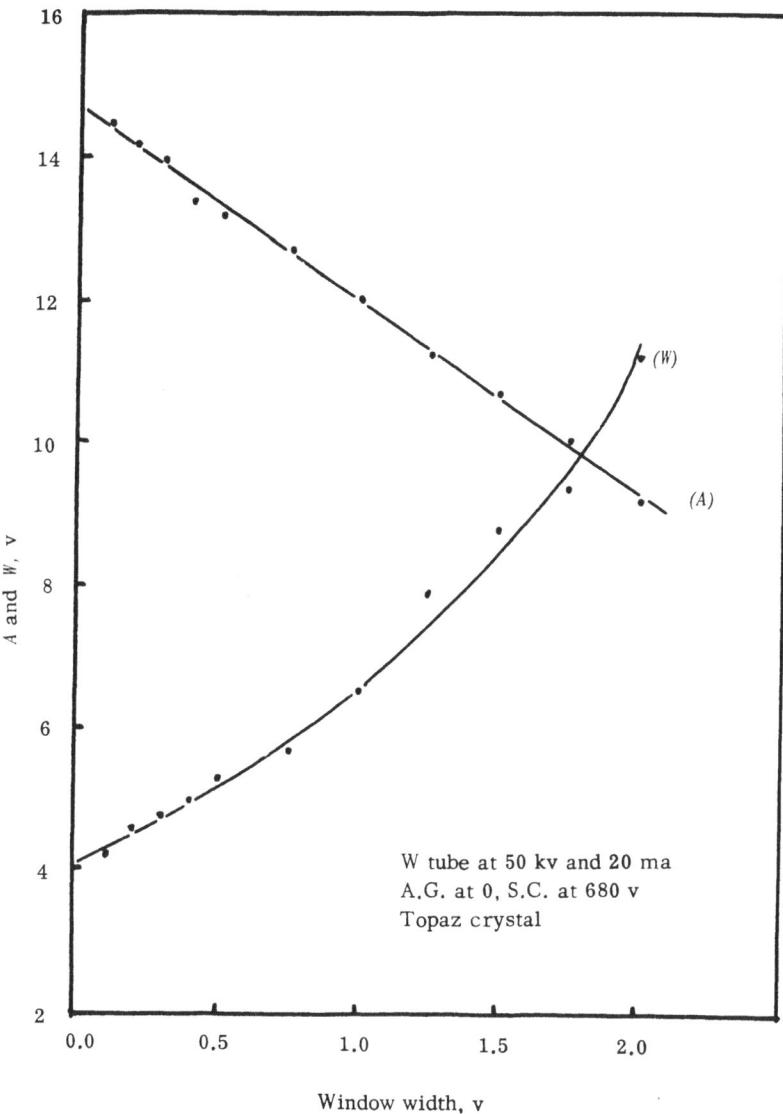

Figure 14. Effect of pulse-height window width on peak amplitude and width at half height at cadmium K edge (topaz crystal).

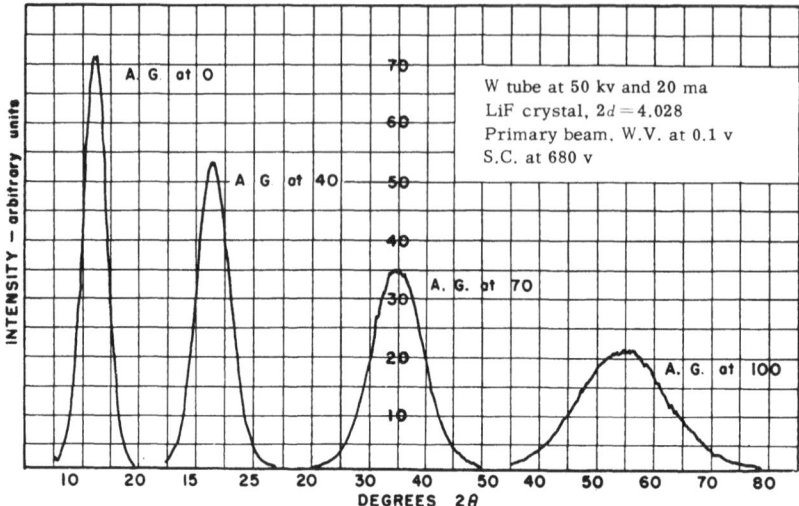

Figure 15a. Effect of pulse-height amplifier gain on form of differential
PAD peaks.

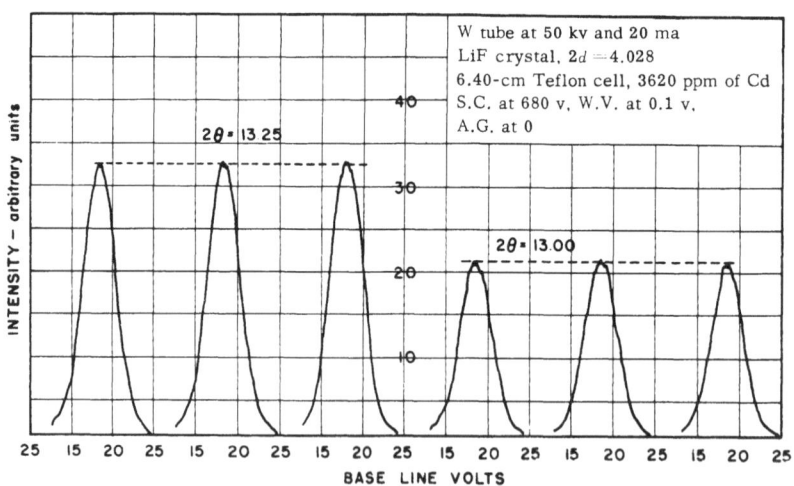

Figure 15b. Differential PAD peak-height analysis for cadmium.

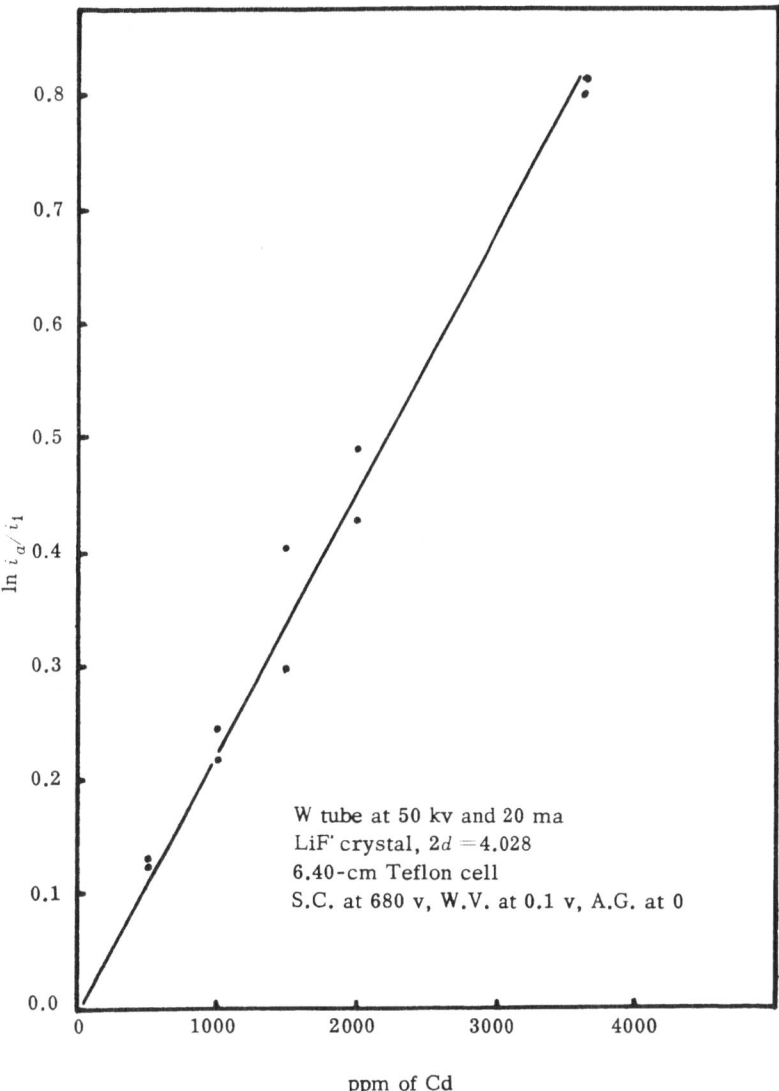

Figure 16. Calibration curve for cadmium solutions as determined by new differential PAD peak-height analysis.

shown in Figure 16. The precision of the new rapid procedure is somewhat less than that of more conventional counting procedures, but the accuracy is probably better and little or no calibration is required because the new method conforms to equation (13).

Cadmium, element No. 48, in the middle of the periodic table is an ideal element for application of absorption-edge methods. Barium, however, has an atomic number of 56 and is more difficult to detect using only a Geiger tube. With the apparatus shown in Figure 1 the writer was unable to determine barium at its K edge. With pulse-height analysis, however, the direct beam scattering into the detector at low-angle settings of the goniometer is eliminated by discrimination. This application of improved X-ray absorption-edge spectrometry to higher-atomic-number elements at low angular settings of the X-ray goniometer is a most fortunate result of the application of the pulse-height analysis. It illustrates a particular advantage of the new method as compared with X-ray fluorescence spectrography. In particular, it also indicates that absorption-edge spectrometry should be applicable to elements of still higher atomic number, as discussed above.

The new procedure, as well as the procedure employed by Barieau, lends itself to the determination of mass absorption coefficients by experimental measurements and the determination of the "jump" at absorption edges. These determinations can be made as a result of the concordance of the procedure with theory. A further advantage is that it is actually unnecessary to calibrate the system ahead of time in order to determine a given element in the range of the periodic table to which the method may be applied. All that is necessary is that the effective absorption-edge jump must be known by prior determination, but no actual calibration curve is required. This can be done simply because the absorption-edge spectrometric method is independent of other elements present in the sample, i.e., there is no matrix effect. This fact alone greatly simplifies and speeds up the analytical procedure and should increase the popularity of X-ray absorption-edge spectrometry.

Further advantages of the new method in comparison with other types of instrumental chemical analysis, and X-ray fluorescence spectrography in particular, is that the method may be installed at a relatively low cost if the laboratory is already equipped for X-ray diffraction work. The only additional equipment required is a pulse-height analyzer and an analyzing crystal holder. It is not necessary to purchase an extra external tube nor an extra set of collimators as required for X-ray fluorescence spectrography. Neither is it necessary to purchase another goniometer in order to avoid the time-consuming chore of re-alignment each time it is moved from the X-ray diffraction tube position to the fluorescence analysis apparatus.

ACKNOWLEDGMENTS

The assistance of David J. Kaup, who carried out most of the experimental work and prepared the figures, and that of the Atomic Energy Commission, who supplied research grant funds for the work, is acknowledged with thanks. Thanks also are due the Continental Oil Company for releasing data from a 1955 laboratory report.

REFERENCES

[1] R. Glocker and W. Frohnmayer, Ann. Physik, Vol. 76, 1925, p. 369.

[2] A. Engstrom, Acta Radiologica, Suppl. LXIII, 1946.

[3] C. G. Dodd, Continental Oil Company (Development and Research Dept.) Report No. 52-55-503-16, June 1955.

[4] R. E. Barieau, Anal. Chem., Vol. 29, 1957, p. 348.

[5] B. L. Henke, R. White, and B. Lundberg, J. Appl. Phys., Vol. 28, 1957, p. 98.

[6] B. L. Henke and J. W. M. DuMond, J. Appl. Phys., Vol. 26, 1955, p. 903.

[7] B. L. Henke, Proceedings of the Cambridge Meeting on X-ray Microscopy and Microradiography, Academic Press, New York, 1957.

A VERSATILE 19-cm-DIAMETER
LOW-TEMPERATURE DEBYE-SCHERRER CAMERA

A. Taylor

Westinghouse Research Laboratory, Pittsburgh, Pennsylvania

ABSTRACT

The camera was constructed to fill the need for a low-temperature instrument capable of yielding high spectral resolution and accurate lattice parameters. Specimen temperatures in the range −190 to 300°C are attainable.

A special feature of the camera is the film holder which is split along the beam axis into two symmetrical halves and which can easily be removed from the base without disturbing the cooling system or the specimen. The specimen holder is mounted on a cross-slide to enable specimens of large radius to be rotated off-center with the beam at glancing incidence to their surface. Finally, diffraction patterns from large metallurgical samples may be obtained by removing one of the film holders and employing the apparatus as a "half camera."

Some applications of the camera to the thermal expansion of SiC and MnSe are described.

INTRODUCTION

While the majority of commercially available Debye-Scherrer powder cameras are perfectly adequate for most routine applications, such as identification or simple lattice parameter work, their shortcomings become immediately obvious when it is required to deal with specimens of unusual shape and size, and when it is required to take diffraction patterns with the specimen at subzero or at elevated temperatures.

The camera described below is designed for use with typical cylindrical powder specimens and also for bulky specimens frequently encountered in metallurgical research. On account of its large radius, it is capable of yielding high spectral resolution and lattice parameters to an accuracy of approximately one part in 50,000 in favorable cases.

CAMERA DESIGN

Basically, the camera, which is shown schematically in Figures 1 and 2, consists of a slit system and beam trap mounted on an open circular frame, the proportions being such that all

41

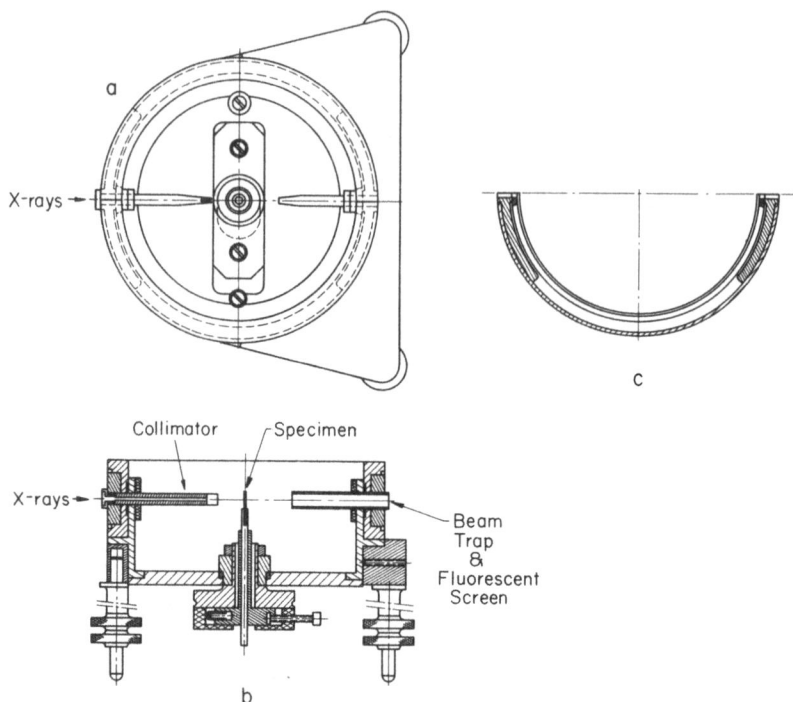

Figure 1. 19-cm-diameter Debye-Scherrer camera.

Debye-Scherrer reflections in the angular range from approximately 3 to 87° in Bragg angle can be accommodated on both sides of the direct beam. This hollow frame, which is made of brass, is mounted in a heavy aluminum table attached to three adjustable legs. The mounting is such that the hollow frame and associated slit system can be rotated for purposes of alignment with respect to the incident beam. A hole, slot, and plane in a wooden block serve kinematically to locate the legs supporting the aluminum table. This ensures that the camera can be removed and replaced without disturbing its accuracy of alignment.

A circular flange on the lower side of the open frame permits the attachment of a round plate. The latter is recessed to carry an adjustable cross-slide drilled to accommodate a thin teflon bearing into which the specimen holder is inserted from below, and retained by means of a brass collar. For normal Debye-Scherrer work the specimen holder consists of a hollow shaft mounted on a pulley wheel and fitted with a central stem which retains the specimen, the stem being capable of movement in

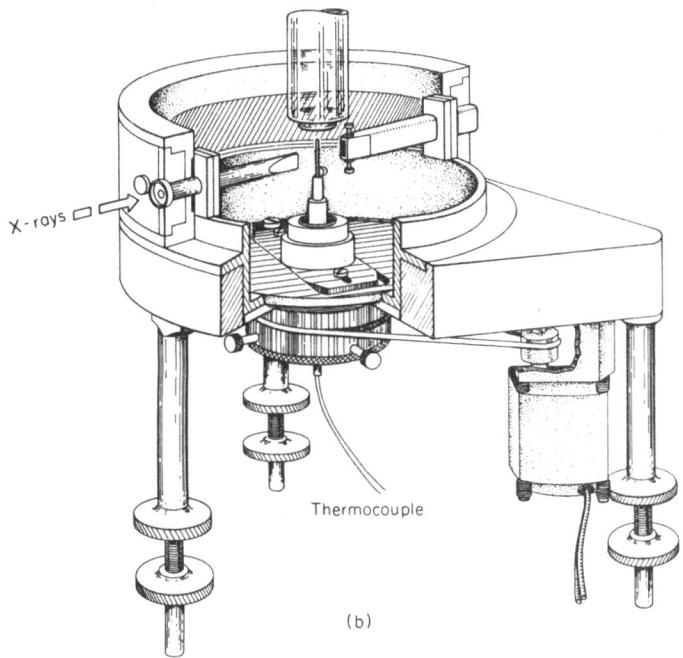

X-rays

Thermocouple

(b)

Figure 2. 19-cm-diameter Debye-Scherrer camera (one film holder removed).

two directions at right angles by means of adjusting screws located in the cylindrical walls of the pulley wheel. These two screws ensure accuracy of rotation around the specimen axis, while the cross-slide ensures that this axis is accurately located at the camera center.

Normally the cross-slide is locked in position by means of its mounting screws, but if it is desired to examine the surface of a cylindrical specimen, up to 1 in. in diameter, the specimen holder may be moved off-center until the X-ray beam just grazes the surface of the specimen, and in this position the specimen may be continuously rotated about its own axis. Since the round base plate may also be rotated on its flanged retaining surface, the displacement of the specimen is not limited to being at right angles to the beam, but may be made at any angle with respect to it.

The camera is not limited to cylindrical specimens, but may be used with X-ray samples in the form of plates. By means of an adjustable reversing switch in the circuit of the electric motor driving the pulley wheel, the specimen can be made to oscillate at any suitable speed through any required angle.

The specimen itself, in Debye-Scherrer work, is held in one of the holes of a thin ceramic thermocouple tube which is conveniently slid from above into the hollow stem of the specimen holder, the other hole carrying a fine copper—constantan thermocouple, the junction of which is held in contact with the specimen just below the point where the beam strikes. To prevent the thermocouple leads from twisting unduly, the direction of rotation of the specimen is reversed every 360°.

On account of the open construction of the camera, which makes the specimen readily accessible, it is possible to employ any one of many available procedures for cooling the specimen. For simplicity, the method developed by Fankuchen et al. was employed.[1,2] Dry gas, usually nitrogen, cooled by passage through a copper coil immersed in coolant in a Dewar flask, is passed down a central silvered vacuum-jacketed glass tube and over the specimen situated just below the orifice, every effort being made to ensure streamline flow. By keeping the gas stream both wide and smooth, the deposition of moisture on the sample is effectively prevented. A stream of dry gas at room temperature is passed through a concentric tube and forms an outer mantle of dry gas about the specimen and the main cooling stream, thus preventing the formation of frost. Temperatures may be controlled to within 5° down to −180°C by varying the rate of flow of cold gas or by admitting some dry gas at room temperature into the cold stream.

The film holder is a special feature of the camera. It is split along the beam axis into two symmetrical halves, either of which can be easily and independently removed from its location on the periphery of the circular frame without disturbing either the cooling system or the specimen. An advantage of this arrangement becomes evident when it is desired to take room-temperature diffraction patterns from a large unwieldy specimen. In such cases, it is possible to dispense with one film holder entirely and employ the apparatus as a "half camera," recording over the range of Bragg angles from 3 to 87° on one side of the beam only.

The camera in its present form has been successfully employed down to −180°C, and by using a long heated tube through which air was passed on to the specimen, it has been possible without difficulty to extend the range up to a modest 200°C. It would be a simple matter to design a special furnace mounting to extend the range still further, and also to incorporate a helium cryostat to get down to 4°K.

---

[1] Superscripts pertain to references at the end of the paper.

Exposure times naturally depend on the nature of the specimen, the radiation employed, and the slit dimensions which, in this case, are approximately 0.5×1.2 mm. Using a copper target in a demountable high-intensity rotating anode X-ray tube at 120 ma and 50 kv, excellent patterns of a metal can be obtained in as little as 20 minutes with filtered radiation, but normally exposure times run anywhere from 30 minutes to 1 hour.

APPLICATIONS

A great advantage presented by a camera of 19-cm diameter is the extra resolution which it affords as compared with smaller commercial instruments. For example, in the case of hexagonal silicon carbide, Mod II, which is based on the stacking of tetrahedral units, the axial ratio, in the case of perfect stacking, would be $c/a = 3 \times 1.632$. However, as shown by the Debye-Scherrer powder pattern taken with Fe K_α radiation, the 11212

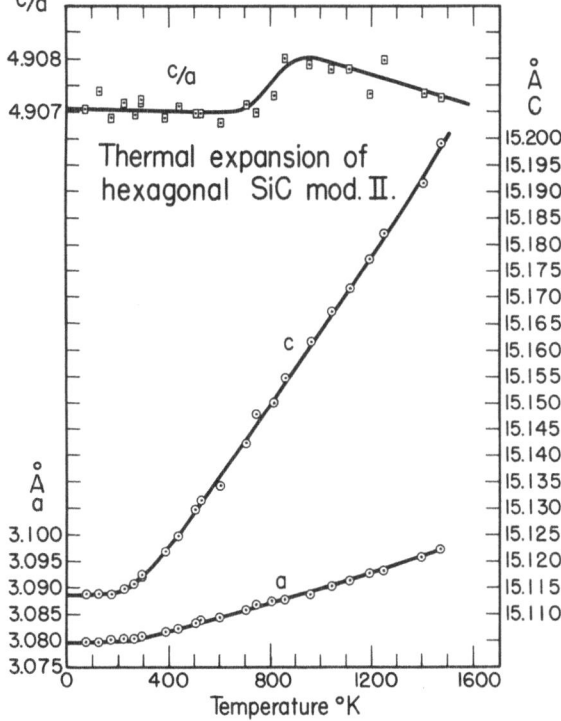

Figure 3. Lattice parameters and axial ratio of hexagonal SiC Mod II.

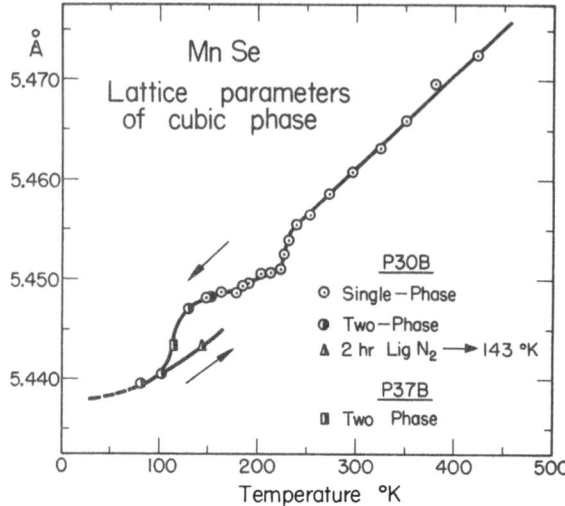

Figure 4. Polymorphic transformation and thermal
expansion of MnSe.

and 21$\bar{3}$4 lines do not overlap each other at $\theta \approx 84°$ as would be
anticipated, but split up into two parts of clearly resolved dou-
blets, measurement of which indicates an axial ratio $c/a = 3 \times$
$\times 1.6357$,[3] thus implying that the bond lengths or bond angles are
slightly different from the accepted values. Low-temperature
measurements show that this c/a ratio remains almost constant
at 4.907_0 down to $-190°C$, while extending the range in a 19-cm-
diameter Unicam high-temperature camera shows a sharp increase in c/a ratio in the region of 600°C to 4.909_0, falling
again to its original value at 1400°C approximately, as shown
in Figure 3.

In addition to thermal expansion work, a low-temperature
camera is useful in observing polymorphic changes taking place
at subzero temperatures. In the case of the cubic structure
MnSe,[4] specific heat and magnetic susceptibility measurements[5]
had indicated that some change was taking place at subzero temperatures, but the data were not in agreement with each other.
Diffraction patterns showed a marked break in the lattice parameter vs temperature curve at 250°K in agreement with the
peak observed in the specific heat curve. However, a second
(hexagonal) form of MnSe was not observed in the diffraction
patterns until the temperature had dropped to 160°K approximately, the specimen remaining two-phase down to 83°K. Apart
from a little hysteresis in the spacing of the cubic constituent,

the phase change is reversible and the material reverts to its original single-phase cubic structure at room temperature. above room temperature, the thermal expansion is almost linear to 450°K, the patterns being taken in the low-temperature camera after the Fankuchen cold-air jet had been replaced by a heater tube above the specimen. Further details will be published elsewhere.

ACKNOWLEDGMENT

The author wishes to thank Miss Brenda Kagle, Richard M. Jones, and Normal Doyle for most of the experimental data presented in this note.

REFERENCES

[1] H. S. Kaufmann and I. Fankuchen, Rev. Sci. Instr., Vol. 20, 1949, p. 733.

[2] B. Post, R. S. Schwartz, and I. Fankuchen, Rev. Sci. Instr., Vol. 22, 1951, p. 218.

[3] A. Taylor and R. M. Jones, Symposium on Silicon Carbide, Pergamon Press, Boston, 1959.

[4] K. K. Kelly, J. Am. Chem. Soc., Vol. 61, 1939, p. 203.

[5] R. Lindsay, Phys. Rev., Vol. 84, 1951, p. 569.

THE UNIVERSAL VACUUM SPECTROGRAPH AND COMPARATIVE DATA ON THE INTENSITIES OBSERVED IN AN AIR, HELIUM, AND VACUUM PATH

D. C. Miller and P. William Zingaro

Philips Electronics, Inc., Mount Vernon, New York

ABSTRACT

Equipment is described whereby comparative measurements were made on the relative absorption of air, helium, and vacuum for the softer radiations which are of interest in furthering the range of sensitivities of the X-ray spectrograph. Data on magnesium, aluminum, and silicon and shorter wavelengths are presented.

INTRODUCTION

Considerable interest has been expressed by users and potential users of X-ray spectrographic equipment in an instrument which would permit the use of a vacuum path all the way from the surface of the specimen to the window of the detector. While somewhat the same results are obtained if a helium path is used, the cost of helium and the recent period of helium scarcity has accentuated the need for a unit which does not depend upon the use of any gas in the X-ray optical path.

After several years of engineering and test, a vacuum path X-ray spectrograph is now commercially available. The first units will come from the Philips P.I.T. group in Eindhoven, Holland, where this development was undertaken. Such a spectrograph has been in test in our laboratory for about eight months and we have found decided advantages not only over the air path unit but also the helium path units. Before going into detail on these points, we shall describe the new spectrograph.

DESCRIPTION

The instrument, which incorporates several novel features in addition to the vacuum path, is shown in Figure 1.

The spectrograph proper, exclusive of the standard basic X-ray generator, electronic circuit panel, and goniometer, consists of:

    1. A specimen chamber housing the X-ray tube, a rotatable disk accommodating four specimen holders, a motor for

rotating two of the sample holders, and a preamplifier for the flow proportional counter. See Figure 2.

2. A drum, attached to the specimen chamber, which contains the main parallel-plate collimator, a two-position double crystal holder, and a flow proportional counter with its collimator and cathode follower. These are clearly shown in Figure 3.

3. An external 2θ scanning arm on which is mounted a collimator and scintillation counter. See Figure 2.

Complete freedom of choice of crystal—counter combination is possible. Change-over from one crystal to another, or from flow counter to the scintillation counter, is accomplished in less than 1 sec. Under vacuum conditions, successive positioning of the four samples for analysis is possible without breaking the vacuum. The vacuum required can be obtained with most commercially available mechanical forepumps. For volatile liquids, provision is made to seal off the specimen chamber at a position close to the specimen so that the vacuum path begins at a point less than $1\frac{1}{2}$ in. from the sample, thereby insuring a minimum of radiation loss due to air absorption. For routine analysis, involving both soft and hard radiations, one crystal and the scintillation counter can be adjusted independently of the second crystal and the flow counter. Change-over from one set to the other takes a fraction of a second.

Figure 1. The universal vacuum spectrograph.

Figure 2. Specimen chamber attached to the goniometer.

Since the unit is designed for vacuum operation, it can be used with a light gas filling in those cases involving the analysis of trace light elements in volatile liquids, and for which an air path of $1\frac{1}{2}$ in. would excessively absorb the excited X-radiation.

PERFORMANCE

In evaluating the performance of the vacuum spectrograph, the most important factors to be considered are:

1. The time required to produce the desired vacuum.
2. The stability of pressure to be maintained for precision measurements of X-ray intensities.
3. The intensities to be realized (as compared to an air or a helium path) for the longer wavelengths.

Time. The time required by both the vacuum and helium systems to reach proper operating conditions starting from an air path is plotted in Figure 4. In addition to the advantage of shorter start-up time, the vacuum instrument permits the use of a vacuum gauge which constantly measures radiation path

Figure 3. Closeup of drum interior.

conditions. Return to normal atmospheric pressure takes about
50 sec. Air is admitted through a carefully designed port so
that the thin collimator plates will not be damaged by the in-
rushing air.

Stability/Vacuum. The stability required depends on the
wavelength of the radiation detected and the required precision
of the intensity measurements. As an example, let us assume
we have a vacuum of 500 μ Hg. Now, if the pressure varies by
$\pm$ 10%, a very large variation, then the intensity of various radiations will vary as listed in Table I.

Table I

Wavelength, A	Element	Pressure range, μ Hg	Variation in intensity, %
4.7	Cl	450-550	0.0
5.36	S	"	0.1
6.14	P	"	0.2
7.11	Si	"	0.3
8.32	Al	"	0.4
9.87	Mg	"	0.7

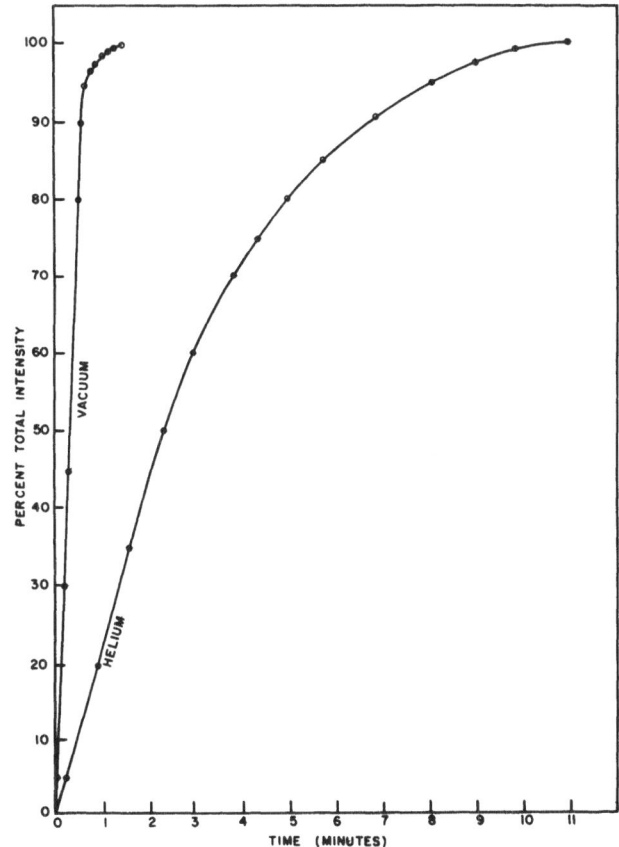

Figure 4. Times required to reach operating condition.

Table II

Wavelength, A	Element	Calculated	I_{vac}/I_{He} Experimental
3.35	Ca	1.00	1.00
5.36	S	1.02	1.00
6.14	P	1.04	1.03
7.11	Si	1.07	1.07
8.32	Al	1.13	1.11
9.87	Mg	1.24	1.18

It can be seen that the effect on intensity of a large variation is almost negligible even for the softest radiation.

The percentage transmittance of the *K* radiations of the elements listed in Table I are shown in Figure 5 as a function of pressure in microns of mercury.

Intensities. The intensities of the radiations from 3- to 10-A wavelength will be greater in a vacuum than in a helium atmosphere unit. The calculated increase as well as the experimental data are shown in Table II.

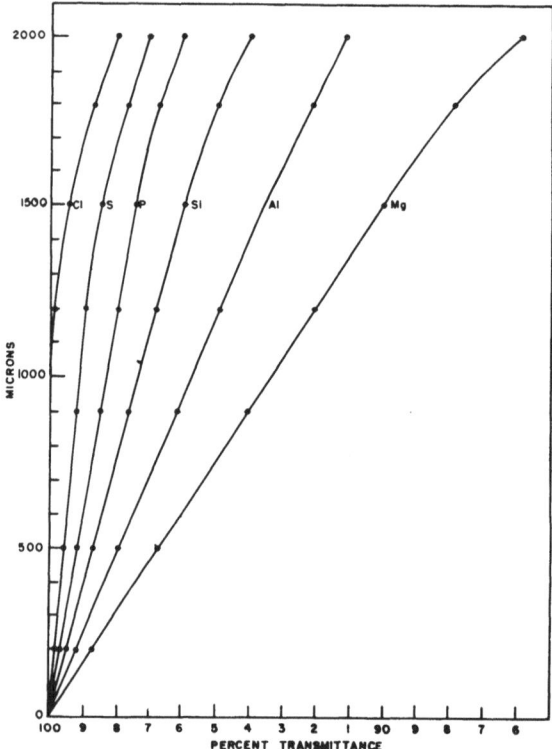

Figure 5. Transmittance of *K* radiations as a function of pressure.

Figure 6 shows a plot of percentage transmittance versus wavelength in angstroms. Since we could not find the absorption coefficients of air for X-radiations from 18 to 44 A, no interpolation was attempted in this range.

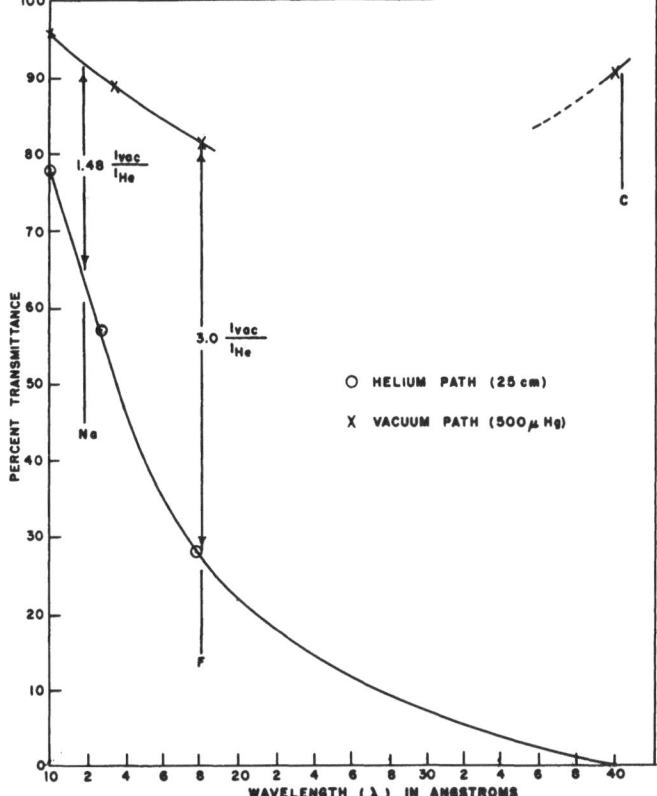

Figure 6. Variation in transmittance with wavelength.

Even the heavier elements, whose characteristic K or L lines are in the 1-A region, will benefit significantly intensity-wise in a vacuum (or helium) system. Table III shows percentage increase of intensity as a function of X-ray wavelength.

CONCLUSIONS

Thus it will be seen that this new vacuum unit offers performance as good or better than the standard helium units as regards intensity and stability. Pump-down time is quicker than a helium unit starting from an air path and is comparable to the time required for a helium system to come to equilibrium after insertion of a new sample. The fact that four samples can be inserted simultaneously means the 2-min pump-down time actually averages about 30 sec per sample, which compares very well with the helium case. When we add the extra intensity available for the harder elements, the ease of changing crystals and the ease of changing detectors, we are sure this new vacuum unit will be a welcome addition to any X-ray spectrographic laboratory.

Table III

Wavelength, A	Element	Atomic No.	Line	% Increase I vacuum or He path over air path
1.04	Br	35	K	20
1.1	Bi	83	$L1$	23
1.4	Zn	30	K	32
1.47	W	74	$L1$	35
1.54	Cu	29	K	38
1.57	Hf	72	$L1$	39
1.66	Ni	28	K	42
1.67	Yb	70	$L1$	43
1.79	Co	27	K	47
1.84	Ho	67	$L1$	50
1.94	Fe	26	K	54
2.04	Gd	64	$L1$	57
2.1	Mn	25	K	62
2.19	Sm	62	$L1$	66
2.29	Cr	24	K	72
2.46	Pr	59	$L1$	82
2.5	V	23	K	87

for the harder elements, the ease of changing crystals, and the ease of changing detectors, we are sure this new vacuum unit will be a welcome addition to any X-ray spectrographic laboratory.

THE NORELCO PORTABLE SPECTROMETER (PORTOSPEC)

D. C. Miller

Philips Electronics, Inc., Mount Vernon, New York

ABSTRACT

A light-weight portable X-ray spectrograph is described in some detail. The analyzing head, weighing only 18 lb, contains a goniometer, ratemeter, and Geiger counter detector. It also contains a special window-type X-ray tube with a gold target which has been developed specifically for this spectrograph, and which provides a very high efficiency of excitation of the sample. An associated power supply contains an rf source providing 40 kv at 60 μa together with the necessary voltages for the rectifier filaments, the X-ray filament, and the Geiger counter voltage. It will detect wavelengths from atomic number 23, vanadium, to atomic number 50, tin, and then using the L series from atomic number 58, cerium, to atomic number 92, uranium. Sensitivity and some applications are discussed.

INTRODUCTION

The latest addition to the Norelco line of X-ray instruments is a junior member of the family, which we have named the Portable Spectrometer, or the Portospec. This unit consists of two components, one an X-ray "reading head" with which the analysis is done, and the other a power supply which provides the voltages required to make the unit work. The unit is shown in Figure 1.

READING HEAD

The heart of the unit is a specially developed tube known as the 2.5 FA-50 X-ray tube. This X-ray tube has a special construction as can be seen in Figure 2. The source of the X-rays is a thin layer of gold on the back surface of the beryllium window target. When the electrons strike this gold film, they generate the gold L spectra as well as an intense white spectrum, both of which are then transmitted through the beryllium window onto the actual specimen to be analyzed. This type of X-ray tube geometry obviously permits a very large solid angle of radiation to be used and as a result provides very efficient X-ray excitation for the power which is used in the X-ray tube.

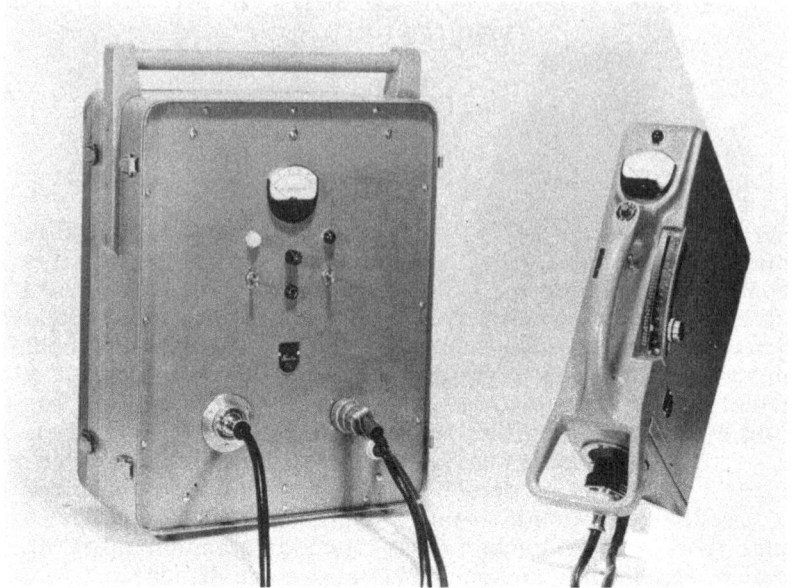

Figure 1. Power supply.

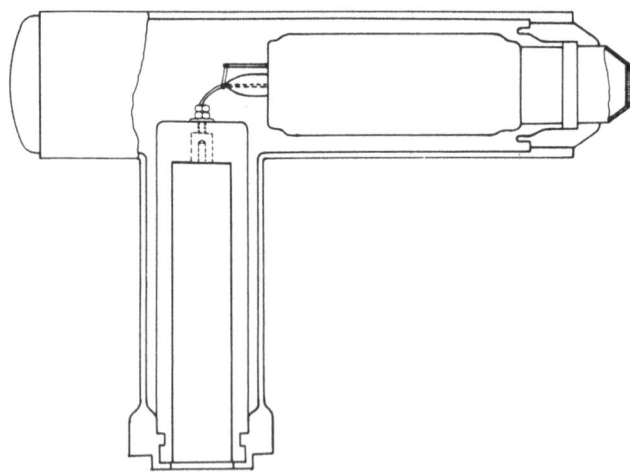

Figure 2. Specially designed X-ray tube.

This tube was specifically developed for this particular equipment and has the clean spectrum required for the spectrographic
X-ray tubes. As can be seen in Figure 3, the X-ray tube fits
into the reading head at a slight angle so that the incident rays
strike a flat specimen surface and the excited fluorescent radiation is taken off at about a 20° angle. Using flat-plate-col-
limator—flat-crystal optics, we can measure the fluorescent
radiation coming from the specimen. The actual X-ray optical
arrangement consists of two collimators and a lithium fluoride
analyzing crystal. The first collimator is $1\frac{1}{4}$ in. long and consists of 2-mil plates with 5-mil spacing. The lithium fluoride
crystal is about 1 in. square and this is followed in the optical
path by a second collimator 1 in. long of 2-mil plates spaced
10 mils apart. The X-rays are detected in an Amperex 100C
argon-filled Geiger-Mueller counter tube mounted on the goniometer arm which carries the second collimator. The goniometer is a simple angle-measuring device where two-to-one
gearing provides the appropriate matching between crystal and
detector motion. A thin metal bar attached to the crystal shift
acts as a lever to move the goniometer through the angular
range which it is able to cover. As is shown in Figure 4, at the
end of the bar is a scale and along the side is a clamp so that
the crystal angle can be read and the bar clamped securely at
a particular crystal angle. The scale can be shifted to allow for
calibration and is labeled so that the goniometer can be set by
elements or by angular position. The counter has its optimum
sensitivity in the copper K region but has reasonable sensitivity
all the way from vanadium to silver radiation.

The ratemeter, fed from the output of the Geiger-Mueller
counter, is a simple current-integrating device which displays
the mean current on a meter calibrated in counts/sec. The ratemeter has two ranges, full scale 1000 counts/sec and full scale
100 counts/sec, and incorporates time constants corresponding
to 1 sec for the 1000 counts/sec scale reading and 5 sec for the
100 counts/sec scale reading.

For radiation protection, microswitches have been provided
on either side of the specimen radiation opening in the bottom
of the reading head. It is necessary that both of these switches
be depressed before the X-rays can be turned on by a third
switch in the handle of the reading head. In addition, there is a
spring-loaded sleeve which comes down on any flat area and
gives protection until the microswitches open. The reading head
is 19 in. long, 13 in. high, and 5 in. wide, and weighs approximately 18 lb.

Figure 3. Reading head.

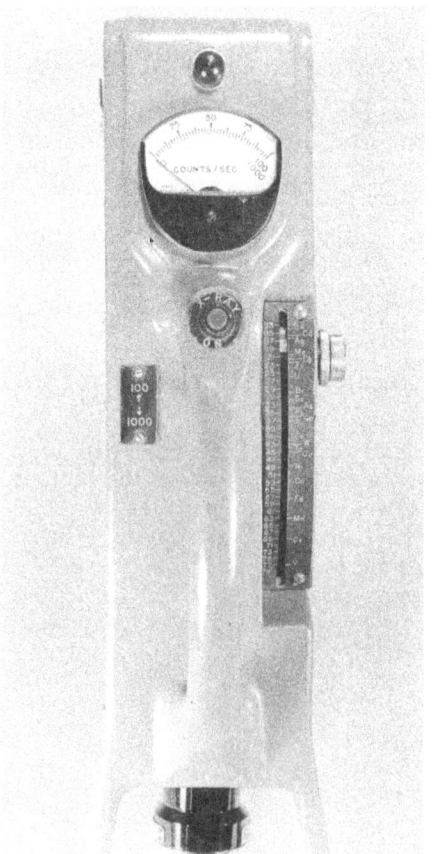

Figure 4. Selector scale.

POWER SUPPLY

The power supply, which provides the voltages for the reading head, consists basically of rf voltage supply and suitable regulators. The rf voltage supply provides, first of all, the high voltage for the X-ray tube. The tube can use up to 40 kv at 60 μa. Incorporated in the supply is a switch which enables one to choose half power (30 μa) for the tube when the ratemeter is driven off scale by the readings obtained at full power. The high voltage is smoothed and kept at constant potential with a ripple percentage of less than 0.1%.

The rf supply also provides the 1200 v dc used on the Geiger-Mueller counter and provides, by means of suitable coupling, filament supplies for the X-ray tube and for the high-voltage rectifier tubes. Included in the same power supply chassis is a voltage regulator which will maintain a voltage of $\pm 1\%$ for 10% change in line voltage and a current regulator having the same specifications. On the front panel of the power supply is the X-ray tube current and the controls, including a power switch, and the X-ray "ON" switch, which also switches between the two ranges of power previously mentioned. The unit will run off a 110 v, 50 to 60 cycle power and has a total power consumption of about 200 w. It can be provided with a battery converter for cases where the unit will be used in a mobile vehicle, in which case it can be run from a normal 6- or 12-v storage battery. The power supply is 17 in. wide, 23 in. high, 11 in. deep and weighs about 48 lb.

It is envisioned that this power supply will normally be used on a special dolly mounted on casters enabling it to be moved around readily. The dolly will also provide space for carrying along suitable standards with the instrument. Shock-proof connecting cables between the power supply and the reading head carry the high voltage and the other voltages necessary for the ratemeter on the reading head.

SPECIMEN REQUIREMENTS

The specimen requirements vary according to the execution of the instrument chosen. For the standard unit, the specimen must be a flat sheet solid and the minimum specimen area which is irradiated is a 1×3 in. rectangle. The standard unit can be provided with adaptors which provide a variety of possibilities as far as specimens are concerned. Two of these adaptors are shown in Figures 5 and 6.

The drawer adaptor, fitting on the underside of the unit, makes it possible to use specimens approximately $1\frac{3}{8} \times 1\frac{15}{16}$ in. and

Figure 5. Drawer adaptor.

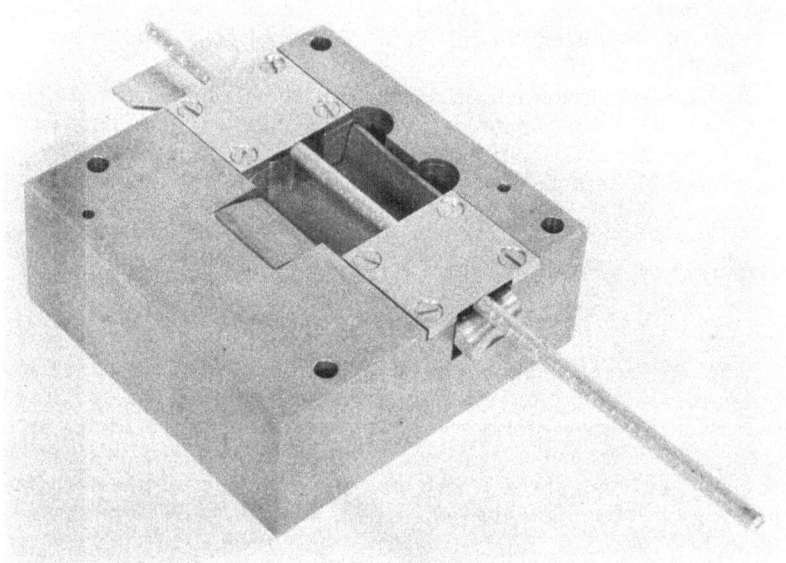

Figure 6. Small rod adaptor.

up to $\frac{1}{2}$ in. thick, including solid powder pellets or even liquids contained in a cup of these general dimensions. The specimen is inserted in a drawer which fits into the side of the adaptor and moves the specimen under the X-ray tube window. Loading of the specimen is always perfectly safe since there is no X-radiation until the button on the top is pressed.

The second type of adaptor is known as the welding rod, or small rod stock adaptor. This device, which permits the choice of holes from $\frac{1}{16}$ to $\frac{5}{16}$ in. in diameter, can be positioned underneath the X-ray tube window and permits rods to be positioned and analyzed at any point along their length without the operator being exposed to a significant amount of radiation. If used with coated welding rods, obviously, the coating on the rods must be scraped at the point where the analysis is going to be made. Any solid rod material can be analyzed in this particular setup, including such things as bolts, nails, drill stock, and any other type of stock less than $\frac{5}{16}$ in. diameter.

The third adaptor is known as the round stock adaptor and depends in size upon the stock which it is desired to measure. It can be made to take stock from $\frac{1}{2}$ in. diameter up to flat sheet and is designed principally to measure solid cylindrical stock material. Other adaptors will be designed and made as required.

RANGE OF ELEMENTS

The goniometer covers the angles from 14 to 78° 2θ. This means that K lines of elements No. 23 to 47, vanadium to silver, and L lines of elements No. 58 to 92, cerium to uranium, can be excited. The detector is not particularly sensitive to the extremes of these ranges and is best suited to handle elements in the midregion of the range.

APPLICATIONS

A most interesting aspect of this instrument is in the applications one can find for it and how it performs in the chosen applications. We, unfortunately have not had the unit long enough to make exhaustive tests as to what it will do, but we have a few representative examples of data which we have obtained using the instrument and they show some of the things which the unit is capable of doing.

Table I shows data obtained with low-level components in steels. As can be seen, the instrument is capable of detecting less than 1% of chromium, and 1% levels of nickel and manganese.

Table I.   Low-Level Components in Steel
On 100 scale - 1 div. - 2.5 counts/sec

Element	%	P/B (div. 100 scale)	Net div.	Net counts/sec	Counts/percent
Mn	1.3	17/4	13	32	24
Ni	1.73	15/5	10	25	14
Cr	0.74	13/5	8	20	27

Table II shows how the instrument can be used to identify major components and thereby sort out complex alloys, such as stainless steels. By measuring two elements in stainless steels, chromium and nickel, a number of types can be sorted out. Types 309, 308, and 301 were sorted out from two samples of Type 302 and two samples of Type 304, which are both 18-8 types. The data show the 302 and 304 intensities to be similar. Type 309 can be identified by the high chrome (22%). Type 301 has about the same chrome as the 302 and 304 but has much lower nickel (6%). Type 308 has high chromium and high nickel. In some cases there is enough variation from the usual 2% manganese to use manganese as the identifying element as well. Obviously, for each set of steels to be sorted, an appropriate pattern of data must be worked out, but it appears to be possible to handle many different types.

Table II.   Identification of Stainless Steel Types

Type	% Cr	Cr counts/sec	% Ni	Ni counts/sec	Div. on graph
301	17.08	310	6.81	66	26.5
309	22.86	385	Not available	Not available	X
308	20.42	350	9.78	90	36
302	17.70	330	8.71	80	32
302	17.92	310	8.86	82	33
304	18.39	320	9.18	79	31.5
304	18.58	335	9.24	80	32

Table III shows the results obtained from three iron ore samples. In this case, we chose samples having ranges of composition from 20 to 60%. The data show that we can detect this variation in composition very easily.

Table III. Data on Iron Ore Pellets

% Fe	Intensity		
	Peak	Bkgd.	Net
50.5	550	10	540
60.5	650	7	643
30.0	425	5	420

In addition to analytical problems, we have tried to use the unit to measure plating thicknesses, and in Figure 7 we see the results obtained with a series of tin plate specimens. We are equally certain that quite satisfactory results can be obtained from chromium, nickel, cadmium, or any other coating on an iron base.

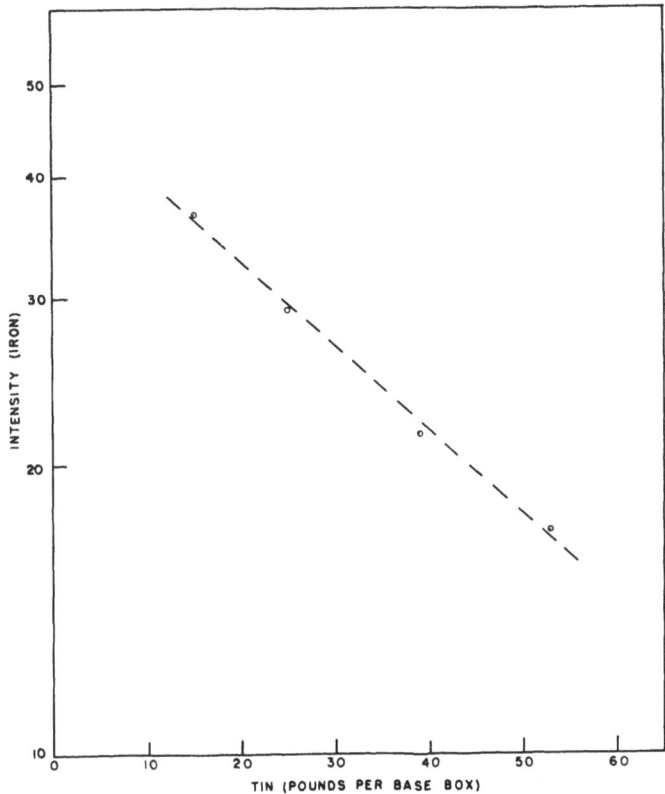

Figure 7. Tin plating thickness data.

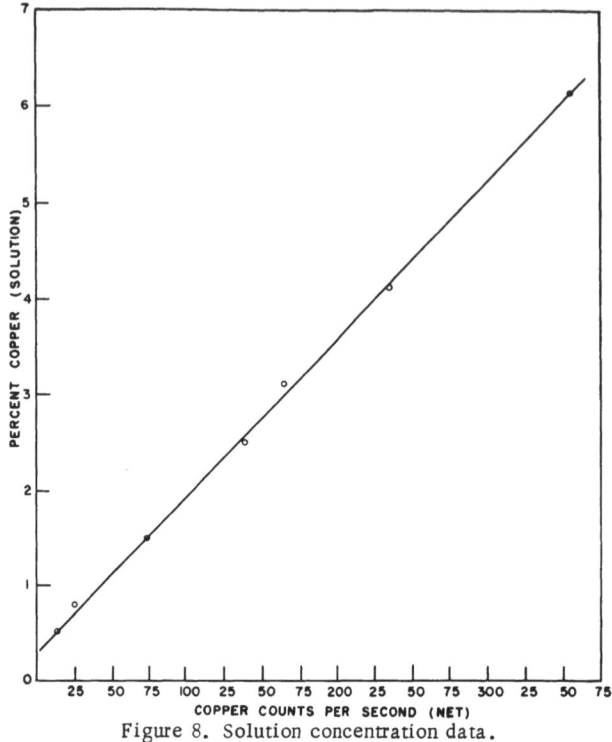

Figure 8. Solution concentration data.

In addition to the work using solid samples, a liquid holder was used with the drawer-type adaptor and curves were made using successively dilute solutions of copper sulfate. As can be seen in Figure 8, using this liquid sample, it was possible to go to less than 1% of copper in the solution, with good peak-to-background ratio. Finally, as shown in Figure 9, some curves were prepared using 0.0005-in. sheets of Mylar placed on an iron plate and the absorption in the Mylar film of the excited iron radiation was measured using an increasing number of sheets of Mylar. The fact that a straight-line calibration is not observed is not due to statistical variation in the counts or inaccuracies in the instrumentation but actually represents variation in thickness of the individual sheets of the Mylar. This we proved to our own satisfaction by repeating the curves placing Mylar sheets in the X-ray beam in the same order and then by changing the order and getting different curves, which once more departed from a linear curve but at different points. This leads us to believe that we can measure the variation of thickness of thin foils of relatively low absorption power.

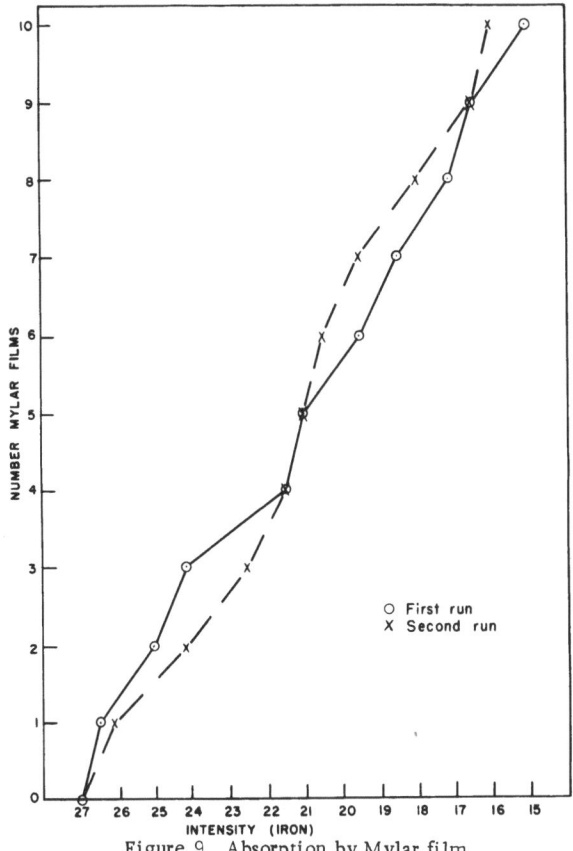

Figure 9. Absorption by Mylar film.

RELIABILITY AND MAINTENANCE

The unit has been developed with the idea of maximum reliability and minimum maintenance. A minimum number of tubes has been used in order to get the most reliable performance. The electron tubes which are used are all standard radio- or television-type electron tubes. The rectifier tubes are inexpensive, and if there is need to replace them, they can be replaced from any television repair shop.

ACKNOWLEDGMENT

I wish to thank Dr. A. A. Sterk, under whose supervision this development was carried out, and Mr. P. W. Zingaro, who assisted me in obtaining the data for this paper.

AN X-RAY CAMERA FOR PRECISION LATTICE PARAMETER MEASUREMENTS

A. Franks

National Physical Laboratory, Teddington, Middlesex, England

ABSTRACT

A back-reflection focusing camera has been constructed in which instrumental errors, including those due to uneven film shrinkage, have been eliminated or very much reduced. This has been achieved by employing metrological techniques both in construction and calibration. The temperature of the camera can be controlled to 0.05°C. The form of the specimen is such that horizontal and vertical divergence cause no line shift. The X-ray source is a demountable semimicrofocus tube with an adjustable focus. The radiation is monochromatized by a bent and ground quartz crystal.

INTRODUCTION

Several techniques are now available for the accurate measurement of lattice parameters and among the more elegant are single-crystal measurements using divergent beam techniques. From the practical point of view, however, the most convenient and usually the only possible method requires the use of a powder specimen. A further choice is open to us: powders can be examined photographically or with diffractometers.

Development of diffractometry during recent years has created a renewed interest in precision and accuracy. The introduction of a second method of measurement has raised the question of the relative merits of the older and newer methods, and it is of interest to determine the limits of precision attainable with a camera. Other reasons which prompted this research are, firstly, the request from the International Union of Crystallography that the Laboratory take part in the general scheme for comparing parameter measurements on a standard material by different laboratories and, secondly, the need for a guide to manufacturers of X-ray cameras. Several British laboratories have had measurements made of the diameter, circularity, and concentricity of their X-ray cameras by the Standards Division of the National Physical Laboratory. Results of the measurements show that there is a definite need to improve on commercial tolerances for work of high precision.

THE SYMMETRICAL BACK-REFLECTION
FOCUSING CAMERA

The back-reflection focusing camera offers one of the best means of eliminating systematic errors in the position of the diffracted lines. The main factors which give rise to errors are (a) errors in measurement of the angle of diffraction, (b) effects of uneven film shrinkage, (c) horizontal and vertical divergence of the beam, (d) missetting and absorption in the specimen, (e) Lorentz and polarization factors, and (f) wavelength distribution of "monochromatic" radiation.

These errors and their elimination are best discussed in conjunction with the focusing camera shown diagrammatically in Figure 1. The camera consists of two stainless steel rings with an internal diameter of 5.7 in. and a wall thickness of 2 in. The height of each ring is 1.5 in. Metrological experience has shown that this type of massive construction is needed to ensure circularity of the bore to an accuracy of $1\,\mu$. Water channels within the walls of the camera provide a means of controlling the temperature to 0.05°C at 20°C, the temperature at which the camera was calibrated. The film is held in position by a spring which forces it centrifugally against the inner bore of the camera — a method advocated by Buerger.[1] There is no measurable gap between the emulsion and the bore.

Under normal conditions of photographic processing, we have shown by printing scales on X-ray film that there is a significant irregularity in film shrinkage which may exceed $30\,\mu$ over 1 cm of film. Extrapolation of measurements based on the supposition of even film shrinkage can thus not be tolerated over lengths greater than 1 cm for work of the highest precision. In order to eliminate this error and to facilitate angular measurement of the diffraction lines, fiducial grooves were cut at 1° intervals in the raised edge of the lower cylinder, as shown in Figure 1. An illuminating system consisting of four external lamps and a lucite light conductor reproduces the grooves on the film, which thus act as a scale; only the emulsion in contact with the camera wall is developed. If sufficient care is taken in the production and calibration of the grooves they provide a scale of accuracy surpassing that of conventional diffractometers. Figure 2 shows the camera in position on a circular dividing table of the highest precision, having an angular accuracy known to be better than 2 sec of arc. This accuracy has been transferred to the camera by means of a cutting tool

---

[1] Superscripts pertain to references at the end of the paper.

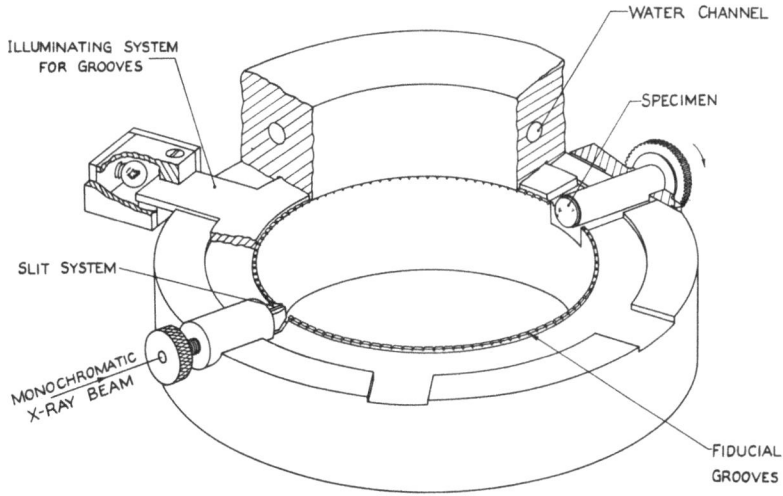

Figure 1. Diagram of focusing camera.

Figure 2. The camera and angular dividing table.

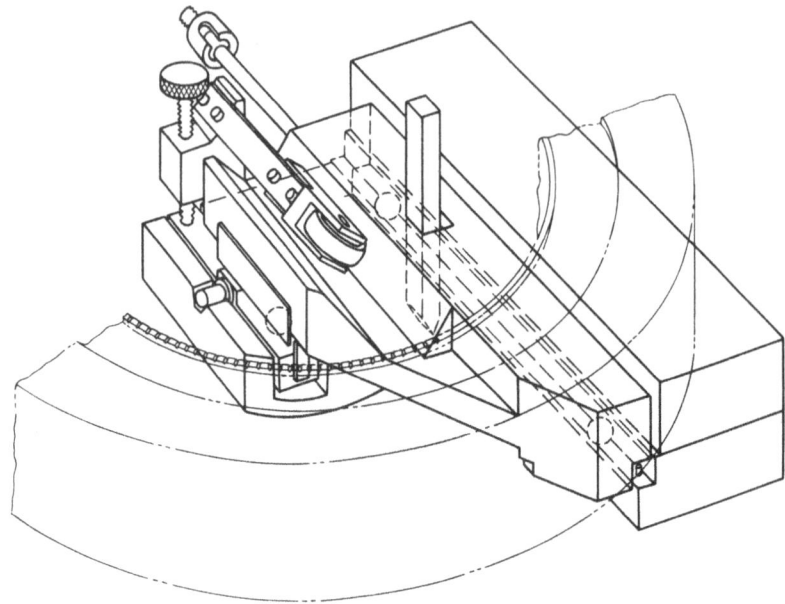

Figure 3. Diagram of cutting tool.

in a kinematic slide. Great care was taken in the design and use
of the cutting tool to ensure that any lateral shifts during the
cutting process did not exceed a fraction of a micron. A diagram
of the tool and its slide is shown in Figure 3. The fiducial lines
are 60° V grooves with a depth of 0.005 in. On the far side of
the rotating table in Figure 2 is an air gauge which controls and
measures the concentricity of the camera and the axis of rotation during the cutting process.

Errors arising from the horizontal and vertical divergence
of the beam have been eliminated by using a specimen, in the
form of a powder compact, whose reflecting surface forms part
of a concave spherical segment, the diameter of the sphere being
equal to the diameter of the camera. Perfect focusing on the
equator of the camera is thus obtained from rays emanating
from a point on the equator. This condition is effectively achieved
by using rays from a semimicrofocus tube reflected from a
bent and ground quartz monochromator. The vertical divergence
is controlled by the width of the adjustable X-ray focus. A vertical and horizontal slit system in the bore of the camera elim-
inates the unwanted radiation and controls the height of the beam.

The specimen surface is correctly set with respect to the
bore by a ball-ended optical indicator. The datum position of

the ball is determined by placing the indicator against the bore of the camera. It is then swung round to contact the specimen which is adjusted by a micrometer movement to give the same indicator reading. The effect on line position due to absorption of the beam could partially be compensated by a small adjustment of the specimen; its effect is eliminated by extrapolation to $\theta = 90°$.

The errors resulting from the Lorentz and polarization factors can be eliminated by extrapolation of the peaks of the reflections[2] but not by the more popular method of extrapolation of centers of gravity.[3]

The effect of the spectral width of monochromatic radiation will be referred to in the discussion.

Temperature control is effected by a mercury-in-glass contact thermometer which regulates a heater in the circulating water system. This method is used by the Thermometry Section at N.P.L. to achieve constancy of temperature to 0.01°C. The camera is enclosed in a wooden box which provides thermal insulation.

DESCRIPTION OF THE APPARATUS

For the sake of mechanical stability the design of the X-ray equipment was conceived as a whole: the X-ray tube, monochromator, and camera forming one unit.

The X-ray tube is fundamentally a fine-focus tube of the Ehrenberg-Spear type[4] and incorporates the modification for a line focus described by Brech and Stansfield.[5] The increased width of the focus results in increased vertical divergency, which is desirable for reasons of intensity. The body of the tube is rectangular and is maintained at room temperature by water cooling. The position of the center of the anode is known with respect to the external faces and the axis of the vertical pumping tube passes through the focal spot. This facilitates the setting up of auxiliary apparatus.

The pumping tube is clamped to the side of a large surface plate which acts as a base for the camera. The surface plate stands on a slab of concrete 4 in. thick, which is in turn supported by an angle iron frame.

A diagrammatic arrangement of the adjustments is shown in Figure 4, the movements being indicated by arrows. The monochromator unit slides in a groove machined in the surface plate giving a take-off angle of about 6°. A fine adjustment (a) is provided in this direction and the distance between source and center of monochromator is indicated on a vernier scale.

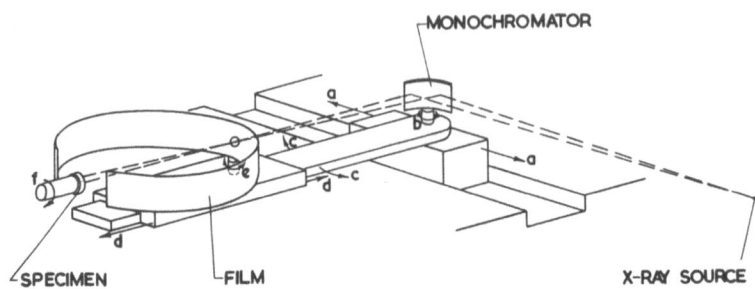

Figure 4. Diagrammatic representation of the adjustments.

Figure 5. The X-ray tube, monochromator, and camera.

A micrometer movement (b) is used for the angular setting. The monochromator can be rapidly set up and the focal position found by observing the image of the focused X-ray beam under high magnification (×200) on a high-resolution fluorescent screen — a single crystal of cesium iodide, activated with thallium.[6] Once the monochromator has been set, the camera is rotated (adjustment c) about the center of the monochromator until the beam passes through the entrance slit; a linear adjustment (d) positions the bore at the focus. Finally the camera is rotated (e) about the focus until the specimen is irradiated symmetrically by the beam. Since the specimen is spherical, a simple rotation (f) suffices for scanning. Figure 5 is a photograph of the apparatus; the wooden box which normally surrounds the camera has been removed for the sake of clarity.

DISCUSSION

If the metrological precision of the camera were the only limiting factor, lattice parameters should be obtained to 1 part in 10^6. This figure does not however take into account difficulties associated with the spectral distribution of wavelengths and the difficulties in measuring the position of the X-ray lines on the film. Two problems are involved in the question of wavelength. Firstly, a range of wavelengths exists in "monochromatic" radiation and, secondly, it is doubtful whether the wavelengths are known to a sufficient accuracy.[7] The difficulties in measuring the line peak positions to 1μ are large but a differential microphotometer based on an electronic scanning technique[8] has recently been developed at N.P.L. for measurements to this accuracy, on line standards, and promises to be useful for X-ray applications.

Examination of lattice parameter measurements reported by different laboratories shows that consistency of results at any one laboratory can be of a high order but comparison of measurements made by different laboratories reveals large unexplained discrepancies. It would appear that this may be due to the customary approach in which it is assumed that all errors are eliminated by suitable extrapolation. This is obviously not the case — the precision which can be obtained in lattice spacings must depend fundamentally on the accuracy of the angular measurement of the diffraction angles and this requires the use of sophisticated metrological techniques.

ACKNOWLEDGMENTS

The work described above has been carried out as part of the general research program of the National Physical Laboratory and this paper is published by permission of the Director of the Laboratory.

The author desires to acknowledge the advice and assistance rendered by his colleagues of Standards and Basic Physics Division.

REFERENCES

[1] M. J. Buerger, "X-ray Crystallography," John Wiley and Sons, Inc., New York, 1942, p. 168.

[2] B. Gale. Private Communication.

[3] E. R. Pike and A. J. C. Wilson, Brit. J. Appl. Phys. Vol. 10, 1959, p. 57.

[4] W. Ehrenberg and W. E. Spear, Proc. Phys. Soc., London, B, Vol. 64, 1951, p. 67.

[5] F. Brech and J. R. Stansfield, Proc. 6th Annual Conference on Industrial Applications of X-ray Analysis, Denver, 1957, p. 17.

[6] A. Franks, Proc. Phys. Soc., London, B. Vol. 68, 1955, p. 1054.

[7] H. S. Peiser, H. P. Rooksby, and A. J. C. Wilson, "X-ray Diffraction by Polycrystalline Materials," The Institute of Physics, London, 1955, p. 652.

[8] A. H. Cook and R. G. Hitchins, J. Sci. Instr., Vol. 36, 1959, p. 337.

THE ESTABLISHMENT OF A Q RATING FACTOR FOR DECORATIVE CHROMIUM PLATE BY X-RAY FLUORESCENCE

William C. Keesaer

Buick Motor Division, General Motors Corporation,
Flint, Michigan

ABSTRACT

Procedures and methods employed to develop an X-ray emission spectrometer technique for determining the quality of decorative chromium plate are presented. Individual plate quality factors and their significance are discussed:
1. Plate thickness
2. Plate density
3. Plate corrosion resistance.

Geometric instrumentation changes to the General Electric XRD-5S spectrometer are detailed. These alterations have made it possible to take more complete advantage of the nondestructive nature of the fluorescent method, and have also yielded an increase in useful radiation intensity of 50%.

The unsuccessful attempts to employ commercially plated Hull-cell-type test panels as "standards" are described, along with the satisfactory application of fabricated standards of vacuum vapor-deposited high-purity chromium, laminated to spectrographically pure nickel foils. The merits and limitations of the resultant "absolute purity — absolute density" standard curve are shown and are related to other measurement techniques and accelerated corrosion test data.

INTRODUCTION

In the highly competitive automobile industry, it is well recognized that each piece of gleaming exterior chromium on our products is a potential source for a future owner's proud satisfaction — or his angry resentment. If his new car's chromium trim remains smooth and bright, withstanding the rigors of de-icing compounds and industrial atmospheres during at least three years of service, then he will (other factors being equal) look with favor on that same manufacturer when he again considers a new car. But if his car's chromium becomes pitted or blistered during his first year of ownership, that customer will not be easily persuaded to return to the same manufacturer for his next new automobile.

Because he has a duly justified regard for the almighty customer, it is necessary for the manufacturer to develop answers to the following questions:

1. How may we reliably predict the actual service performance for a given plated exterior part?
2. How may we certify that the desired degree of plate protection is present without insulting economic propriety?

To date, the above questions have only been partially resolved. Perhaps the most acceptable approach, now widely employed among General Motors' automotive divisions, is the 16-hour, copper chloride modified, acetic acid salt spray test (CASS test).[1] This accelerated corrosion test yields a fair measure of correlation with one year's environmental exposure in Detroit, Michigan. The CASS test was developed by General Motors Research Staff with the aid of several G. M. Divisions. Many current specifications require plated parts to be corrosion-free after 16 hours exposure to the CASS test fog spray at 120°F.

In addition to specifying acceptance requirements based on the accelerated corrosion test, it is also necessary to rely on some measurement process to insure that a minimum thickness value for each plate and a minimum total thickness value for the combined copper—nickel—chromium plates are provided. These measurement techniques usually require part destruction, whether the nature of the procedure is wet chemical, metallographic, or by microinterferometer technique.[2] From the standpoint of accuracy the interference microscopic method is most desirable for chromium plate since it affords little opportunity for human error. Wet chemical processes are not generally applicable to quality control thickness evaluations of tri-plated specimens because they are either impossibly slow or of debatable reliability. Some chemical techniques for single plate deposits, such as Metal and Thermit Company's dissolution and titration procedure for determining chromium thickness, have shown excellent reliability.

The metallographic microscope technique for plate thickness measurement is normally conducted at magnification factors of 750-1000× and employs a filar micrometer. The accuracy of this method is dependent on careful sample preparation, and is not usually considered satisfactory for the chromium determination when the plate thickness is less than 20 μ in. Plate evaluation by metallographic examination does offer some unique advantages, however. Stress cracking and plate structure may be apparent and possibly, contaminants and defects in the base

---

[1] Superscripts pertain to references at the end of the paper.

material may also be visible although sample area is normally quite limited.

X-ray fluorescence has been applied to the measurement of plate or cladding thicknesses for many years, and this approach holds certain advantages over the afore-mentioned techniques. Most significant among these advantages are: The X-ray methods are basically nondestructive and are potentially fast enough to provide an economical and continuous plating quality control process.

It is clear that a great volume of information is contained in the stream of characteristic and scattered radiation coming from a tri-plated specimen under irradiation. It is not clear how this information should best be sorted, classified, and interpreted in terms of plate quality. With commercially available X-ray spectrometers, only a portion of the available information is directly observed. In the case of a tri-plated steel sample, the individual and combined absorbing thicknesses will define the limitations of the X-ray technique. When these physically restricting factors are added to those imposed by instrumental

Figure 1. XRD-5 fluorescent spectrometer before modification.

limits, the X-ray method is unfortunately retarded from widespread commercial application. At present, the X-ray fluorescent technique for decorative tri-plate appears limited to the measurement of the chromium and the underlying nickel, providing their combined thickness does not exceed 0.00083 in. Although the theoretical upper limit for nickel measurement by the absorption technique has been reported as 0.0013 in.,[3] the method appears to lose sensitivity for thicknesses greater than about 0.00083 in.

Since commercial specifications for nickel plate thickness frequently exceed 0.001 in., the external chromium survives as the only fit subject for X-ray fluorescent evaluation on our tri-plated specimen if a nondestructive method is mandatory. Fortunately, the chromium layer has recently achieved greater significance as an inhibitor of corrosion due to the duplex chromium plating process development.[4] Now chromium plate can no longer be regarded as simply a bright, glamorous cover-up for the true corrosion protector – the nickel deposit. CASS test results on plated specimens which had received the duplex chromium treatment have exhibited a remarkable increase in corrosion resistance over similar specimens with a single chromium deposit of equivalent thickness. With this added incentive for carefully inspecting chromium plate, the X-ray fluorescence method was investigated.

EQUIPMENT

Because of the sample size restrictions imposed by commercially available X-ray fluorescent equipment, and with due consideration to the cost involved in labor and material for sample preparation, a more flexible arrangement for sample acceptance was highly desirable. Therefore, Buick designed and built a bulk sample holder suited to our particular laboratory requirements. Our design aims were as follows:

1. The facility must be radiation-safe in operation under existing policy.
2. No reduction in fluorescent intensity must be involved, and an increase in intensity should be obtained if possible.
3. The construction and installation should be kept as uncomplicated as possible.
4. Better cooling of sample and X-ray tube should be provided.
5. The facility should be capable of handling, with a high degree of adaptability, all sizes and shapes of commonly submitted analytical samples with a minimum of sample preparation.

To acquaint those who may not be familiar with the commercial XRD-5 spectrometer. Figure 1 shows the basic physical arrangement. The small sample drawer, located directly beneath the Machlett X-ray tube port. is capable of accommodating a maximum specimen size $\frac{3}{8}$ in. thick by $1\frac{1}{2}$ in. wide by 3 in. long. This size limitation imposes a burden of labor involved in sample preparation.

The obvious space limitation beneath the X-ray tube dictated the rotation of the X-ray tube 180° so that the space above the tube could be utilized for the large samples. Figure 2 illustrates

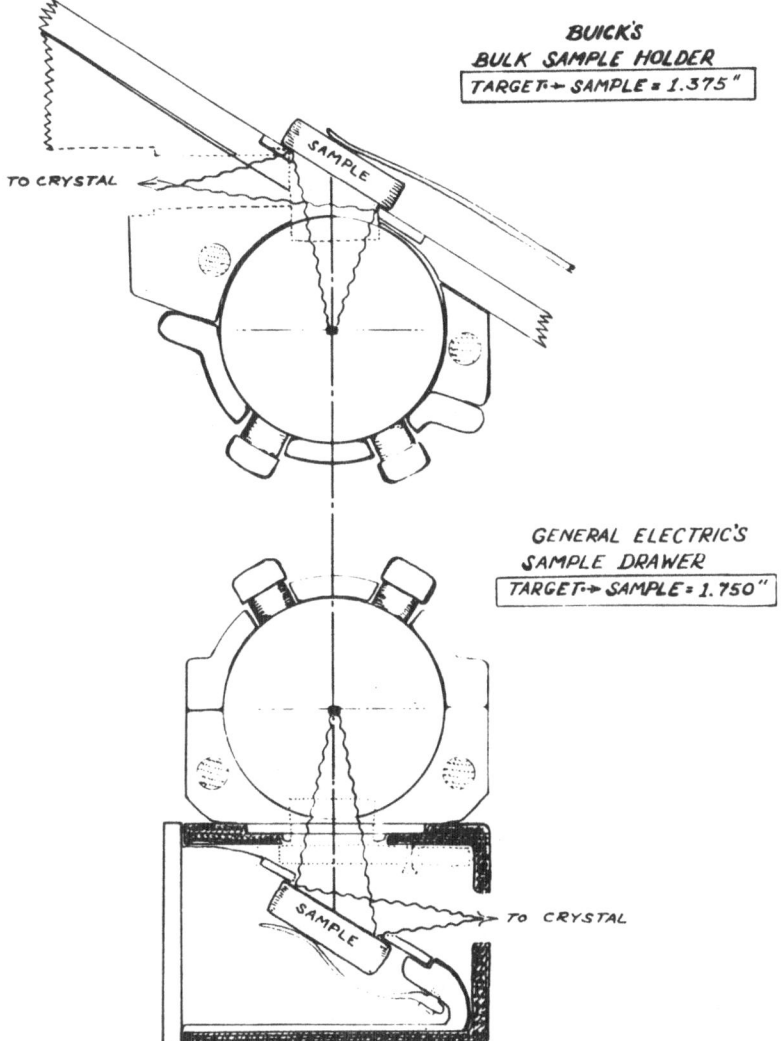

Figure 2. Comparison of target-to-sample relationships.

the comparative geometric relationship between these two concepts. The mechanical shutter arrangement was discarded and
radiation protection achieved by interrupting the high-voltage
supply through two lid-closure-operated microswitches, wired
in series. In this manner, a decrease in target-to-specimen
distance was possible, and a considerable gain in useful
fluorescent radiation resulted. Figure 3 shows the bulk specimen
holder fixed in place on the XRD-5S goniometer table.

While using the small sample drawer arrangement, very low
count rates (those requiring counting intervals of several minutes duration) left the irradiated sample literally too hot to
handle. This condition prevailed even with the coolant flow rate
adjustment at the upper limit of the recommended range. To
eliminate excessive sample heating in the bulk specimen holder,
a separate tap-water cooling supply is employed to circulate
through the tube cooling girdle, while the closed system (using
distilled water) is employed to cool the tube anode only. In this
manner, we have gained control of the sample temperature and
can handle even volatile liquids. The achieved gain in intensity

Figure 3. View of Buick bulk specimen holder.

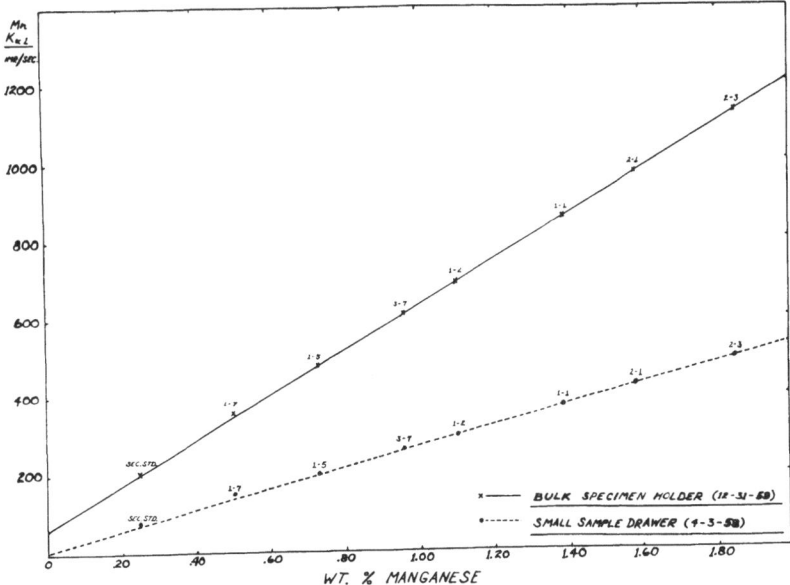

Figure 4. Graphic comparison of intensities observed on Mn K_α.

is illustrated by Figure 4. Depicted are the relative intensities for manganese K_α radiation obtained with the small sample drawer and the bulk specimen holder. The same General Motors Spectrographic Standards were used in both instances with a $\frac{1}{4}$-by $\frac{3}{4}$-in. sample aperture mask and the tube operated at peak capacity. Oldsmobile Division in Lansing, Michigan, has also purchased a GE XRD-5 with bulk specimen holder, and its reported increases in intensity are depicted in Figure 5.

A maximum sample size acceptance of 3 in. thick by $3\frac{7}{8}$ in. wide by $8\frac{3}{4}$ in. long is provided with the radiation-proof cover closed on the bulk specimen holder. There is virtually no upper limit as to size if the cover is left open. This procedure is radiation-safe only for objects of ferrous material thicker than $\frac{1}{8}$ in. and having a flat area 2 in. square centered over the X-ray tube port. The installation was approved by the General Motors Industrial Hygiene Department.

Working with the bulk specimen holder requires little change in operating procedure, and the installation is very simple, involving only ten bolts. The sample mask construction is illustrated in Figure 6. Use of the G.E.-supplied helium path apparatus is possible with the bulk specimen holder. One restriction is imposed by the bulk holder which limits high-angle goniometer

GAIN IN USEFUL FLUORESCENT INTENSITY WITH
THE BULK SAMPLE HOLDER REPLACING THE SAMPLE DRAWER

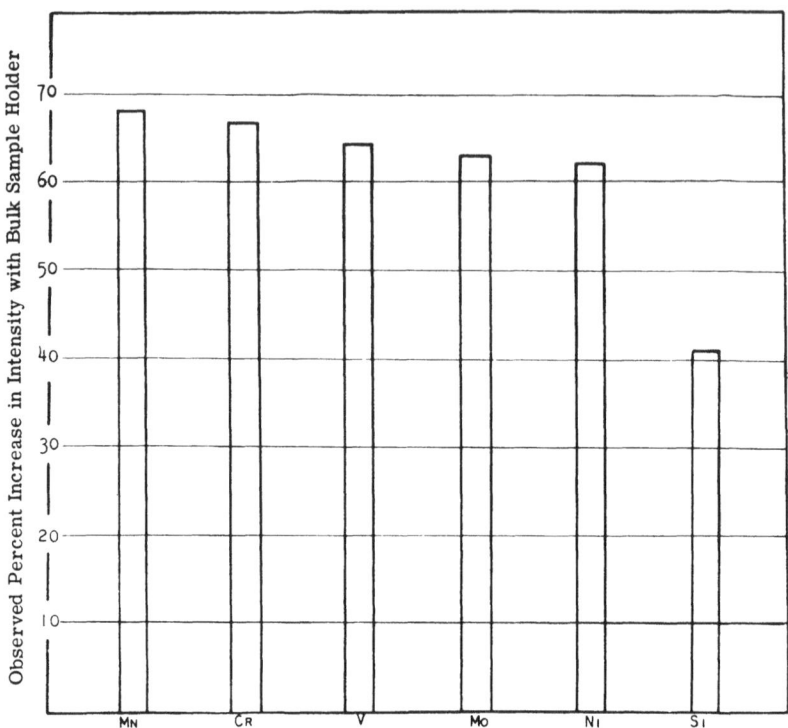

NOTE: The indicated intensity improvements were observed with each element at
approximately 1.0 wt. % in ferrous material.

Figure 5. Observed percent increase in fluorescent intensity for six elements.

travel to 138.00°. This limitation has had little effect on the routine analysis work performed at Buick, however, and could be eliminated by contouring the specimen holder.

PREPARATION OF "STANDARD" PLATED SPECIMENS

Initially, twelve 3 × 4 in. polished copper Hull panels were plated by Buick's Standards Research Department. The purity of the copper panels was verified prior to plating by an emission spectrographic analysis. Varying quantities of nickel and chromium were plated on each panel, using optimum conditions of current density, bath temperature, electrode contact, etc., and

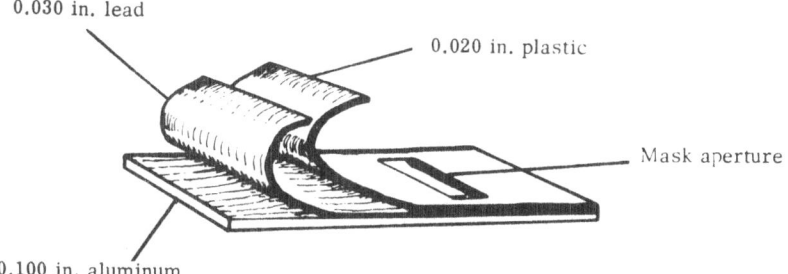

0.030 in. lead

0.020 in. plastic

Mask aperture

0.100 in. aluminum

TECHNIQUE EMPLOYED:

Machlett AEG 50-SW X-ray tube
Operated at 50 kvp and 23 ma
LiF crystal. $2\theta - 69.34°$. Cr K_{α_1}

0.010-in. soller
No. 2 SPG detector
Fixed-time method, 1000 sec
Average background 26 counts/sec

Useful sizes for sample mask apertures, in.	Chromium K_{α_1} counts/sec from a 20-μ in. plate*
0.500 × 0.750	2280
0.250 × 0.750	1265
0.090 × 0.750	480
0.250 diam.	525

* Microinterferometer measurements of vacuum vapor-deposited chromium.

Figure 6. Sketch of sample mask construction.

only the factor of time was changed to provide different plate thicknesses. These plated specimens were analyzed for chromium plate thickness by three procedures: Microexamination at 750 with filar micrometer measurement, X-ray fluorescence measurement of the chromium and nickel K_α intensities, and a wet chemical analysis for total weight chromium by dissolution and titration over an area 1 cm square with the appropriate calculation to translate the observed weight of chromium to a thickness value. Plotting microscopic thickness determinations versus chromium K_α intensities yielded very poor correlation. The wet chemical determination similarly plotted against chromium intensity provided a precise linear correlation. It should be noted however, that the same plated area used for the chemical measurement was also used previously for determining the average chromium K_α intensity. The microexamined specimen was sectioned from the immediately adjacent area toward the edge of the panel. Since the observed thicknesses under 750× magnification disagreed nearly 100% with the wet chemistry report, plate uniformity was highly suspect. The calculated chromium absorption curve for characteristic Ni K_α radiation based on the equation

$$I_x = I_0 \exp\left(-\frac{\mu}{\rho}\,\rho x\right) \ ,$$

where I_x is the intensity of transmitted X-ray beam, I_0 is the intensity of incident X-ray beam, μ is the linear absorption coefficient, ρ is the density, and x is the thickness, cannot be applied since it does not simultaneously consider the reduction in nickel excitation due to the absorption by chromium of the "white" radiation from a W target X-ray tube. Moreover, this is not an easily calculable effect due to the polychromatic beam generated by the W tube at 50 kvp. This discrepancy in calculation is only the first of several which arise in a theoretical determination of chromium thickness by absorption. Obviously the problem would be somewhat simplified if a monochromatic exciting radiation were employed; but having only a W tube available, this change was not possible. An empirical calibration procedure appeared necessary.

Considering the fact that many different proprietary chromium plating baths are currently popular, it was advisable to create a standard high-purity, uniform-density chromium, which would be independent of the unavoidable plating process variables, to serve as a basis for comparison.

General Motors Research Staff kindly provided eleven vacuum vapor-deposited high-purity chromium standard specimens having a chromium layer deposited on thin Mylar. The thickness of the chromium was determined by microinterferometer measurements reporting the best average observed along the entire length of the deposit. The density of this chromium was calculated at 7.0 g/cm^3 ($\pm 1\%$). These films were laminated to spectrographically pure nickel 0.00210 in. thick and mounted on glass slides for ease of handling. The specimens were then analyzed for fluorescent intensity of the chromium and nickel K_α radiation using the dispersive system of the XRD-5 with LiF crystal. Average intensities were obtained on the basis of both fixed-time and fixed-count procedures and the resultant data were plotted against the chromium thicknesses reported by G. M. Research. The resulting graph is shown in Figure 7. An acceptable degree of repeatability was established using a 100-sec fixed-time counting procedure. Percent standard deviation for 10 μ of chromium was ± 0.64 as calculated by

$$\% \ \sigma = \frac{100}{C} \sqrt{\frac{\Sigma (C - C_i)^2}{n - 1}} \ ,$$

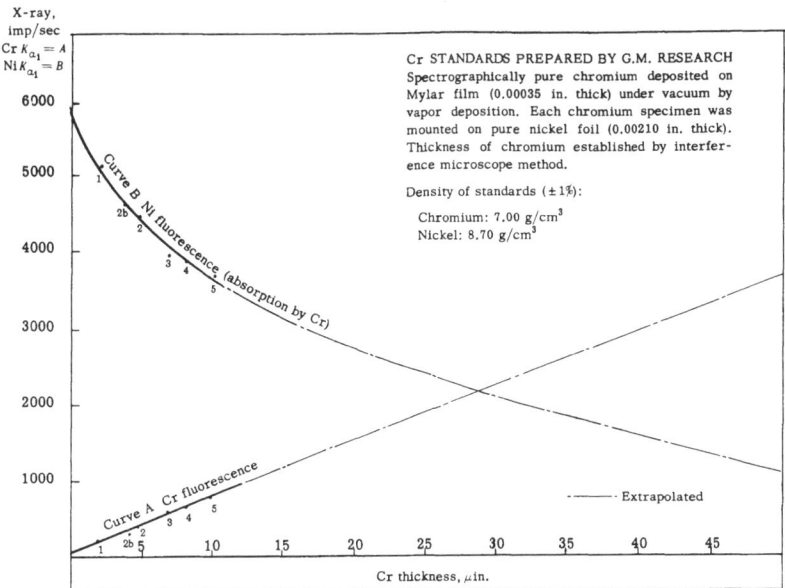

Figure 7. Working curve for chromium thickness based on vacuum vapor-deposited Cr.

where $\% \sigma$ is the percent standard deviation; $\Sigma(C - C_i)^2$ is the summation of squares of the deviation. C is the average or mean thickness. C_i are the individual values, and n is the number of determinations.

Thirty-one standard polished steel Hull panels were then prepared by Buick's Standards Research Department. All panels were plated with 0.0005 in. pyrophosphate copper as a basic substrate. Varying quantities of nickel and chromium plates were applied using optimum conditions as previously described, and changing only the factor of time to produce differences in plate thickness. The plating data are given in Table I.

The 31 standard Hull panels were individually analyzed for their characteristic chromium K_α and nickel K_α intensities. An example of the observed variations is shown in Figure 8. Using the curve developed with the vapor-deposited standards, the effective chromium plate thickness of each panel was determined. Since the area involved in each determination was $\frac{1}{4} \times \frac{3}{4}$ in., a total of ten different area locations on each Hull cell panel was counted, all on the face opposite the soldered electrode. The panels were then trimmed to eliminate factors of edge build-up,

Table I. Plating Technique for Standard Test Panels
Copper—Nickel—Chrome Panels
All groups have 0.0005 in. pyrophosphate copper for base.

	Thickness of nickel (plating aim), in.	Plating time, min.	Thickness of chromium (plating aim), in. $\times$ 10^{-6}	Plating time
Group 1				
B	0.0001	3	5	10 sec
C	0.0001	3	10	1 min
D	0.0001	3	20	2 min
E	0.0001	3	30	3 min
F	0.0001	3	50	5 min
Group 2				
A	0.0002	6	70	7 min
B	0.0002	6	80	8 min
Group 3				
B	0.0003	9	5	10 sec
C	0.0003	9	10	1 min
D	0.0003	9	20	2 min
E	0.0003	9	30	3 min
F	0.0003	9	50	5 min
Group 4				
A	0.0004	12	30	3 min
B	0.0004	12	50	5 min
Group 5				
B	0.0005	15	5	10 sec
C	0.0005	15	10	1 min
D	0.0005	15	20	2 min
E	0.0005	15	30	3 min
F	0.0005	15	50	5 min
Group 6				
A	0.0005	18	50	5 min
B	0.0005	18	60	6 min
Group 7				
B	0.00075	23	5	10 sec
C	0.00075	23	10	1 min
D	0.00075	23	20	2 min
E	0.00075	23	30	3 min
F	0.00075	23	50	5 min
Group 9				
B	0.001	30	5	10 sec
C	0.001	30	10	2 min
D	0.001	30	20	4 min
E	0.001	30	30	6 min
F	0.001	30	50	10 min

Nickel plate - Udylite - 130°F at approximately 35 amp/ft^2

Chrome plate - conventional bath
       32 oz/gal CrO$_3$ 100-1 ratio
       125°F at 150 amp/ft^2

OBSERVED VARIATION IN CHROMIUM THICKNESS
ON
STANDARD HULL PLATING TEST PANEL
BY
X-RAY FLUORESCENCE METHOD

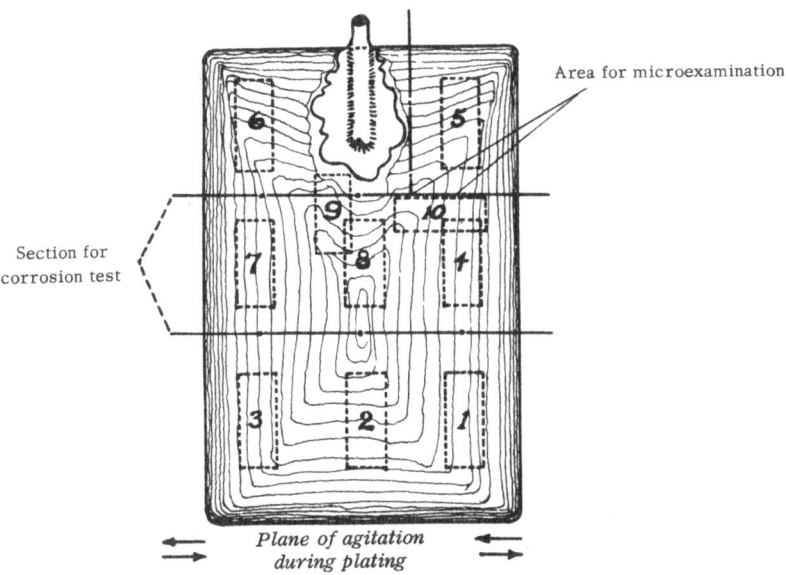

Area* of excitation	Chromium thickness, in.	Summary
1	110	Maximum variation in
2	95	chromium thickness:
3	99	
3	89	
4	93	36 μin. or 32.7%
5	82	
6	80	
7	75	
8	74	
9	73	
10	76**	

* All measurements taken on side opposite the soldered electrode.
** The measured value reported by microexamination was 73 μin.

Figure 8. Recorded plate thickness variations on standard Hull cell panel.

Table II.   CASS Test Results

Specimen	Plate thickness		Hours in CASS test	Final rating for degree of failure	Acceptable after 16 hours
	nickel (in. $\times 10^{-5}$)	chromium (in. $\times 10^{-6}$)			
1B	8	1	16	3	No
1C	10	12	4	1	No
1D	6	27	4	2	No
1E	5	37	8	3	No
1F	10	65*	16	1	No
2A	9	39	36	2	Yes
2B	10	42	36	2	Yes
3B	14	3	8	3	No
3C	16	12	8	3	No
3D	18	24	18	3	No
3E	15	34	18	3	Yes
3F	16	66*	18	3	Yes
4A	22	19	18	4	Yes
4B	23	33**	36	4	Yes
5B	26	1	16	2	No
5C	26	7	16	3	No
5D	27	19	18	3	Yes
5E	32	33	36	2	Yes
5F	24	51***	36	4	Yes
6A	27	19	36	2	Yes
6B	32	27***	36	4	Yes
7B	38	1	16	2	Yes
7C	38	6	18	3	Yes
7D	37	12	36	4	Yes
7E	44	22	36	4	Yes
7F	49	42***	36	5	Yes
9B	49	1	36	3	Yes
9C	49	2	36	3	Yes
9D	46	12	36	3	Yes
9E	47	20	36	3	Yes
9F	47	47****	36	4	Yes

* Specimens 1F and 3F exhibited macrocracking and surface haze as received.
** Specimen 4B was considered satisfactory after 36 hours in the CASS test.
*** Specimens 5F, 6B, 7F, and 9F were all considered satisfactory after 36 hours in the CASS test.

bath position, and electrode "shadowing," and the approximately uniform-thickness remainder was sectioned into three pieces:

1. Specimen for metallographic microscope plate thickness determination at 750×.
2. Specimen for accelerated corrosion testing by CASS method.
3. Specimen as permanent standard for future X-ray fluorescence comparison.

The corrosion tests and metallographic thickness measurements were completed in the Buick Metallurgical Laboratory. Following the exposure of the CASS test specimens, each was rated according to the General Motors method employing numbers "5-4-3-2-1" as representative of the percent failure of total area. No failure is rated "5," and a part exhibiting slight failure affecting no more than 10% of total area is rated "4." One which has failed in excess of 10%, but less than 30%, is rated "3." Severe failure involving from 40 to 70% of the total visible area is "2," and "1" indicates a failure exceeding 70% of total area. The data developed for each of the 31 test specimens are given in Table II.

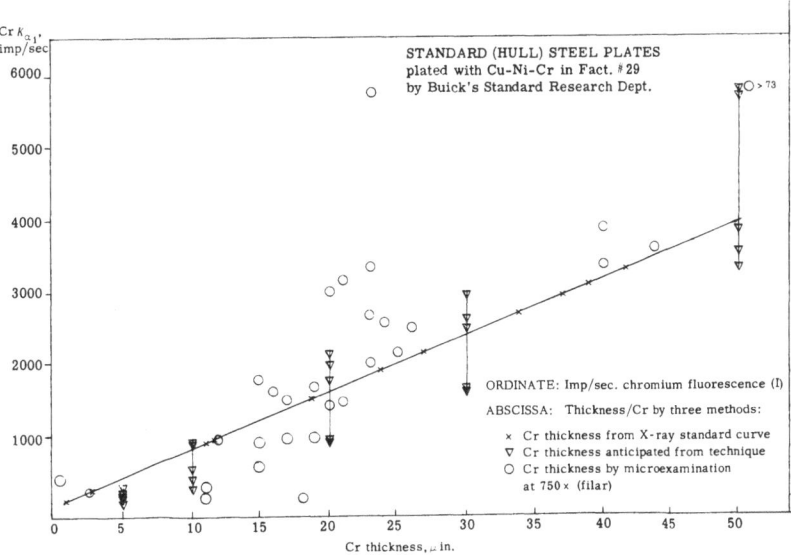

Figure 9. Plated panel curve for Cr thickness showing scattered results.

DISCUSSION AND CONCLUSIONS

Our somewhat oversimplified investigation has pointed out several significant facts:

A. An X-ray fluorescent analysis for chromium plate quality by a dispersive technique to measure chromium K_α intensity alone will not provide sufficient information. Measurements of the scattered radiation and the characteristic radiation from the nickel substrate must also be employed.

B. Variables in commercial plating practice and the widespread use of proprietary bath compositions create the necessity for an arbitrary "absolute density—absolute purity" basis for a working curve. Figure 9 illustrates the magnitude of thickness—density differences and the observed deviations from plate thicknesses anticipated from the standard plating technique employed.

C. The metallographic measurement for chromium thickness does not account for density differences in plate. Hence the large degree of scatter shown in Figure 9. The principal advantage possessed by the metallographic technique is the unrestricted choice of sample area for plate examination. Figure 10 depicts a sample headlamp door positioned for fluorescent analysis. It is obviously impossible to align the small internal radii for analysis (where plated chromium is likely to be thinnest).

D. Fluorescent analysis of duplex chromium plated parts will indicate the presence of double or triple the chromium thickness observed by the microscopic method. This is believed due to the initial chromium deposit being nearly free from porosity, while the external layer exhibits a high degree of microcracking. The two layers in combination will tend to increase the observed chromium K_α intensity; the first by providing more chromium atoms via density, and the second by increasing effective sample area.

E. New concepts in instrumentation and equipment design are both possible and necessary to extend the commercial application of X-ray fluorescence for plate quality determination. The specialized continuous plate measurement spectrometer developed by Applied Research Laboratories must certainly be considered as a step in the right direction. This instrument records the ratio of scattered to emitted characteristic radiation from a plated element. Continuous strip plating operations should now achieve optimum process control from this type spectrometer. (The analysis for plate thickness in small internal radii is still virtually impossible without part destruction.)

ACKNOWLEDGMENT

The writer wishes to thank W. L. Grube and S. R. Rouze of General Motors Research Staff for furnishing the vacuum vapor-deposited chromium standards; R. P. Durham and R. M. Peterman who supplied the plated test panels; R. A. Martin who performed the metallographic measurements and corrosion testing; L. A. Campana for his photographic assistance; and Mrs. Etta Jennings for typing the paper.

Figure 10. Headlamp trim positioned for Cr analysis.

REFERENCES

[1] Harold A. Kahler, "Automotive Trim: Performance and Testing," Products Finishing, Vol. 23, No. 7, 1959, p. 22.

[2] J. D. Thomas and Stanley R. Rouse, "Using the Interference Microscope for Thickness Measurements of Decorative Chromium Plating," Plating, Vol. 42, 1955, p. 55.

[3] Paul S. Goodwin and Charles L. Winchester, "Continuous X-ray Measurement of Plating Thickness," Plating, Vol. 46, 1959, p. 41.

[4] Edgar J. Seyb, "Improved Chromium Plate," Metal Progress, Vol. 76, No. 1, 1959, p. 113.

COUNT DISTRIBUTION AND PRECISION IN X-RAY FLUORESCENCE ANALYSIS

Kurt F. J. Heinrich

E.I. du Pont de Nemours and Company, Wilmington, Delaware

ABSTRACT

The statistical fluctuations of photon counts are discussed as a factor limiting the precision of the analytical result. Assuming a Poisson distribution, the theoretical standard deviation of the result can be calculated. While this prediction does not consider causes of variation other than the count statistics, it is useful in developing methods and checking instrument reliability. Practical examples using experimental results are given.

INTRODUCTION

The two basic assumptions made in X-ray fluorescence are: a) that under defined conditions of excitation a sample emits X-ray photons of wavelengths characteristic of the elements that constitute the sample at a constant rate measured by the instrument, and b) that the amount of photons of a particular wavelength, as registered by the instrument in a given time, is a function of the concentration in the sample of the emitting element.

The accuracy of the result is thus limited both by the errors committed in determining the counting rate of the sample under supposedly standardized conditions (errors of count), and those committed in deducing the analytical result from the counting rate obtained (error of evaluation).

The errors of count are produced by variations of the conditions of excitation, by errors committed in the measurement of the radiation emitted by the sample, and by the fact that the X-ray photon emission, as a sum of individual processes subject to the laws of probability, is affected by a statistical distribution function.

The errors of evaluation include, for example, errors of standardization, interference (due to lines of other elements), interelement effects (absorption and enhancement), background variations, inhomogeneity of sample, particle size, etc.

Frequently, this type of error is the principal factor in limiting the accuracy of the analytical result, and therefore the errors of count can be neglected. This is not the case, however, when the net counting rates are low, or the signal-to-background ratio

95

is unfavorable. Here, in order to obtain information regarding these types of error, the statistical error must be considered first.

Knowledge of the distribution function applicable to the photon counts permits the prediction of the statistical distribution of a repetition of counts under identical conditions. Therefore, the statistical variations affecting the analytical result, when no other errors are committed, can be calculated, and the experimental deviations from the expected values, under selected conditions, can then be used to estimate other types of errors. Furthermore, this procedure permits the prediction of the limits of detection of a particular method and selection among alternative methods, according to their predicted precision, with a minimum of experimental work.

THEORETICAL COUNT DISTRIBUTION

The repetition of X-ray photon counts for equal time intervals, under identical conditions of excitation, results in a distribution practically identical with the Poisson distribution.[1] Thus, for a given set of counts under identical conditions

$$\sigma_n^2 = \bar{n} \simeq n, \tag{1}$$

where $\bar{n}$ is the average number of counts, σ_n^2 is the variance of the number of counts, and n is the number of counts obtained in a single count.

This can be verified experimentally over wider ranges of and counting rates than those usually employed in analytical work.[2]

If r is the counting rate

$$r = \frac{n}{t}, \tag{2}$$

where n is the number of counts registered, and t is the time; it follows that the variance σ_r^2 of the counting rate is:

$$\sigma_r^2 = \frac{\sigma_n^2}{t^2} = \frac{n}{t^2} = \frac{r}{t}, \tag{3}$$

and the coefficient of variation:

$$V_r = \frac{100\,\sigma_r}{r} = 100\,(r\,t)^{-\frac{1}{2}}. \tag{4}$$

If a constant-time method is used $(t = K_t)$, it follows that

$$\sigma_r = r^{\frac{1}{2}}t^{-\frac{1}{2}} = K_t^{-\frac{1}{2}}r^{\frac{1}{2}}, \tag{5}$$

---

[1] Superscripts pertain to references at the end of the paper.

and

$$V_r = 100 \ r^{-\frac{1}{2}} K_t^{-\frac{1}{2}} \ . \tag{6}$$

If a constant-count method is used $(n = rt = K_n)$,

$$\sigma_r = r^{\frac{1}{2}} \left(\frac{K_n}{r}\right)^{-\frac{1}{2}} = r K_n^{-\frac{1}{2}} \ ,$$

and

$$V_r = 100 \ r^{-\frac{1}{2}} \left(\frac{K_n}{r}\right)^{-\frac{1}{2}} = 100 \ K_n^{-\frac{1}{2}} \ . \tag{7}$$

In the constant-count method the standard deviation of the count is independent of the counting rate, and so is the coefficient of variation of the counting rate. The standard deviation of the counting rate, however, is proportional to the counting rate. This is not necessarily desirable for a method covering a wide range of counting rates.

Comparison of the variance obtained experimentally with the predicted variance permits testing for sources of random variation other than the statistic distribution, such as imprecision of the timing device, or mechanical and electronic sources of error. By application of analysis of variance to a planned experiment, their magnitude can be estimated.

The manner in which the various possible errors contribute to the final error must be considered in order to estimate the error of the analytical result.

The effect of the statistical count fluctuation upon the precision of results derived from two countings (e.g., ratio of counting rates) has been dealt with in several publications.[3]

For quantitative analysis, however, implicitly or explicitly, a result, expressed as concentration of the element to be determined, is obtained by plotting the count of the sample on a standard curve, defined by the counts of two standards so that three counting errors must be taken into account. While these curves frequently are nonlinear, the use of corrections for nonlinearity essentially permits the reduction of the problem to plotting the count of the unknown on a linear standard curve as defined by the counts of two standards. See Figure 1.

The analytical result is thus obtained by

$$X_1 = R + X = R + C \ \frac{w - u}{v - u} \ , \tag{8}$$

where u, v, and w are the counting rates of the lower standard, the higher standard, and the sample, respectively. X_1 is the concentration of element to be determined in the sample, and R is the concentration of element to be determined in the lower standard.

If C and R are known exactly, while u and w are estimated from the number of counts $U = ut_u$, $V = vt_v$, and $W = wt_w$ obtained in the time intervals t_u, t_v, and t_w according to (1), the variance of X is obtained by

$$\sigma_x^2 = \sigma_u^2\left(\frac{\partial x}{\partial u}\right)^2 + \sigma_v^2\left(\frac{\partial x}{\partial v}\right)^2 + \sigma_w^2\left(\frac{\partial x}{\partial w}\right)^2 ;$$

$$\sigma_u^2 = ut_u^{-1}; \quad \sigma_v^2 = vt_v^{-1}; \quad \sigma_w^2 = wt_w^{-1};$$

$$\frac{\partial x}{\partial u} = C(w-v)(v-u)^{-2} ; \quad \frac{\partial x}{\partial v} = C(w-u)(v-u)^{-2} ; \quad \frac{\partial x}{\partial w} = C(v-u)^{-1} ,$$

$$\sigma_x^2 = C^2(v-u)^{-4}\left[ut_u^{-1}(w-v)^2 + vt_v^{-1}(w-u)^2 + wt_w^{-1}(v-u)^2\right]. \quad (9)$$

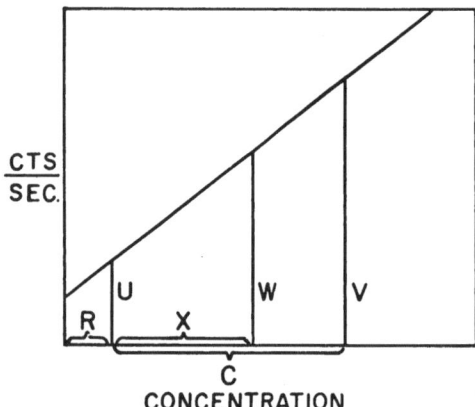

U   COUNTING RATE OF LOWER STANDARD
V = COUNTING RATE OF HIGHER STANDARD
W = COUNTING RATE OF SAMPLE

R   = CONCENTRATION OF UNKNOWN IN LOWER STANDARD
C   = CONCENTRATION OF UNKNOWN ADDED TO HIGHER STANDARD
X + R = CONCENTRATION OF UNKNOWN IN SAMPLE

Figure 1.

For a fixed-time counting method of the type

$$t_u = t_v = t_w = const,$$

equation (9) is reduced to

$$\sigma_x^2 = C^2 (V-U)^{-4} \left[U(W-V)^2 + V(W-U)^2 + W(V-U)^2 \right]. \tag{10}$$

LOWER LIMIT OF DETECTION

The lower limit of detection of a method can be defined as the minimum amount of the unknown that can be detected with a given degree of confidence. If the variance of the result at the low end of the range of the method $(\sigma_{x,0})$ is known, the range of concentration that will not be exceeded in one direction, with a 95% confidence, is given by $X + 1.65\,\sigma_{x,0}$ or $X - 1.65\,\sigma_{x,0}$, depending on the direction.[4] Thus, if $\sigma_{x,0}$ is known, $1.65\,\sigma_{x,0}$ can be assigned as the lower limit of detection (95% confidence).

Comparison of these lower limits of detection obtainable with different methods is an important criterion for selecting a method for determination of low concentration.

In the case of X-ray fluorescence, for the low end of the range, W in equation (10) will be practically equal to V, so that

$$\sigma_x^2 = 2C^2 U(V - U)^{-2}, \tag{11}$$

and the theoretical limit of detection can be defined as

$$LD = 2.33 CU^{1/2}(V - U)^{-1}. \tag{12}$$

While other types of error might in practice raise this limit, it constitutes a theoretical limit of sensitivity particularly useful for comparing different possible instrumental setups. From equation (12) it follows that in varying t

$$LD_1 : LD_2 = t_1^{-1/2} : t_2^{-1/2}. \tag{13}$$

COMPARISON OF COUNTING TECHNIQUES FOR THREE-POINT DETERMINATION

Equation (9) can be transformed using the parameters:

$$K = \frac{v - u}{C} = \text{slope of standard line,}$$

$$A = \frac{w - u}{u} = \text{net signal-to-background ratio of unknown,}$$

$$B = \frac{w - u}{v - u} = \text{ratio of net signals of unknown to higher standard,}$$

$$r_u = \frac{3t_u}{t_u + t_v + t_w} \; ; \; r_v = \frac{3t_v}{t_u + t_v + t_w} \; ; \; r_w = \frac{3t_w}{t_u + t_v + t_w} \; .$$

We obtain

$$\sigma_x^2 = \frac{3u}{K^2 \Sigma t} \left[r_u^{-1}(B-1)^2 + r_v^{-1}(AB + B^2) + r_w^{-1}(A+1) \right] . \quad (14)$$

Normalizing with respect to t, K, and w, we define

$$\sigma_N^2 = \sigma_x^2 \frac{K^2 \Sigma t}{3w} \; ;$$

therefore,

$$\sigma_N^2 = \left[r_u^{-1}(B-1)^2 + r_v^{-1}(AB + B^2) + r_w^{-1}(A+1) \right](A+1)^{-1}. \quad (15)$$

The time distribution yielding the smallest σ_N^2 for a given pair of values of A and B is found by

$$\frac{\partial \sigma_N^2}{\partial r_u} = -r_u^{-2}(B-1)^2(A+1)^{-1} + (3 - r_u - r_v)^{-2} = 0 \; ,$$

$$\frac{\partial \sigma_N^2}{\partial r_v} = -r_v^{-2}(AB + B^2)(A+1)^{-1} + (3 - r_u - r_v)^2 = 0 \; ,$$

and we obtain:

$$r_u = \frac{3(1 - B)}{D} \; ; \; r_v = \frac{3(AB + B^2)^{\frac{1}{2}}}{D} \; ; \; r_w = \frac{3(A+1)^{\frac{1}{2}}}{D}$$

where

$$D = 1 - B + (A+1)^{\frac{1}{2}} + (AB + B^2)^{\frac{1}{2}} \; ; \quad (16)$$

and

$$\sigma_{N\,min}^2 = 3^{-1}(A+1)^{-1}D^2 . \quad (17)$$

The values of $\sigma_{N\,min}^2$ as a function of A and B are shown in Figure 2. In order to compare some more conventional methods with the least-variance method described above, the values of the normalized variance and of the time ratio are derived for them.

A) *Fixed-Time Method.* (Same time is used for each counting.)

$$r_u = r_v = r_w = 1 \; ,$$

$$\sigma_N^2 = B + 1 + \frac{(B-1)(2B-1)}{A+1}$$

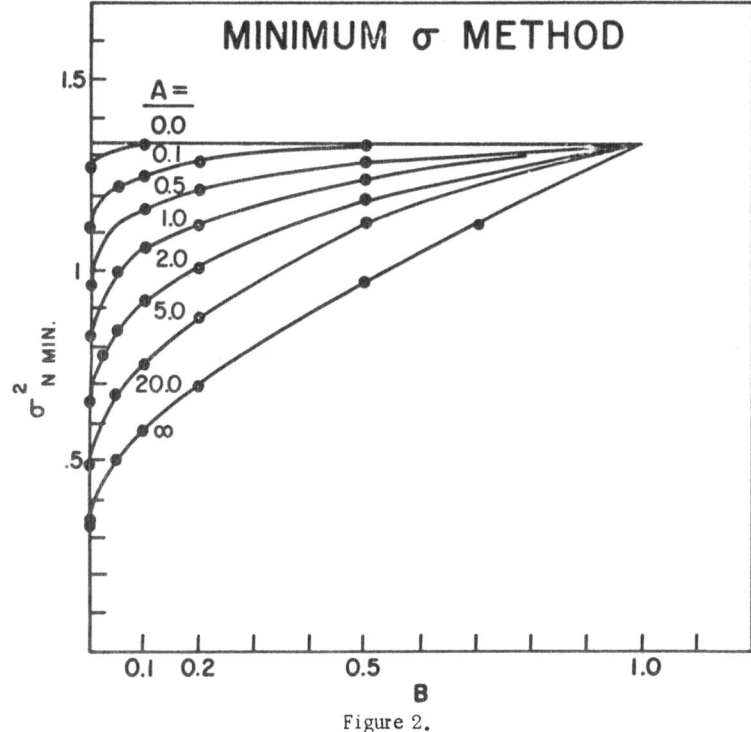

Figure 2.

B) Modified Fixed-Time Method. (Twice as much time is used for the sample as for each standard.)

$$r_u = r_v = 0.75 ; \quad r_w = 1.50 ;$$

$$\sigma_N^2 = \frac{4}{3}\Big[B(2B - 3) + 1\Big](A + 1)^{-1} + \frac{2}{3}(2B + 1).$$

C) Fixed-Count Method. (Same amount of count used for each counting.)

$$r_u = \frac{3(A + 1)(A + B)}{(A + 2B)(A + 1) + (A + B)}; \quad r_v = \frac{3(A + 1)B}{(A + 2B)(A + 1) + A + B};$$

$$r_w = \frac{3(A + B)}{(A + 2B)(A + 1) + A + B};$$

$$\sigma_N^2 = \frac{(A + 2B)(A + 1) + A + B}{3(A + 1)^2(A + B)} \left[(B - 1)^2 + (A + B)^2 + (A + 1)^2 \right] .$$

D) *Modified Fixed-Count Method.* (Twice as many counts are used for the sample as for the standards.)

$$r_u = \frac{3(A + 1)(A + B)}{(A + 2B)(A + 1) + 2(A + B)} \; ; \; r_v = \frac{3(A + 1)B}{(A + 2B)(A + 1) + 2(A + B)} \; ;$$

$$r_w = \frac{6(A + B)}{(A + 2B)(A + 1) + 2(A + B)} \; ;$$

$$\sigma_N^2 = \frac{(A + 2B)(A + 1) + 2(A + B)}{6(A + 1)^2(A + B)} \left[2(B - 1)^2 + 2(A + B)^2 + (A + 1)^2 \right]$$

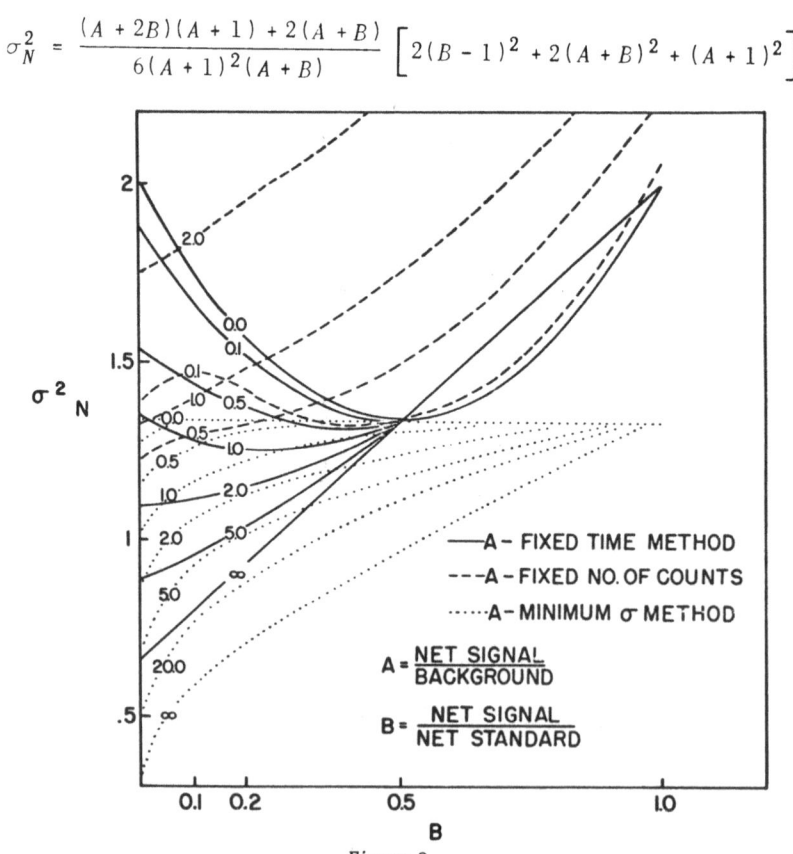

Figure 3.

Figure 3 shows that while in many cases the differences of the variances obtainable with different counting methods are trivial, the fixed-count methods give high variances when A is larger than 1 (low background). Therefore, the fixed-time method (in particular the modified fixed-time method) seems more recommendable for values of A below 1 in cases where extreme time economy is important. Table I gives the percentages of total time to be used for each count, as a function of A and B.

These figures also permit, by observing the values of σ_N^2 as a function of B, the selection of the optimum value of the concentration of the higher standard. It can be observed that for an ideal time distribution (and also for the fixed-time methods, when $A < 2$), the precision of the determination increases with increasing concentration of the element to be determined in the higher standard.

EXPERIMENTAL

Determination of Iron in Niobium Pentoxide. Three samples of niobium pentoxide were fused with nine times their weight of borax, and iron is determined as described elsewhere,[2] using a General Electric XRD-5 instrument. The instrumental conditions are as follows:

Line used: First-order Fe K_α
Excitation: W target, 50 kvp, 50 ma
Crystal: Lithium fluoride
Collimator: 0.020 in.
Counter: Flow proportional counter
Counter voltage: 2000 v
Sample holder covered with silver foil
Counting time: 100 sec
Average counts obtained: (approximate figure)
  Lower standard (50 ppm Fe):   12,200
  Higher standard (10,000 ppm Fe): 100,000
  Sample A: 12,400
  Sample B: 17,400
  Sample C: 33,400

Each determination was made 16 times. The results are given in Table II. The experimental value of σ_x was determined and the theoretical value σ_t was then obtained by using equation (10).

Table I.   Optimum Time Distribution in %

A		0.00	0.01	0.02	0.05	0.10	0.20	0.30	0.50	0.70	0.90	1.00
							B					
0.00	U	50	49	49	47	45	40	35	25	15	5	0
	V	0	1	1	3	5	10	15	25	35	45	50
	W	50	50	50	50	50	50	50	50	50	50	50
0.05	U	49	48	48	47	44	39	34	24	15	5	0
	V	0	1	2	3	6	11	16	26	35	45	50
	W	51	51	50	50	50	50	50	50	50	50	50
0.20	U	48	47	46	44	42	37	32	23	14	5	0
	V	0	2	3	5	8	13	18	27	36	45	50
	W	52	51	51	51	50	50	50	50	50	50	50
0.50	U	45	43	42	41	38	33	29	21	13	4	0
	V	0	3	5	7	10	16	20	29	37	46	50
	W	55	54	53	52	52	51	51	50	50	50	50
0.70	U	43	42	41	39	36	32	27	19	12	4	0
	V	0	4	5	8	12	17	22	30	38	46	50
	W	57	54	54	53	52	51	51	51	50	50	50
1.00	U	41	40	39	37	34	30	26	18	11	4	0
	V	0	4	6	9	13	18	23	31	39	46	50
	W	59	56	55	54	53	52	51	51	50	50	50

Table II

Determination	Results (ppm Fe)		
	a	b	c
1	51	614	2471
2	57	659	2493
3	28	631	2434
4	63	632	2453
5	81	650	2453
6	55	617	2455
7	35	596	2470
8	66	630	2514
9	71	644	2470
10	23	601	2476
11	78	634	2487
12	51	633	2499
13	33	614	2488
14	90	644	2505
15	41	626	2464
16	60	616	2495
Average	55.2	629.4	2483
σ_x (experimental)	19.0	16.9	20.7
σ_t (calculated)	17.9	19.2	24.5

Determination of Aluminum Oxide in Titanium Oxide.
Aluminum oxide was determined in two samples of uncompacted titanium dioxide powder. The same instrument was employed. The third-order Ti K_α radiation was eliminated using a pulse-height analyzer. A chemically pure titanium oxide was used as lower standard. Two higher standards were employed: a titanium dioxide powder containing 1.69% Al_2O_3, according to wet analysis, and an alloy of high aluminum content, whose equivalent expressed in Al_2O_3 had been previously determined by repeated counting. All oxide samples were counted 400 sec, and the alloy standard was counted 100 sec. The equivalent of the alloy with such counting times was determined as 110% Al_2O_3. The instrumental conditions were as follows:

    Line used: First-order AlK_α
    Excitation: W target, 50 kvp, 50 ma
    Crystal: EDDT
    Collimator: 0.020 in.
    Counter: Flow proportional counter
    Counter voltage: 1900 v

Sample holder covered with silver foil
Helium path
Pulse-height analyzer set:
 Base line: 0.0 v
 Channel width: 7.5 v

The results given in Table III were obtained by repetition of the determination, and by calculation, using equation (10).

Table III

Average counts:				
Lower standard				6,400
Higher standard I (1.69% Al_2O_3)				10,400
Higher standard II (110% Al_2O_3)				267,000
Sample A				8,800
Sample B				11,400

Results:				
Sample	a	b	a	b
Standard	I	I	II	II
Values obtained:	1.01	2.06	1.03	2.09
	1.10	2.11	1.05	2.03
	0.99	2.12	0.99	2.13
	1.10	2.26	1.03	2.13
	0.89	1.94	0.92	2.01
	0.97	2.08	0.98	2.12
	0.94	2.03	0.96	2.08
			0.99	2.18
			1.06	2.16
			1.06	2.18
Averages:	1.000	2.086	1.007	2.111
σ_x (experimental)	0.073	0.094	0.044	0.056
σ_t (calculated)	0.049	0.071	0.052	0.056

In both experiments the check between predicted and obtained standard deviation is fairly close. This confirms the usefulness of equation (10) in predicting the precision obtainable by the respective methods.

SUMMARY

While in X-ray fluorescence analysis the random error due to the count distribution is frequently negligible as compared with the systematic errors involved in the method, this is not true when the counting rate is unusually low. Therefore, in trace analysis, microfocus work, and other operations where low net counting rates are obtained, especially when the background counting rates are relatively high, consideration of the effects of count distribution in photon counting is useful. As shown experimentally, the statistical precision of a method can be predicted by these means.

REFERENCES

[1] R. M. Brissey, H. A. Liebhafsky, and H. G. Pfeiffer, ASTM Special Technical Publication, No. 157, 1954, p. 50.

[2] K. F. J. Heinrich and T. D. McKinley, Paper presented at the Pittsburgh Conference on Analytical Chemistry, March, 1959.

[3] N. Parrish, Philips Technical Review, Vol. 17, No. 7-8, Jan., 1956, p. 206.

[4] O. L. Davies, Statistical Methods in Research and Production, Chapter 4; Hafner Publishing Co., New York, 1957.

INTENSITIES OF THE *K*, *L*, AND *M* SPECTRAL
LINES FOR THE ELEMENTS WITH
ATOMIC NUMBERS 16 TO 92*

William J. Campbell

Bureau of Mines, U.S. Department of the Interior
College Park, Maryland

ABSTRACT

Line intensity and background measurements were made on the *K* lines for the elements with atomic numbers 16 to 60, *L* lines for the elements above atomic number 42, and *M* lines for elements above atomic number 80. Three general classes of samples were investigated: (1) infinitely thick, (2) microgram deposits, and (3) thin layers.

These studies show that longer-wavelength *L* radiation may be preferable to the *K* series lines from the same element in the range of elements with atomic numbers 42 to 60. In particular the L_α lines are more intense than the *K* series lines from Class 2 and 3 samples. With Class 1 samples the L_α lines are weaker than the *K* series but their line-to-background ratio is superior to the *K* series.

M series lines show little promise for spectrochemical analysis except for elements with atomic numbers 90 to 92; for example, with uranium samples in Class 2 and 3, the very high line-to-background ratio of the U M_{β_1} line may have limited application.

Elements with atomic numbers from 16 to 22 are more sensitive than expected due to the very high line-to-background ratios and the reduced collimation requirements in this long-wavelength region.

INTRODUCTION

This investigation was undertaken to determine the optimum X-ray spectral lines to use for the analyses of a wide variety of samples received by the College Park Laboratory. Until several years ago the fluorescent X-ray spectroscopist was limited to the *K* series for elements with atomic numbers 22 to 60 and the *L* series for higher-atomic-number elements. Upon

---

* This paper was presented at the Seventh Annual Conference. Approval for publication was received too late for inclusion in the Proceedings of that conference.

commercial development of the flow proportional counter used in conjunction with a pulse-height analyzer the longer-wavelength radiation of the K, L, and M series of elements with atomic numbers 16 to 22, 42 to 60, and higher atomic numbers, respectively, were investigated by the author for possible use as analytical lines.

Theoretical calculations and experimental studies were made on the line intensities and line-to-background ratios for three classes of samples: (1) infinitely thick, e.g., powders, solutions, and metals, (2) intermediate thickness, e.g., ion exchange membranes, and (3) very thin, e.g., oxidized layers such as antimony on molten lead.

THEORY

The fluorescent X-ray intensity from element A in the volume $S\,dx$ can be derived as follows: Let I_0, shown in Figure 1, be the intensity of the incident radiation of wavelength λ_i at the surface of the sample. All radiation between $\lambda_{\text{D.H.}}$ and $\lambda_{\text{C.E.}}$ is effective in exciting element A, where

$$\lambda_{\text{D.H.}} = \frac{hc}{Ve}$$

is Duane-Hunt's law,[1] and $\lambda_{\text{C.E.}}$ is the critical absorption wavelength for A (can be K, L, or M edge, depending on atomic number of A).

The intensity of the incident radiation is reduced by the factor $\exp(-\lambda_i \mu x / \sin\theta_1)$ when passing from the surface of the sample to the layer $S\,dx$, where $\lambda_i \mu$ are the linear absorption coefficients of the matrix for the incident radiation, θ_1 is the angle between the incident radiation and the surface of the sample (beam divergence is neglected), S is the effective excited sample area, limited to approximately 0.75 by 1.0 in. by a parallel-plate collimator, and x is the perpendicular distance from the surface of the sample to the excited atom.

If W_A is the weight fraction of A, then there are

$$\frac{W_A \cdot \rho \cdot S\,dx \cdot N}{\text{Atomic weight A}}$$

atoms of A in the layer $S\,dx$, where ρ is the density of the sample, and N is Avogadro's number.

The probability of atom A emitting a particular line is related to the photoelectric absorption coefficient, r, the fraction of the exciting radiation which results in raising the atom to the required energy state, F, the fluorescent yield, and the proba-

---

[1] Superscripts pertain to references at the end of the paper.

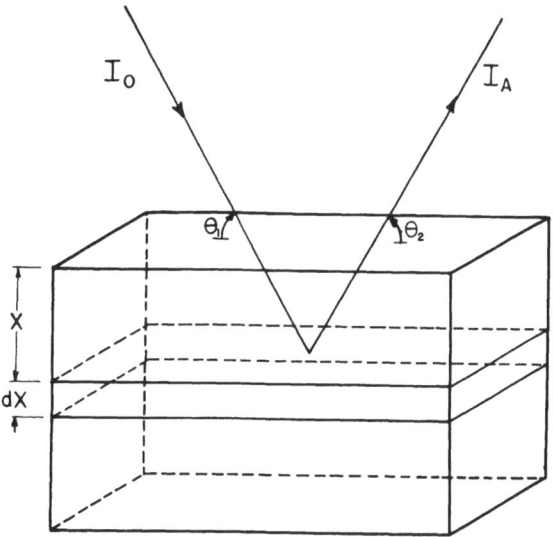

Figure 1. Schematic diagram of path of exciting
radiation, I_0, and fluorescent radiation, I_A.

bility P that a given electronic transition, e.g., $L_{III} \to K$ emitting K_{α_1} radiation, will occur. With all these factors considered, the fluorescent intensity of a characteristic line of A can be expressed as follows:

$$dI_A' = \sum_{i=D.H.}^{i=C.E.} \frac{\lambda_i I_0 \; \lambda_i \gamma r F P W_A \rho S N \exp\left(-\lambda_i \mu \dfrac{x}{\sin\theta_1}\right) dx}{\text{Atomic weight A}} . \quad (1)$$

The intensity of the fluorescent radiation F is also reduced by the absorption term $\exp(-\lambda_A \mu \; x/\sin\theta_2)$ where $\lambda_A \mu$ is the linear absorption coefficient of the matrix for radiation λ_A. Also the measured intensity is reduced by blocking in the collimator B, diffraction efficiency D, percent transmission T of X-radiation through air or helium, and detector efficiency E. For a given kilovolt and milliampere setting the intensity dI_A can be expressed as follows:

$$dI_A = \sum \frac{\lambda I_0 \; \lambda \gamma r F P W_A \rho N S B D T E \; \exp\left[-\left(\dfrac{\lambda_i \mu}{\sin\theta_1} + \dfrac{\lambda_A \mu}{\sin\theta_2}\right)\right] dx}{\text{Atomic weight A}} . \quad (2)$$

Equation (2) is difficult to evaluate because of the interdependence of x and λ. However, for the special cases where the samples are thin layers or microgram deposits, $\exp(-x)$ terms are approximately unity as

$$\lim_{x \to 0} \exp(-x) = 1 - x + \frac{x^2}{2!} - \frac{x^3}{3!} + \ldots \cong 1 \ . \qquad (3)$$

The T term can be considered as independent of the element being determined since the transmission coefficient for longer-wavelength radiations, 1-5 A through helium, is approximately unity, as shown in Table I. Also, the B term is constant if the same collimator is used for all measurements. With these considerations, equation (2) is reduced to the following expression:

$$I_A = \sum_{i=D.H.}^{i=C.E.} {}_{\lambda_i} I_0 \ {}_{\lambda_i} \gamma r FPDEA \ , \qquad (4)$$

where A is the number of grams of atom A and ${}_{\lambda_i} \gamma$ is now the mass absorption coefficient. If ${}_{\lambda_i} \gamma$ is expressed as the atomic absorption coefficient, then A is the number of A atoms present.

Table I. Transmission of X-Rays in a
27-cm Path through Air and Helium

Wavelength, A	$I/I_0 \cdot 100$	
	Helium	Air
1	100	92
2	100	56
2.5	100	33
3	100	15
3.5	100	5
4	99	1
7	92	$\ll 1$
10	79	$\ll 1$

Evaluation of Terms.

1) $\sum_{\lambda_i} I_0$. The wavelength distribution of the continuous radiation from a thick-target X-ray tube, operated with a full-wave rectified power supply, can be expressed by the following relationship:[2]

$$I_\lambda = \frac{1}{\lambda^2} \left(\frac{1}{\lambda_0} - \frac{1}{\lambda} \right) T_1 \ ; \qquad (5)$$

where
$$\lambda_0 = \frac{12.35}{V_t \; (kv)},$$

$$V_t = V_{Peak} \sin 2\pi ft$$

and T_1 is the transmission coefficient, calculated for a 1-mm beryllium window.

A series of intensity values for various wavelengths were calculated for nine values of V_t, where $2\pi ft$ was varied in 10° increments up to 90°. The average intensity for each wavelength is given in column 2 in Table II.

These values correspond to the relative intensity distribution of the continuous radiation inside the X-ray tube. The longer-wavelength X-rays are absorbed preferentially in the X-ray tube window as shown by the transmission coefficients listed in column 3. Column 4 then gives the relative intensity distribution reaching the surface of the excited sample.

From examination of column 4 it is apparent that very little long-wavelength radiation is incident on the surface of the sample. Therefore, all of the scattered radiation observed in the

Table II. Relative Intensity Distribution from
an X-Ray Tube Operated at 57 kvp

Wavelength, A	I target *	I/I_0 tube ** window	I sample surface
0.25	2.66	0.97	2.58
0.30	5.38	0.97	5.22
0.35	6.21	0.96	5.96
0.40	6.25	0.96	6.00
0.45	5.90	0.96	5.66
0.50	5.47	0.96	5.25
0.60	4.52	0.96	4.34
0.75	3.41	0.95	3.24
1.00	2.21	0.90	1.99
1.50	1.12	0.75	0.84
2.00	0.67	0.55	0.37
3.00	0.32	0.20	0.064
4.00	0.18	0.035	0.0063
5.00	0.12	0.0015	0.00018

* Contribution of characteristic tungsten radiation is neglected.
** Calculated for 1-mm beryllium window.

long-wavelength region must be higher-order short-wavelength radiation which can be discriminated against by pulse-height analysis.[3]

2) $\lambda_i \gamma$. Photoelectric absorption coefficients are related to the atomic number of the atom and the wavelength of the incident radiation by the following expressions:[1]

$$\mu_a = C_a Z^4 \lambda^3 + b_a \ , \quad \lambda < K_{edge} \ ,$$

$$\mu_a = C_a' Z^4 \lambda^3 + b_a \ , \quad K_{edge} > \lambda < L_{edge} \ ,$$

(6)

where μ_a is the atomic absorption coefficient equal to $\dfrac{\mu/\rho}{N/A}$; when λ is in centimeters,

$$C_a = 2.25 \cdot 10^{-2} \ ,$$

$$C_a' = 0.33 \cdot 10^{-2} \ ,$$

and b_a terms are neglected in the following calculations.

For a given element the relative dosage rate can be calculated by multiplying the intensity of the incident radiation by the appropriate value of μ_a. A plot of these terms is shown in Figure 2 for the K and L excitation of iodine (atomic number 53). In all the theoretical calculations and experimental studies 57 kvp was used, as that value approximates the highest operating voltage obtainable with commercially available equipment.

The characteristic radiation, W L series, of the X-ray tube was neglected in the calculations. However, in every instance the characteristic tungsten radiation would ionize only those elements that emit radiation longer than 1.2 A.

There are two very important facts contained in Figure 2, which are essential to this discussion: (1) There is a much greater X-ray dosage available for exciting elements emitting long-wavelength radiation (approximately 22/1 for Iodine$_L$/Iodine$_K$) and (2) most of the radiation which raises the atom to the L or M ionized states can be effectively discriminated against by the pulse-height analyzer.

3) r. The r terms represent that fraction of the radiation absorbed which raises the atom to the K or L ionized state.[4] These values are computed by measuring the ratio of the absorption coefficients across the critical absorption edge; e.g., if the absorption coefficients across the K absorption edge were 100 and 20, respectively, then 80% $[100 \cdot (100-20)/100]$ of the

---

* A limited number of 100-kv constant potential sources are available in the Philips Electronic Autrometer.

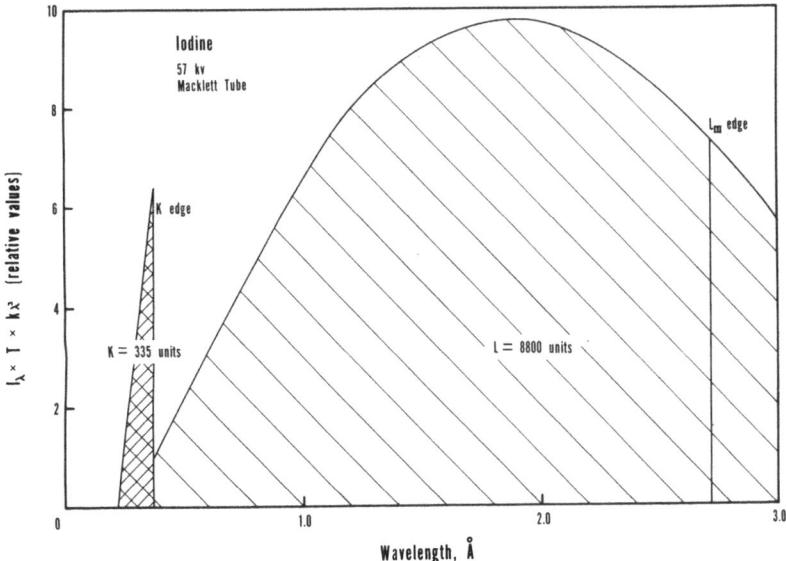

Figure 2. Relative X-ray dosage available to excite K and L spectra of iodine.

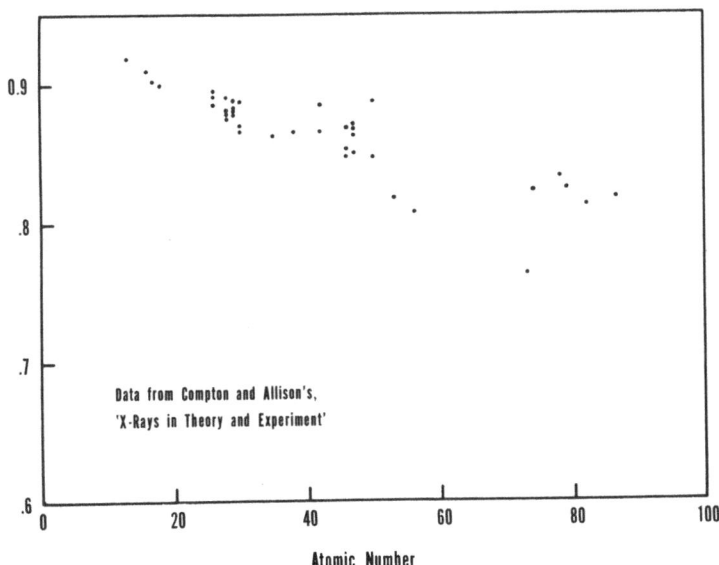

Figure 3. K absorption jump ratios.

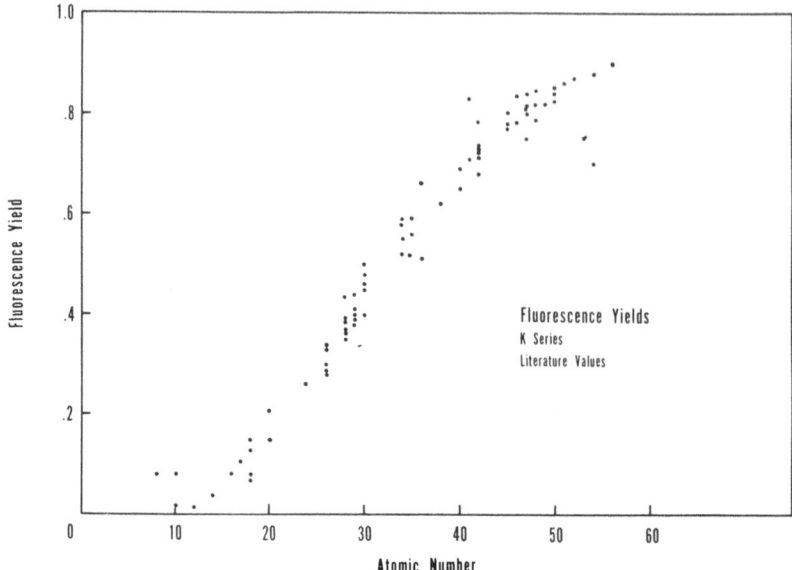

Figure 4. Fluorescent yields for the K series.

radiation resulted in a K ionized atom. Figure 3 shows the r_K values for selected elements between atomic numbers 13 and 90, while Table III lists the r_L and $r_{L_{111}}$ values for certain elements with atomic numbers between 47 and 82.

4) F. The fluorescent yield factor, which varies systematically with atomic number, is defined by Stephenson[5] as the ratio of the number of fluorescence K quanta emitted to the number of atoms ionized in the K shell per unit of time. The definition for the L and M ionized atoms would be similar.

Fluorescence yield data for the K series are presented in Figure 4. No attempt was made by the author to evaluate the different values reported by various investigators for the same element. A theoretical relationship between W_K and atomic number was given by Stephenson:

$$W_K = \frac{bZ^4}{1 + bZ^4}, \text{ where } b = 1.127 \cdot 10^{-6}. \tag{7}$$

Fluorescence yields for the L series were difficult to locate, however; the few values found are given in Table IV. No attempt was made to find fluorescence yields for the M series.

5) P. Transition probabilities for the K_α doublet and the L_{α_1} line are summarized in Table V. These values were determined by dividing the intensity of the spectral lines $(K_\alpha$ or $L_{\alpha_1})$ by the

sum of the intensities for all lines originating from the same energy level, e.g.,

$$P_{K\alpha} = \frac{I_{K\alpha_1} + I_{K\alpha_2}}{I_{K\alpha_1} + I_{K\alpha_2} + I_{K\beta_1} + I_{K\beta_3} + \dots} . \tag{8}$$

The transition probability for the L_{γ_1} line was calculated for the lines originating from the L_{111} ionized atom.

6) I. The diffraction efficiency is proportional to the coefficient of reflection R_C. This coefficient is the fraction of a monochromatic beam which is effectively diffracted at the Bragg angle. In this paper the prime interest is how R_C varies with wavelength since lithium fluoride is used for the wavelength range 0.3 to 3.0 A. Compton and Allison[4] reported calculated R_C values of 1.98 and $4.00 \cdot 10^{-5}$ for wavelength 0.71 and 2.94 A, respectively, for single crystals of calcite. The corresponding values measured experimentally were 2.31 and $4.70 \cdot 10^{-5}$. Theoretically, R_C is related to the square of the wavelength for mosaic crystals; therefore, a much higher diffraction efficiency would be expected for longer-wavelength radiation. The R_C values given on page 404 of Compton's and Allison's book were used for all calculations.

As the analyzing crystal may not accept the entire flux from the collimator at small theta angles, a second term must be included:

$$\% \; E.F. = \frac{l \sin \theta \cdot 100}{0.75 \text{ in. (width of collimator)}} ; \tag{9}$$

where $\% \; E.F.$ is the percentage of effective flux from the collimator, l is the length of crystal in inches, and θ is the Bragg angle.

Since $n\lambda = 2d \sin \theta$, equation (10) can be expressed more usefully as

$$\% \; E.F. = \frac{n\lambda}{1.5d} \cdot 100 . \tag{10}$$

7) E. The scintillation counter (Norelco No. 52245) is rated to have a quantum counting efficiency in excess of 95% for the wavelength range 0.3 to 1.0 A. In addition this counter intercepts the total flux of the diffracted radiation.

The counting efficiency of the gas flow proportional counter (Norelco No. 62033) varies with wavelength as indicated by the gas absorption data in Table VI. Since this counter has an opening absorption data $\frac{3}{8}$ by $\frac{5}{8}$ in., only one-half of the total flux from the collimator enters the counting chamber.

Table III. L Shell and L_{111} Subshell Absorption
Jump Ratios for Elements with
Atomic Numbers 47 to 82[*]

Element	$\dfrac{r_{L-1}}{r_L}$	$r_{L\,111^{-1}}$
Ag 47	0.84	0.50
Ba 56	0.82	0.67
Ce 58	0.81	0.65
W 74	0.80	0.60
Pt 78	0.82	0.62
Au 79	0.81	0.61
Hg 80	0.80	0.59
Tl 81	0.79	0.58
Pb 82	0.80	0.58

[*] Values calculated from data given by Compton and Allison.[4]

Table IV. L Fluorescence Yields

Element	Fluorescence yield[*]
krypton, 36	$W_L = 0.1$
xenon, 54	$W_L = 0.25$
lead, 82	$W_{L\,111} = 0.32$
thorium, 90	$W_{L\,111} = 0.42$
uranium, 92	$W_{L\,111} = 0.44,\ 0.41$

[*] Literature values.

Table V. Transition Probabilities

Element	P_{K_α}	$P_{L_{\alpha_1}}$
23 V	0.88	
26 Fe	0.89	
29 Cu	0.90	
32 Ge	0.87	
35 Br	0.86	
40 Zr	0.86	
45 Rh	0.84	
47 Ag		0.73
50 Sr	0.80	
74 W		0.74
68 Pt		0.72
90 Th		0.70

Table VI. Calculated P-10 Gas Absorption Data

Wavelength, A	$\frac{\mu}{\rho}(P\text{-}10)$	% Absorbed,[*] $I\text{-}I/I_0$
1.0	31.2	10.0
1.24	56.5	17.3
1.54	103	29.3
1.93	212	50.8
2.29	321	66
2.50	429	76.3
2.74	542	83.8
3.03	687	90.0
3.35	917	95.4
3.59	1100	97.5
3.87	138	26.9
3.93	143	27.1
5.17	304	63.9
6.97	693	90.2
8.32	1094	97.5

[*] 2-cm path length in 90% argon—10% methane mixture.
$\rho = 1.677 \cdot 10^{-3}$ STP.

Sample Calculation. One method to check the agreement between theoretical calculations and experimental results is to compare two lines, e.g., Ca K_α and I L_1, of approximately the same wavelength. Using these spectral lines the diffraction efficiency D and detector efficiency E terms cancel out, giving the following expression for the intensities from equal weights of calcium and iodine:

$$\frac{I_{CaK_\alpha}}{I_{IL_{\alpha_1}}} = \frac{\left(\sum_\lambda I_0 \gamma r FP\right)_{Ca}}{\left(\sum_\lambda I_0 \gamma r FP\right)_I} \tag{11}$$

<u>Calcium</u>

$\lambda = 3.35$A
$K_{edge} = 3.07$A
$\lambda_i I_0 \cdot \gamma = 0.57$
$r = 0.90$
$F = 0.20$
$P = 0.88$

<u>Iodine</u>

$\lambda = 3.14$A
$L_{111edge} = 3.72$A
$\lambda_i I_0 \cdot \gamma = 1.0$
$r = 0.55$
$F = 0.25$
$P = 0.74$

Placing these values into equation (11), the ratio of $J_{CaK\alpha}/J_{IL\alpha_1}$ of approximately unity is calculated, whereas the experimental value is closer to 2/1. Since the samples consisted of microgram deposits of iodine and calcium on filter paper there was a finite sample thickness. Therefore the shorter-wavelength $I_{IL\alpha}$ would be preferentially transmitted. The author considers this to be a satisfactory agreement, considering the uncertainty in many of the terms. For example, refer to equation (6), where C'_a value is given as $0.33 \cdot 10^{-2}$, when λ is expressed in centimeters. Using published mass absorption coefficients, C'_a values of 0.0028, 0.0024, and 0.0018 were calculated for wavelengths of 0.88, 1.43, and 2.29 A, respectively. When determining the optimum choice of spectral line for analysis, factors of ten or more will be common if comparing line intensities and line-to-background ratios of the K and L series, so that factors of two are not significant.

EXPERIMENTAL

K and L Intensities from the Same Element. These tests are designed to compare the K_α and L_α line intensities and line-to-background ratios for elements around atomic number 50. The first sample consisted of a 0.25-mil Mylar film supporting approximately 100 μg of iodine. This sample was prepared by lightly brushing a square inch of the Mylar with an iodine swab. Figure 5 illustrates the excellent intensity and line-to-background ratio obtained with the long-wavelength $I_{L\alpha}$ radiation as compared to the high-energy $I_{K\alpha}$ line.

The calculated ratio of $I_{IL\alpha_1}$ to $I_{IK\alpha}$ was obtained as follows:

	$J_{IK\alpha}$	$J_{IL\alpha_1}$
$\lambda_i I_0 \cdot \gamma$	1.0	22.0
r	0.85	0.67
F	0.85	0.25
P	0.80	0.74
D	$2.2 \cdot 10^{-5} \cdot 0.43$	$6.6 \cdot 10^{-5}$
E	1.0	$0.92 \cdot 0.5$

Calculated $I_{IL\alpha_1}/I_{IK\alpha} = 16/1$ (for infinitely thin sample).

Observed $I_{IL\alpha_1}/I_{IK\alpha} = 5/1$.

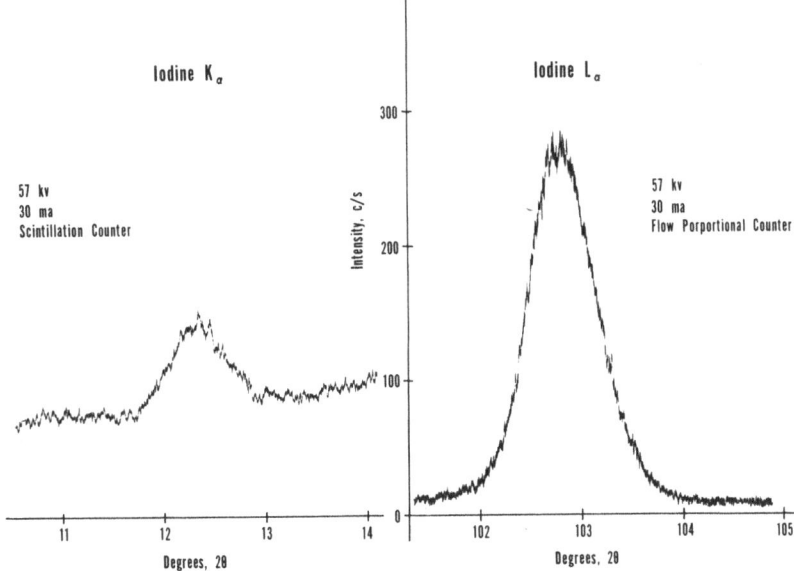

Figure 5. Iodine K_α and L_{α_1} line intensities from microgram deposits.

As expected from the previously derived theoretical expressions the longer-wavelength lines are relatively enhanced over the shorter-wavelength radiation when extremely thin samples are used. Under these conditions only a small fraction of the higher-energy radiation is absorbed in the thin layer while a high percentage of the longer-wavelength radiation ionizes the L shell. Also the transmission coefficient (escape probability) through the thin layer is favorable, resulting in effective excitation of the L lines.

The second series of samples studied consisted of a selectively oxidized layer of antimony on the surface of high-purity lead.[6] The shorter-wavelength antimony K_{α_1} is not very effectively excited since most of the high-energy primary radiation is not absorbed in the antimony layer, as indicated by Table VII. These data show a possible application of the L series radiation for analyses in the parts per million range by preconcentrating the impurity element on the surface of the sample.

Other methods for the determination of impurities in high-purity materials involve preconcentration of the desired elements by precipitation and collection on filter paper or by the use of ion exchange membranes. Table VIII summarizes the K_α and L_α intensity data for elements with atomic numbers 42 to 60, using

Table VII. Antimony K_α and L_α Intensities
on Selectively Oxidized Samples

Antimony, weight %	Intensity, counts/sec	
	Sb K_α	Sb L_α
0.004	–	210.0
0.01	136.0	623
0.02	304	1170
0.02 (unheated)	n.d.	9
0.5	1300	1630
Bg*	120	7

* Background.

Table VIII. Line Intensity and Line-to-Background Ratios
of K and L Lines from Microgram Deposits

Element	K_α , counts/sec			L_α , counts/sec		
	Line	Bg	$\dfrac{\text{Line}}{\text{Bg}}$	Line	Bg	$\dfrac{\text{Line}}{\text{Bg}}$
Mo	2690.0	442.0	6.1	39.0*	1.6	24.4
Cd	308	350	0.9	158**	11.1	14
I	125	342	0.4	354	2.6	136
Ba	19	147	0.1	321	5.3	61
Nd	10	150	0.1	384	21	18

100 μg on filter paper.
L_α 57 kv, 25 ma, flow proportional, LiF.
K_β 57 kv, 25 ma, scintillation, LiF..
*L_{β_1}, NaCl
**L_{β_1}, LiF

filter paper as the supporting medium. These samples were prepared by placing 0.100 ± 0.005 ml of solutions containing 1 g/liter of the desired element on a good-grade ashless filter paper.

Comparable results are obtained with very thin ion-exchange membranes but ratio of the intensities K_α/L_α increased rapidly with thickness of the absorbing layer. With increasing sample thickness equation (4) is no longer applicable since exp $(-x) \neq 0$. Samples such as ion-exchange membranes fall into the intermediate thickness class whereby

$$I_A = \frac{\sum_{i=D.H.}^{i=C.E.} \lambda_i I_0 \,_{\lambda_i} \gamma rFPDEA \left[1 - \exp(-Mx) \right]}{M} \tag{12}$$

where $M = \dfrac{\lambda_i \mu}{\sin\theta_1} + \dfrac{\lambda_A \mu}{\sin\theta_2}$.

As the absorption coefficients of the elements are proportional to λ^3, M will increase rapidly with increasing wavelength. Therefore the longer-wavelength radiation is very highly absorbed in thick ion-exchange membrane. Figure 6 shows the very rapid decrease in transmission with increasing wavelength

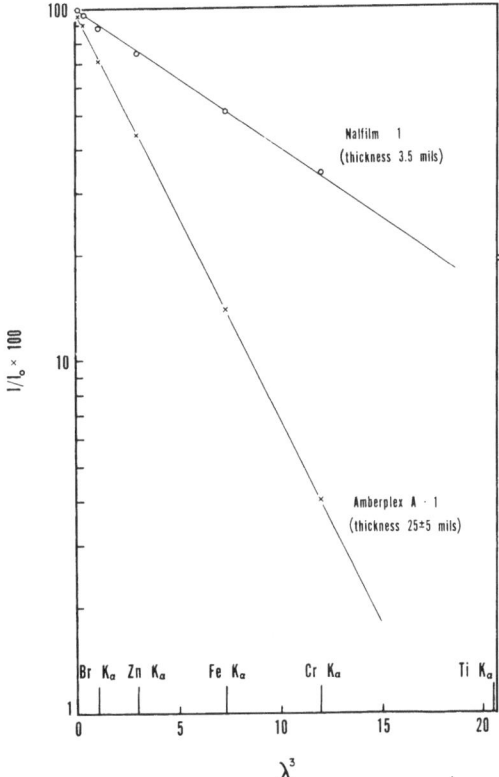

Figure 6. Transmission coefficients for ion-exchange membranes.

Table IX. Line Intensity and Line-to-Background Ratios of
K and L Lines from Infinitely Thick Samples

Element	K_α , counts/sec			L_α , counts/sec		
	Line	Bg	$\dfrac{\text{Line}}{\text{Bg}}$	Line	Bg	$\dfrac{\text{Line}}{\text{Bg}}$
Mo	30,000	2,500	12.0	92[*]	11.0	8.4
Cd	12,000	3,000	4.0	370[**]	8.1	44
I	4,600	2,100	2.2	1160	13	89
Ba	4,000	2,400	1.7	1360	23	59
La	2,600	1,900	1.4	1380	29	48

5 g/liter concentration.

L_α 57 kv, 25 ma, flow proportional, LiF.

K_α 57 kv, 7.5 ma, scintillation, LiF.

[*] NaCl crystal L_{β_1} line.

[**] LiF crystal L_{β_1} line.

Table X. Limits of Detectability of Tin
in Silicate Matrix

Line	Counts/sec		Limit of detectability[*]
	I_{Line}	I_{Bg}	
Sn	928.0	465.0	0.02
Sn	123	6	0.002

[*] Line intensity equal to 10% of background.
Sample consisted of 0.39% Sn added to tin-free base.
Sn K_α, 54 kv, 25 ma, 0.005 in. collimator, LiF.

Sn L_α, 50 kv, 45 ma, 0.02 in. collimator, LiF.

using the 25-ml Amberplex membranes, whereas the thinner
Nalfilm membranes can be used in the long-wavelength region.
Thicker membranes have the advantage of greater exchange
capacity and greater rigidity in the sample holder, but have
limited application for radiation longer than 2 A.

When dealing with infinitely thick samples, the depth of sample analyzed with the short-wavelength radiation makes the K_α
line intensity many times stronger than the L series radiation.

However, the background also increases rapidly for the K series; thus the line-to-background ratio is favorable for the relatively weaker L lines in many instances. Table IX lists the intensity data for several elements between atomic numbers 42 and 60, using solutions as infinitely thick samples.

A practical application of the L series to the analyses of infinitely thick samples is illustrated by the data in Table X. The problem was the determination of tin in very low-grade materials. These data indicate use of the L_α radiation increases the sensitivity of tin by approximately a factor of ten. Also, the background is irregular in the tin K_α region, whereas a uniform background is observed in the long-wavelength region. However, it is important to realize that sample preparation is very critical in the long-wavelength region because of the small depth of sample examined.

L and M Intensities from the Same Element. The elements above atomic number 79 emit M_α and M_β spectral lines having wavelengths shorter than 6 A. Therefore, some consideration was given to the possibility of using the M series for these elements in preference to the commonly used L series radiation. Preliminary investigations show that the M series lines of all elements below uranium required the use of high-spacing crystals which have low reflectivity. This low reflectivity combined with other considerations limits the application of M series radiation to elements above thorium. The data in Table XI indicate that the uranium M_{β_1} line can be used for both microgram deposits and infinitely thick samples.

Table XI. Comparison of Uranium L_{α_1} and M_{β_1} Line Intensities and Line-to-Background Ratios

Sample	$U L_{\alpha_1}$, counts/sec			$U M_{\beta_1}$, counts/sec		
	Line	Bg	$\frac{\text{Line}}{\text{Bg}}$	Line	Bg	$\frac{\text{Line}}{\text{Bg}}$
1	594.0	269.0	2.0	197.0	1.5	131.0
2	6590	1010	6.5	432	4.4	98

1) 100 μg uranium on filter paper.

2) 10 g/liter uranium.

Since the M series lines fall in the wavelength range of the K series of elements below atomic number 20 and the L series below elements of atomic number 50, the possibility of spectral interference exists. Published X-ray spectral tables giving the 2θ angles for the various analyzing crystals do not include the M series; therefore these lines should be included in any future publications.

K Spectral Lines for Lower-Atomic-Number Elements.

Because of the low fluorescence yield of the elements below titanium and the small depth of penetration of their characteristic K lines these elements are considered to be insensitive. However, these studies indicate that these elements are not necessarily insensitive since (1) the K lines of adjacent elements are widely dispersed, (2) their K_β lines are quite weak as shown in Table XII, and (3) the backgrounds are very low for reasons discussed in the theoretical section. The wide dispersion of spectral lines in conjunction with weak β lines permits the use of a minimum of collimation with the resulting increase in line intensity. Therefore parts per million of these elements in infinitely thick samples can be detected under favorable conditions, such as shown in Table XIII.

In the microgram deposits or thin layers, the K lines of elements such as calcium are $\frac{1}{4}$ the intensity of the most sensitive elements, e.g., nickel and the line-to-background ratio of calcium is higher. Therefore, limits of detectability do not decrease as rapidly as suggested early in the field of X-ray spectroscopy.

Table XII.   Variation of IK_α to IK_{β_1} Ratio with Atomic Number

Element	Ratio $\dfrac{IK_\alpha}{IK_{\beta_1}}$
19 K	7.0
17 Cl	11
16 S	18
16 P	24
14 Si	30
13 Al	88

Table XIII. Limits of Detectability in Feldspar Matrix

Element	Weight %	Line, counts/sec	Background, counts/sec	Limit of detectability, weight %
K	0.290	400.0	4.0	0.0003
Ca	0.257	290	3.5	0.0003
Ti	0.0102	25	10	0.0004
Fe	0.047	544	124	0.001

*Line intensity equal to 10% of background.
Bureau of Standards Sample No. 99 (feldspar).
55 kv, 35 ma

SUMMARY

Table XIV lists the limits of detectability for elements with atomic numbers 16 to 90 in solutions (infinitely thick samples). From these data it can be concluded that the limit of detectability is somewhat independent of wavelength. Although the shorter-wavelength fluorescent radiation is more intense (because a greater depth of sample is effectively excited) the line-to-background ratio is about the same over the 0.5 to 3.5 A wavelength region. A few elements, e.g., tungsten and copper, have low sensitivity because of the high background resulting from $W L_\alpha$ and $Cu K_\alpha$ lines present in the primary X-ray beam.

The longer-wavelength L lines are preferable to the shorter K lines for most elements between atomic numbers 42 and 60, particularly for Class 2 and 3 samples.

M series lines show little promise for application in X-ray spectrochemical analysis. However, the M lines can interfere with the determination of longer-wavelength K and L lines so that future 2θ tables should include the M series.

The major analytical disadvantage of longer-wavelength radiation is the small depth of penetration into the sample, thus making sample preparation very critical. It is suggested that longer-wavelength lines should not be used for powder mixtures since particle sizes less than 1 μ would generally be required. However, if solution techniques or borax fusion methods are used this particle size problem is eliminated.

ACKNOWLEDGMENT

The author wishes to acknowledge the technical assistance of Mr. John W. Thatcher, analytical chemist, of the Eastern Experiment Station.

Table XIV. Limits of Detectability under Optimum Conditions

Element	Spectral line	Instrumentation-current, ma	Line intensity, counts/sec	Background counts/sec	Detectability, g/liter
16 S	S K_α	A − 25	2.8	2.1	0.14
17 Cl	Cl K_α	A − 25	6.8	2.4	0.062
19 K	K K_α	B − 25	320	14.5	0.0032
20 Ca	Ca K_α	B − 25	311	6.8	0.0023
23 V	V K_α	B − 25	772	19	0.0015
24 Cr	Cr K_α	B − 25	1030	370	0.0051
26 Fe	Fe K_α	C − 25	2000	170	0.0018
29 Cu	Cu K_α	C − 10	2100	350	0.0024
33 As	As K_α	C − 10	3140	460	0.0018
38 Sr	Sr K_α	C − 10	6400	1300	0.0015
42 Mo	Mo K_α	C − 10	8000	3300	0.0020
42 Mo	Mo L_{β_1}	A − 25	18.5	11	0.049
48 Cd	Cd K_α	C − 10	3200	4000	0.0054
48 Cd	Cd L_{β_1}	B − 25	75	8	0.010
53 I	I K_α	C − 10	1200	2800	0.012
53 I	I L_α	B − 25	232	13	0.0042
56 Ba	Ba K_α	C − 10	1070	3200	0.014
56 Ba	Ba L_α	B − 25	275	23	0.0048
57 La	La K_α	C − 10	695	2540	0.02
57 La	La L_α	B − 25	275	29	0.0053
62 Sm	Sm K_α	B − 25	440	45	0.0041
62 Sm	Sm L_α	C − 25	210	62	0.010
70 Yb	Yb L_α	C − 25	675	285	0.0068
79 Au	Au L_α	C − 10	690	2400	0.019
82 Pb	Pb L_α	C − 10	1060	460	0.0055
90 Th	Th L_α	C − 10	1220	850	0.0065

Concentration = 1 g/liter.
A EDDT crystal, helium, 0.02 in. collimator, flow proportional with PHA. 57 kv.
B LiF crystal, helium, 0.02 in. collimator, flow proportional PHA. 57 kv.
C LiF crystal, 0.02 in. collimator, scintillation counter with PHA. 57 kv.

Limit of detectability = that concentration which results in a line intensity equal to three times the square root of the background for two minutes counting times.

REFERENCES

[1] F. K. Richtmyer and E. H. Kennard, "Introduction to Modern Physics," McGraw-Hill Book Company, 4th Edition, 1947, p. 451.

[2] W. A. Jennings, "A Theoretical Study of Radiation Outputs and Qualities from a Beryllium Window Tube Operated at Low Kilovoltages (10-50 kvp)," Brit. J. Radiology, Vol. 26, 1953, p. 193.

[3] W. J. Campbell, M. Leon, and J. W. Thatcher, "Flat Crystal X-ray Optics," Proceedings 6th Annual X-ray Conference, Denver Research Institute, 1957, p. 193.

[4] A. H. Compton and S. K. Allison, "X-Rays in Theory and Experiment," D. Van Nostrand Company, 2nd Edition, 1935, pp. 394-425, 511-540.

[5] R. J. Stephenson, "X-Ray Fluorescence Yields," Phys. Rev. Vol. 51, 1937, p. 637.

[6] W. J. Campbell and M. Leon, "Preliminary Studies of Selective Oxidation in Molten Metals by Fluorescent X-Ray Spectrography," Pittsburgh Conference on Analytical Chemistry and Applied Spectroscopy, Paper 70, March 1958.

THE DIRECT DETERMINATION OF VANADIUM AND NICKEL IN CRUDE OILS BY X-RAY FLUORESCENCE

Victoria R. Lopp and C. G. Claypool

Jersey Production Research Company, Tulsa, Oklahoma

ABSTRACT

A rapid method has been developed for the determination of vanadium and nickel in crude oils in concentrations above 5 ppm. Duplicate analyses for both nickel and vanadium require only one hour, and excellent precision is obtained. At a concentration level of 100 ppm of vanadium, the standard deviation is 3.1 ppm, and at a level of 30 ppm of nickel, the standard deviation is 1.6. The results are comparable to those obtained by spectrophotometric and optical emission spectrographic analysis.

Conversion of the crude oils to greases by the addition of aluminum stearate minimizes the usual difficulties of handling liquid samples in the conventional-geometry fluorescent X-ray spectrometer. Scattered tungsten emission lines from the X-ray tube serve as internal standards to correct for instrumental and matrix variations.

A new sample tray that permits the insertion of four samples into the instrument at a time has been constructed. Each sample can be moved into position for analysis without opening the sample chamber, making it necessary to wait for the helium to sweep the trapped air from the X-ray path only once for each four samples.

INTRODUCTION

The presence of vanadium and nickel in crude oil has always been a great problem to the petroleum industry, and there has been a continued search for rapid methods for the determination of these trace elements. A preliminary examination of crudes by means of X-ray fluorescence revealed that concentrations above about 2 ppm can be detected in the untreated oil. Since many of the crude oils analyzed by Jersey Production Research have higher concentrations of vanadium and nickel, a direct method of analysis was developed.

131

INSTRUMENTATION

The instrument used is a Norelco fluorescent X-ray spectrometer with a helium path and conventional geometry, that is, with the sample positioned below the X-ray tube. The source of primary X-rays is a Machlett OEG-50 tungsten-target tube. The polychromatic fluorescent X-ray beam is analyzed by a lithium fluoride crystal; a gas-flow proportional counter is used as the detector.

When using the standard helium attachment, considerable time is lost in waiting for the atmosphere in the X-ray path to reach a constant composition. Each time the sample chamber is opened to insert a sample, 5 to 10 min are required to sweep the trapped air from the path. A new sample tray (shown in Figure 1) which permits the insertion of four samples at a time has been constructed to reduce this lost time. Each sample can be moved into position for analysis without opening the sample

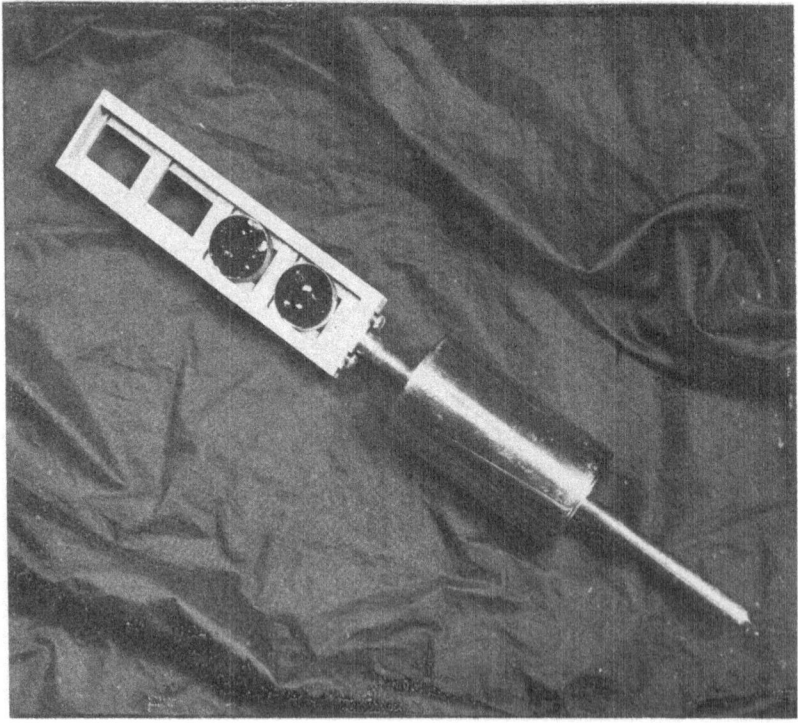

Figure 1. Sample holder.

chamber. The sample tray is designed so that none of the samples are in the X-ray beam where they can become overheated or altered by the X-rays during the waiting period. The sample chamber has been modified to accommodate the longer tray.

The sample holder supplied with the instrument has been replaced by holders cut from linen-Bakelite gear stock. A circular recess 3.5 cm in diameter and 3 mm deep requires only a gram or two of sample.

PROCEDURE

The analysis of crude oils presents several problems in addition to those usually involved in handling liquid samples, since crudes vary considerably in viscosity, volatility, and organic composition. The heat from the X-ray tube causes evaporation of the more volatile crudes and partial loss of sample through creeping. The intensities of three different spectral lines are measured for each sample, and it is necessary that the sample level be the same for each measurement because the intensity of the fluorescent radiation varies with the distance between the sample and the X-ray tube. To avoid the difficulty of loss of sample between and during measurements, we decided to remove the more volatile components before the analysis, and then convert the oil to grease, since a grease will not flow even though it softens with heating.

A weighed 5- to 10-g sample of the oil is heated to approximately 130°C for two hours to remove the lower-boiling constituents. After the oil has cooled and has been weighed, a portion of aluminum stearate, approximately equal to 10% of the weight of the oil after evaporation, is stirred into the oil. The mixture is then heated, with stirring, until it begins to thicken. When the grease has cooled, it is placed in a sample holder and smoothed with a glass slide.

With the X-ray tube at 45 kv and 45 ma, the intensities of the nickel K_α, vanadium K_α, and tungsten L_1 lines are measured. Scattering of the primary X-ray beam differs with variations in crude oil composition. This necessarily affects the measured intensities of the nickel and vanadium lines. To compensate for this matrix effect, as well as instrumental variations and possible differences in sample-to-tube distance, the scattered radiation from the tube is used as an internal standard.[1] The tungsten L_1 line at 49.24° 2θ was chosen because of its proximity

---

[1] Superscripts pertain to references at the end of the paper.

to the nickel line, and because its intensity is of the same order of magnitude as that of the vanadium and nickel lines. The ratios V K_α/W L_1 and Ni K_α/W L_1, rather than absolute intensities, are then used to determine concentrations.

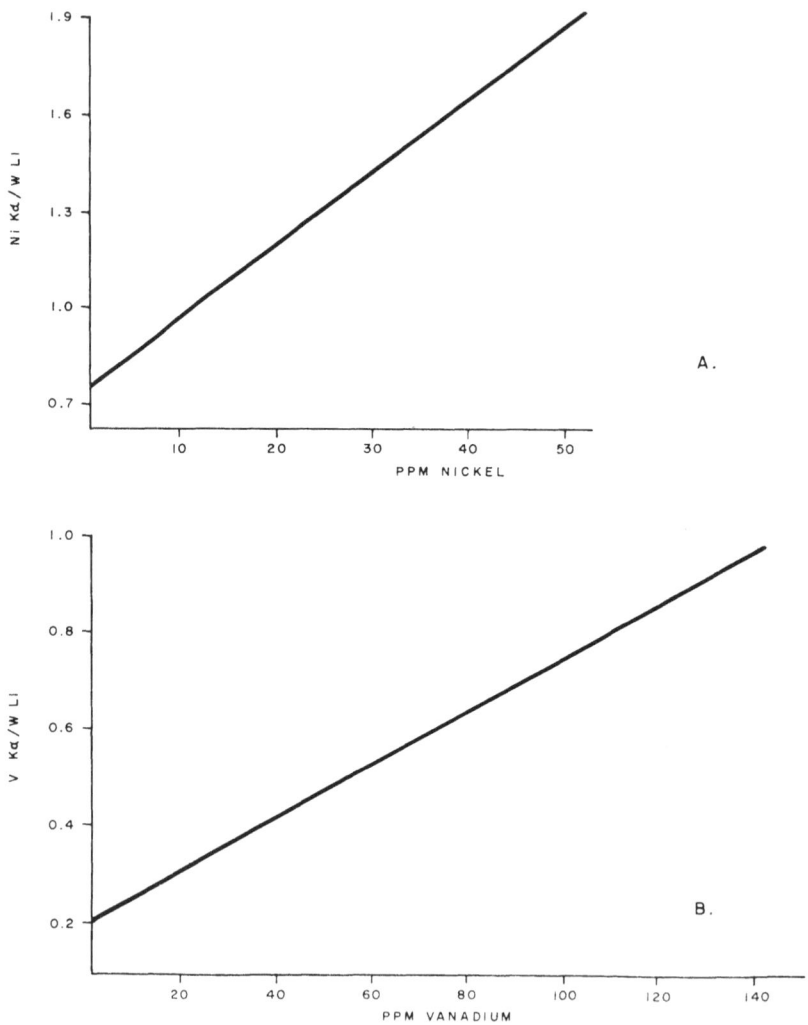

Figure 2. Calibration curves for determination of nickel and vanadium in crude oils.

CALIBRATION

Because of the difficulty in preparing standards of composition similar to that of crude oils, oils which had been analyzed chemically and spectrographically were used in constructing the calibration curves. These oils contained up to 50 ppm of nickel and 150 ppm of vanadium. Linear curves were obtained by plotting V K_α/W L_1 and Ni K_α/W L_1 against concentration, as shown in Figure 2.

RESULTS AND DISCUSSION

In calculating the results, the ratio of the weight after evaporation to the weight before evaporation is multiplied by the value read from the calibration curve to correct back to the original composition. No correction is made for the addition of aluminum stearate since the same percentage is added to all samples. Experiments have shown that varying the quantity of aluminum stearate between 7 and 10% does not significantly alter the intensities of the lines.

At the higher levels of concentration of vanadium and nickel with which we are concerned, excellent precision can be obtained by this method of analysis. Six greases prepared from one crude oil were analyzed on two different days a month apart. As shown in Table I, the standard deviation for the twelve analyses was 3.1 ppm for vanadium, and 1.6 ppm for nickel.

Table I. Precision of Measurement of Vanadium and Nickel

Sample No.	Date of Analysis			
	2/11/58		3/12/58	
	ppm Ni	ppm V	ppm Ni	ppm V
17922-1	25.6	102	25.8	110
17992-2	25.9	108	24.0	106
17922-3	26.8	101	27.6	104
17922-4	29.2	103	28.9	108
17922-5	27.0	106	28.6	110
17922-6	27.6	103	29.2	107
Average (12 analyses)			27.2	105.7
Standard deviation (σ)			1.64	3.05

As a further evaluation of precision, eight analyses were carried out on each of four crude oils varying in composition from 54 ppm of vanadium and 9 ppm of nickel, to 111 ppm of vanadium and 23 ppm of nickel. The results of these analyses are shown in Table II. The analyses were performed by four analysts. The standard deviation calculated from the combined results of these analyses was 2.8 ppm for vanadium and 1.3 ppm for nickel.

Table II. Standard Deviation Based on Multiple Analyses of Four Crude Oils

Sample No.	Average ppm V	Average ppm Ni
1	54	9
2	91	19.5
3	94	17
4	111	23

$$\sigma_V = 2.82 \qquad \sigma_{Ni} = 1.32$$

X-ray fluorescence results obtained compare favorably with those found by wet chemical and optical emission spectrographic analysis for samples containing more than 5 ppm of vanadium and nickel, as shown in Table III.

Because of the relatively few manipulations involved in the analysis, only 1 man-hour is required for duplicate analyses.

Table III. X-Ray Fluorescence Results Compared with Results from Other Methods of Analysis

Sample No.	Emission spec.		Absorption spec.		X-ray fluorescence	
16680	ppm V 44	48	44	46	48	
	ppm Ni 12.5	15	13.4	14	14	
18210	ppm V		297	306	308	
	ppm Ni		74		74	
16693	ppm V	80			82	
	ppm Ni	25			26	
22488	ppm V	4.1			4	3.9*
	ppm Ni	4.4			5.2	4.45*

* Ashed sample analyzed by X-ray fluorescence.

PROCEDURE FOR LOW CONCENTRATIONS

Crudes containing less than 5 ppm of vanadium and nickel are coked with concentrated sulfuric acid in a 250-ml Vycor beaker, and ignited at 500°C in a muffle furnace to concentrate the inorganic portion of the sample. One milliliter of a solution containing 25 ppm of cobalt is added as an internal standard to each oil before ashing is begun. After all the carbon has burned off, the ash is dissolved in hot hydrochloric acid. The volume of the solution is reduced to a few milliliters, and the remaining solution is absorbed in a 30-mm disk of heavy chromatographic paper (Eaton-Dikeman, Grade 652)[2] and dried in an oven at 60°C. The intensities of the nickel, cobalt, and vanadium K_α lines are measured, and the ratios Ni K_α/Co K_α and V K_α/Co K_α are used in determining the concentrations from a calibration curve. These values are then adjusted according to the weight of sample used. A 10-g sample yields enough ash to be analyzed if the crude contains more than 0.5 ppm of nickel and vanadium. This procedure, of course, requires more time than the direct method described for samples with higher concentrations of these elements.

REFERENCES

[1] J. W. Kemp and G. Andermann, Pittsburgh Conference on Analytical Chemistry and Applied Spectroscopy, February 1956.
[2] W. M. MacNevin and E. A. Hakkila, Anal. Chem., Vol. 29, 1957, p. 1019.

A HIGHLY SIMPLIFIED MULTIELEMENT CALIBRATION SYSTEM FOR SEMIQUANTITATIVE X-RAY SPECTROGRAPHIC ANALYSIS

Merlyn L. Salmon

FLUO-X-SPEC Laboratory, Denver, Colorado

ABSTRACT

Semiquantitative results are adequate for the satisfactory solution of many problems involving mineral analyses, and fluorescent X-ray spectrography is gaining more recognition as a satisfactory method for performance of these analyses.

Successful applications of the method in various instances are discussed to demonstrate a system involving minimum sample preparation and the use of instrumental factors in establishing multielement calibration curves.

INTRODUCTION

Mineral analyses by fluorescent X-ray spectrography can be very simple in cases where the over-all composition varies only slightly in a series of samples. The problem becomes more difficult when the matrix varies in composition or when it is necessary to check for the concentration of an element in different mineral systems.

Quantitative analyses, in general, require a minimum of matrix variation from sample to sample for direct analysis by simple calibration procedures; a conversion of variable systems to some standard variety by chemical or physical treatment and subsequent analysis on the basis for calibration of the standard system resulting from the treatment; indirect reference techniques where an internal standard is added to the original sample to provide a means for compensation of matrix variations; or other indirect reference procedures based on measurement of physical properties (i.e., absorption coefficients) of the sample and calibration of the variable systems with physical property parameters, or other parameters available from instrumental factors that can be measured in the course of an analysis by fluorescent X-ray spectrography.

In the latter category above, the value of scattered radiation as an aid in compensating for matrix variations in uranium mineral systems was discussed in reporting results obtained in the Metallurgy Division Laboratories, Denver Research Institute,

139

University of Denver.[1,2] Many similar systems based on reference to background measurements are used and show different ways of evaluating the background or using the values as parameters in the calibration systems. Kemp and associates indicate the value of using scattered X-rays[3,4] in ore, oil, and alloy samples.

There appears to be a greater recognition of the value of fluorescent X-ray spectrography in mineral exploration programs, and this can be attributed to the relative ease and low cost of checking large numbers of samples with reasonably accurate results. In many programs the mineralization within an area is fairly uniform, and it is possible to use simple procedures to obtain results within the desired range of accuracy. In many more programs, however, the accuracy requirements are more stringent or the over-all mineralization varies considerably, and it is not possible to use a direct relationship of intensity and concentration unless the sample is converted, or there is some measurable factor to use as a parameter.

The scattered radiation as a parameter for intensity-concentration calibration is very useful for mineral exploration samples and the accuracy of the results can be improved with very little additional effort compared to the direct correlation of intensity and concentration. The procedure is based on the use of loose-powder samples and this requires little or no additional sample preparation beyond that required to obtain representative portions of the gross exploration sample. The experimental measurements of scattered background radiation intensities and peak intensities for different elements are taken from an automatic chart recording of the type normally run for qualitative analysis of a sample for elements with atomic numbers greater than 21 (above scandium in the periodic table). Studies of several systems for individual elements indicated a fundamental relationship of scattered radiation intensities and mass absorption coefficients of the sample at different wavelengths in the X-ray spectral region. Consolidation of these correlations of mass absorption coefficients and background intensities at different wavelengths revealed a general expression of background intensity as a function of mass absorption that is independent of wavelength within the range of 0.3 to 2.8 A, and this is considered as a basis for the use of background intensities as parameters of absorption behavior by the sample matrix.

---

[1] Superscripts pertain to references at the end of the paper.

Concentration of individual elements in different types of matrices can be correlated with peak and background intensities to compensate for absorption effects by the matrix. A general correlation for several elements provided a multielement calibration system based on K spectra for elements No. 22 to 56, and another system based on L spectra for elements No. 56 to 92.

These multielement calibration curves make it possible to obtain semiquantitative results by evaluation of the information contained in the automatic chart recording normally run for qualitative analysis of the sample. The systems are simple, and yield accuracies within 50% of the actual concentrations with few exceptions.

EXPERIMENTAL PROCEDURE

Data reported were obtained with a 100-kv Norelco spectrograph equipped with a tungsten-target FA 100 X-ray tube, a constant-potential power supply, a scintillation detector, a lithium fluoride analyzing crystal, a 4×0.005 in. spacing collimator between the sample and the crystal, a 2×0.005 in. spacing collimator between the crystal and the detector, and associated electronic and recording mechanisms. The X-ray tube was operated at 50 kv and 40 ma and the scintillation detector at 900 v for all tests. Scanning was done at 4 deg/min (2θ), a scaler time constant of 1 sec, and the indicated scale factor.

This spectrograph has the inverted geometry similar to the three-position spectrograph and the sample holders have been adapted to the use of CaPlug EP-14 sample cups with a 0.00025-in. Mylar window cover held in place with a Kirsch 1404E or Judd 142 plastic curtain ring.

All samples were analyzed as loose powders that had been prepared by grinding in a standard vertical plate pulverizer, a mechanical mortar, or a ball mill agitated in a paint mixer. Samples varied in mesh size according to visual examination; however, no actual measurements of mesh size were attempted. The adequacy of preparation was checked by experimental evaluation of peak intensities as functions of grinding time, and grinding was continued until there was no appreciable change of intensity. The synthetic standards were prepared with commercial oxide preparations in minus 325 mesh silica except in cases such as cesium, rubidium, and strontium. Sodium salts were used for bromine and iodine. Mixing was completed with 15 min of grinding in an alundum ball mill agitated in a paint mixer.

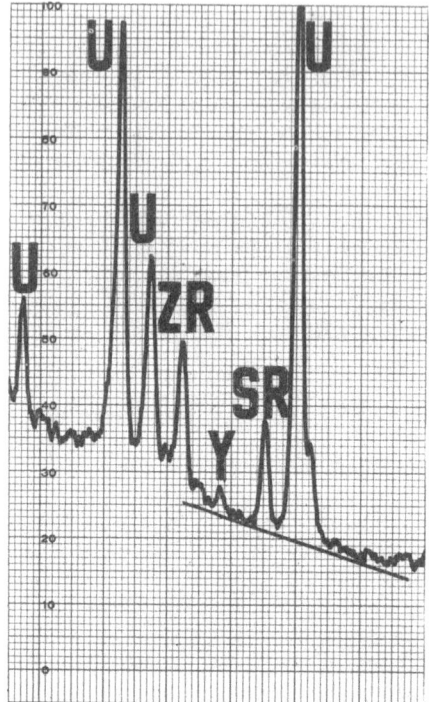

Figure 1. Uranium mineral spectra. 3200
counts/sec full scale.

EXPERIMENTAL RESULTS

Uranium minerals such as carnotite found in the Colorado
Plateau present a simple chart recording of the X-ray spectra,
and as reported in studies of these systems,[1,2] it was possible
to select a standard position in the vicinity of the uranium $L_{\alpha 1}$
peak for measurement of scattered radiation (background). Another alternative as shown in Figure 1 is to use a straight-edge
interpolation of the background on both sides of the peak to measure the scattered radiation intensity at the position of the peak.

As more complex minerals were studied, the interpolation
of background levels in the vicinity of the peak offered the better
basis for measurement of scattered radiation at the peak position
because there could be many interfering lines in the vicinity of
the peak. A curved line was necessary in some cases to inter-
polate the background and the comparison shown in Figure 2

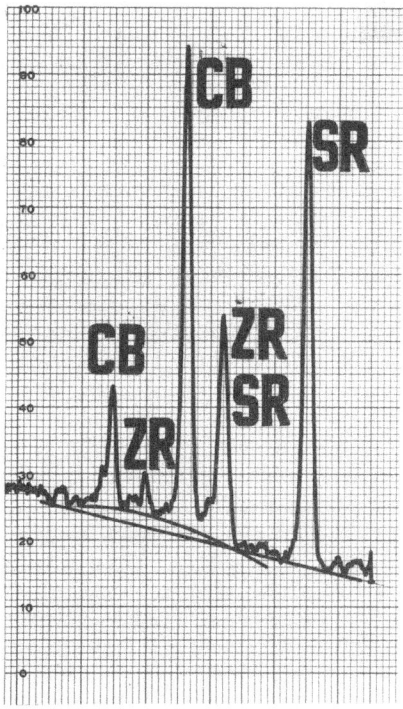

Figure 2. Columbium mineral spectra.
3200 counts/sec full scale.

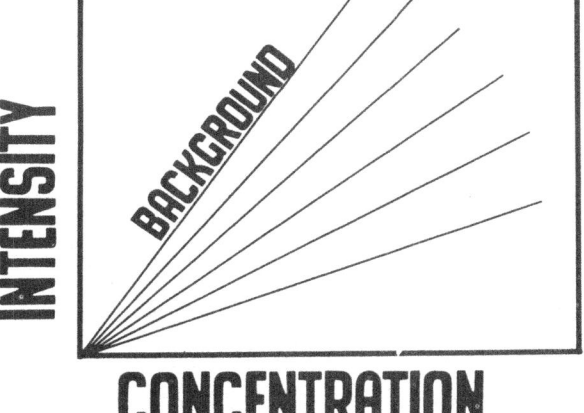

Figure 3. Calibration curve for individual elements.

indicates the difference between the values obtained with a curved line and the straight line. In a study of columbium systems, the curved-line interpolation offered a more consistent correlation of intensity-concentration calibration curves with background parameters.

Studies of several systems with chemically analyzed standards and synthetic standards revealed a series of calibration curves assuming the general form shown in Figure 3, and the use of these curves for individual elements greatly improved the results for systems with variable matrices compared to the results obtained with no compensation for matrix effects.[5] These studies indicated that results could be obtained within $\pm 10\%$ of the actual concentration if proper evaluations of the data were made on the basis of a specific calibration curve for that element. Low-grade samples (less than 10%) can generally be directly evaluated and in some systems it is possible to evaluate some higher-grade samples directly. Higher-grade samples can generally be more accurately analyzed with a simple dilution of the original sample with silica. This method is essentially the same as that used for the preparation of synthetic standards. Ratios of 9 parts or 99 parts of quartz to 1 part of sample are convenient, depending on the range of concentration of the original sample. Nine-to-one dilutions will generally require some attention to the background parameters, but the 100-to-1 dilutions can generally be handled with direct reference to simple calibration curves for synthetic standards prepared with silica matrices (99% plus) and the element in question.

Preparation of diluted samples and synthetic standards requires an adequate mixing of the materials, and the extent of mixing greatly affects the accuracy in establishing calibration curves. A recommended method is grinding in a ball mill agitated by a paint mixer (same basic approach as used in the commercially available Pica and Spex machines). Better results were obtained by this method than by mixing in a mechanical mortar or by mixing in an agitated container without the drastic grinding action of the ball mill. The validity of this approach was checked by evaluation of peak intensity as a function of grinding action until a plateau of intensity level was reached when additional grinding no longer affected the intensity. An additional check of the method was made by the comparison of chemically analyzed standards with synthetic standards. In general, 10 g of sample are ground for 15 min in an alundum ball mill (Spex 8003) with three $^3/_4$-in. and three $^3/_8$-in. porcelain balls agitated in a paint mixer (Red Devil Model 30).

Synthetic standards are very important in the establishment of intensity-concentration curves with background parameters because it may be difficult to obtain series of chemically analyzed samples which exhibit variations in matrix composition as well as a range of concentration of the element in question. In some cases it is also difficult to obtain chemically analyzed samples that warrant consideration as standards.

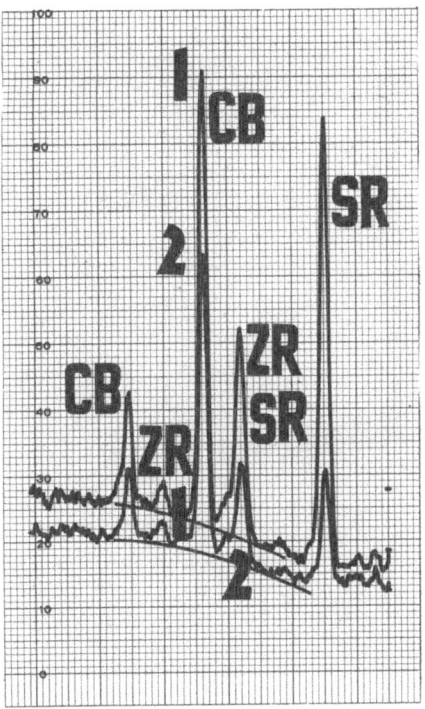

Figure 4. Columbium mineral spectra. 3200 counts/sec full scale. Sample No. 1 contains 0.24% Cb. Sample No. 2 contains 0.26% Cb.

In Figure 4 the comparison of the X-ray spectra for two samples with reliable and nearly equal chemical values for columbium is shown to indicate the correlation of scattered radiation with peak intensity. Synthetic standards are compared in Figure 5 for samples with equal concentrations of columbium in a silica matrix and in a matrix of 60% silica plus 40% iron oxide.

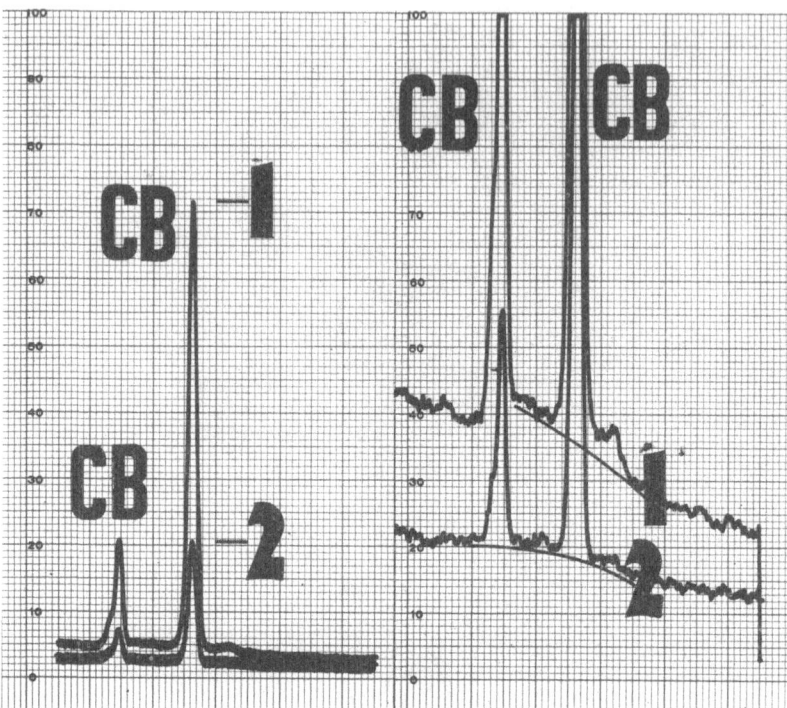

Figure 5. Synthetic columbium standard spectra. Left—25,600 counts/sec full scale. Right—3200 counts/sec full scale. Sample No. 1 contains 1% columbium pentoxide in a silica matrix. Sample No. 2 contains 1% columbium pentoxide in a matrix of 60% silica and 40% Fe_2O_3.

A convenient series of matrix compositions for preparation of synthetic standards can be compounded with various ratios of iron oxide and silica where the iron presents no problem in the analysis. Blank spectra in the region of 3 to 28° 2θ are shown in Figure 6. The ratios of iron oxide and silica are indicated in the range of 0 to 40% iron oxide. The scattered radiation from these samples brackets the range of background parameters for normal samples of minerals with elements exhibiting X-rays in this region. The zirconium and barium indications are from contamination of the silica and the indications of cadmium in the blank result from the coloring agent in the CaPlug EP-14 sample cups.

The variation of columbium peak intensities for a series of samples with equal columbium concentrations in the different

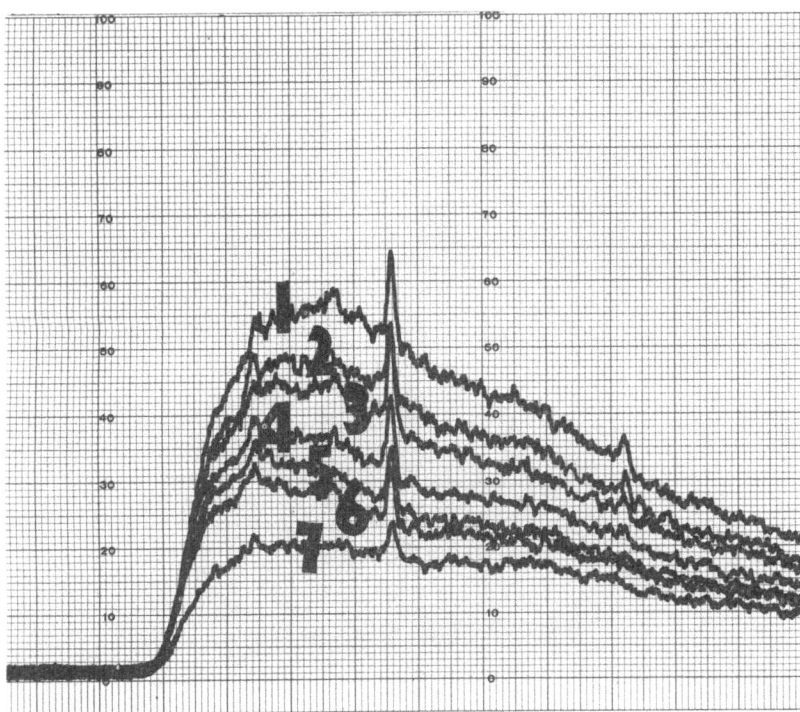

Figure 6. Spectra of iron-silica mixtures. 3200 counts/sec full scale. Range of 0.3 A to 1.0 A. The percentages of silica are shown for the various samples and the balance is Fe_2O_3. No. 1—100, No. 2—95, No. 3—90, No. 4—80, No. 5—70, No. 6—60, and No. 7—0.

matrices of iron oxide and silica is shown in Figure 7. The variation of scattered radiation intensities for the columbium peak position can be noted for the matrices of iron oxide and silica containing no columbium in Figure 6. Low concentrations of columbium will cause no changes in the background intensities exhibited by these matrices in the columbium peak position with the determination of intensities by the interpolation of the average background in the vicinity of the columbium peak. Using the scattered radiation intensities as background parameters, and the columbium peak intensities from Figure 7, a few of the points necessary to define a family of curves such as shown in Figure 3 are available. Study of sufficient samples will clearly define the family of calibration curves for an element in a variety of matrices.

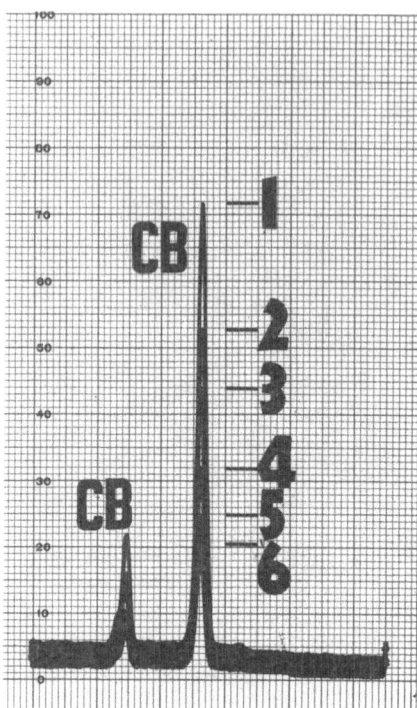

Figure 7. Synthetic columbium standard spectra. 25,600 counts/sec full scale. 1% columbium pentoxide in the matrices as indicated by number in Figure 6.

The iron oxide—silica mixtures provide a convenient series of synthetic matrices; however, it is sometimes advisable to use materials that more closely duplicate the composition of natural minerals. Common host materials are lime, potash, and oxides of other light elements. Blank spectra for some of the common host materials are shown for the region of 3 to $28° 2\theta$ in Figure 8. The variations in scattered radiation intensities for this region are clearly evident and additions of yttrium oxide to these host materials were studied to determine the correlation of yttrium peak intensities with concentration on background parameters. Yttrium peak intensities for 1% yttrium oxide in some of the host materials are indicated in Figure 9. The correlation is similar to that previously discussed for columbium in a series of iron oxide—silica mixtures. This indicates the importance of background as a measurable quantity

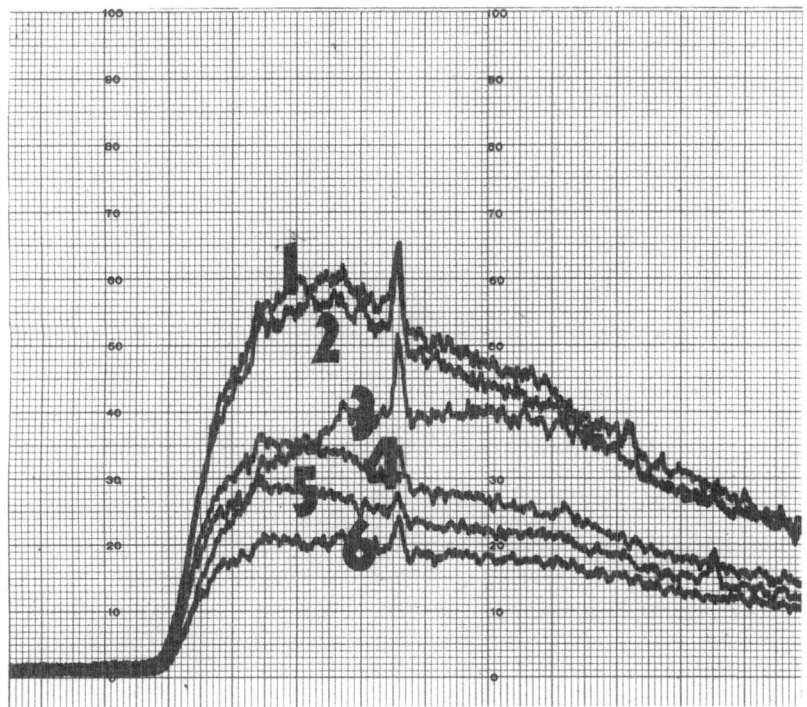

Figure 8. Spectra of common host minerals. 3200 counts/sec full scale. Range of 0.3 A to 1.0 A. No. 1—MgO, No. 2—Al_2O_3, No. 3—SiO_2, No. 4—K_2CO_3, No. 5—CaO, and No. 6—Fe_2O_3.

that can be correlated with concentration vs intensity calibration information and provide a basis for consideration of matrix effects.

Studies of many systems of natural mineral standards and synthetic standards revealed calibration curves for several elements with the general form shown in Figure 3. Especially in the study of synthetic standards, another pattern became evident and that is the general correlation of background intensity with the over-all mass absorption coefficient of the matrix. Calculated mass absorption coefficients based on values from Victoreen[6] for the iron oxide—silica mixtures and common mineral host materials are compared in Table I with the observed background intensities at a wavelength of 1 A (28.75° 2θ for LiF). An inverse relationship of the mass absorption coefficients and scattered radiation intensities is evident and a log-log plot of the values in Table I indicates that this relationship can be expressed as follows:

$$I_b = 1480\mu_m^{-0.37}, \tag{1}$$

where I_b is the scattered radiation intensity in counts per second, and μ_m is the mass absorption coefficient at the wavelength of the scattered intensity measurement.

Table I

Matrix	Mass absorption coefficient (g/cm^2)	Background intensity (counts/sec)
MgO	8.0	640
Al_2O_3	9.0	640
SiO_2	10.1	640
95% SiO_2 5% Fe_2O_3	13.1	570
90% SiO_2 10% Fe_2O_3	16.1	510
80% SiO_2 20% Fe_2O_3	22.2	480
K_2CO_3	26.1	480
70% SiO_2 30% Fe_2O_3	35.2	390
CaO	37.7	390
60% SiO_2 40% Fe_2O_3	41.3	350
Fe_2O_3	70.4	320

The correlation of scattered radiation and absorption per equation (1) is significant, and indicates some validity in the use of scattered radiation intensities as parameters in families of curves relating peak intensity and concentration for different elements. Previous studies of the experimental measurement of absorption by the sample by using a thin layer of the sample as an absorption filter in the monochromatic X-ray beam for the element in question have shown the use of absorption (or transmittance) values as parameters for the correlation of peak intensity and concentration. These experimentally measured absorption coefficients provided a basis for improved analytical results with compensation for matrix variations in a series of samples.[1,2,7,8]

The correlation of scattered radiation and absorption explains the similarity of the calibration curves using background parameters as shown in Figure 3, and the curves using absorption (or transmittance) values as shown in Figure 1 of reference 8.

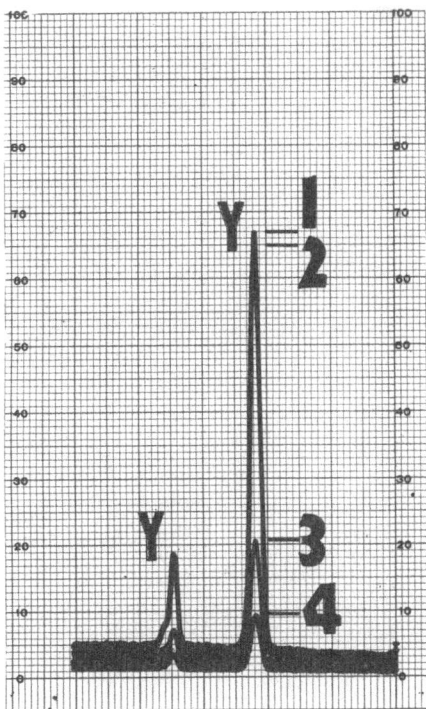

Figure 9. Spectra of synthetic yttrium
standards. 25,600 counts/sec full scale.
1% yttrium oxide in the host materials
indicated by number in Figure 8.

Since equation (1) is based on an intermediate value of wavelength for the spectral region of interest for elements with atomic numbers greater than 21, the possibility of using this equation
in other parts of the region was checked. Calculated values of
scattered radiation intensity are compared with experimentally
observed values in the region up to 2.5 A in the tabulation of
Table II. Comparisons of experimental and calculated background
intensities in the region from 0.4 to 1.5 A indicate a general
agreement within 30% of the calculated values. At longer wavelengths of 2.0 and 2.5 A, the experimental value is approximately
one-third of the calculated value. Absorption effects by the air
path of the instrument are the most probable cause for deviation
at the longer wavelengths and if helium is used for the region of
2.5 A, the intensity of vanadium K_{α_1} radiation is increased by a
factor of three or four compared to the intensity in an air path.

Table II

Wavelength (A)	Background intensity (counts/sec)	
	observed	calculated
2.50	72	240
2.10	96	
2.00		300
1.93	120	
1.54	280	
1.50		410
1.43	280	
1.25	407	
1.20		520
1.17	456	
1.04	586	
1.00		630
0.92	713	
0.83	873	
0.80		790
0.78	996	
0.71	1226	
0.60		1080
0.53	1576	
0.51	1604	
0.50		1290
0.49	1706	
0.40	1626	1590

In view of the results indicated in Table I and Table II, it is apparent that some correction of absorption effects by the sample matrix can be considered on the basis of peak intensity and scattered radiation intensity (background) measured at the peak position. The mechanism for these measurements is summarized in Figure 10, which shows the chart record for a columbium sample. It is necessary to scan a region to include checks for possible interfering elements in the columbium peak vicinity and to scan initially at a sensitivity that allows for good measurement of the background level. It may be necessary to rerun the peak at a decreased sensitivity to measure the peak intensity.

Experimentally measured peak intensities and background intensities for a series of columbium samples are correlated in Figure 11. Each sample contains 1% columbium and the ma-

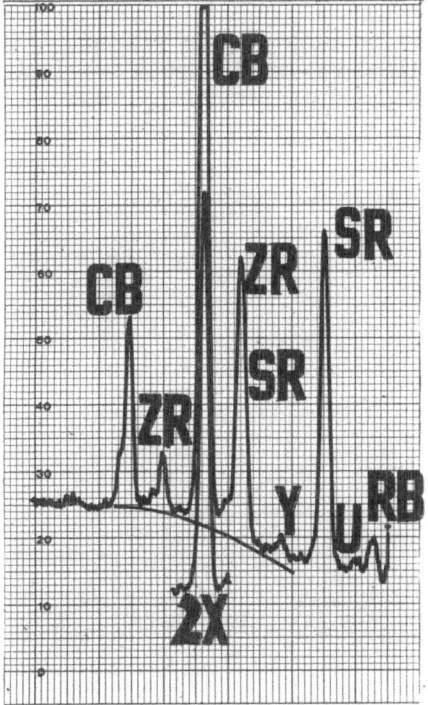

Figure 10. Columbium mineral spectra.
3200 counts/sec full scale for scan and
6400 counts/sec for peak rerun.

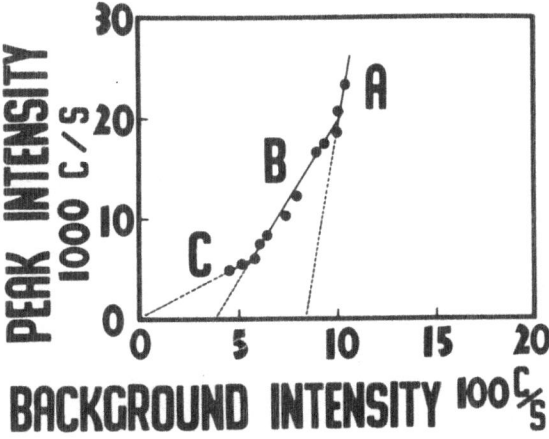

Figure 11. Columbium calibration curve.

trix is different for each sample. Three regions of the curve can each be defined by simple linear relationships of peak intensity and background intensity as follows:

Region A with background intensities above 1000 counts/sec;

$$\% \text{ Cb} = \frac{I_p}{93(I_b - 840)}. \tag{2}$$

Region B with background intensities between 600 and 1000 counts/sec;

$$\% \text{ Cb} = \frac{I_p}{30(I_b - 380)}. \tag{3}$$

Region C with background intensities below 600 counts/sec;

$$\% \text{ Cb} = \frac{I_p}{10 I_b}, \tag{4}$$

where I_p is the peak intensity in counts per second, and I_b is the scattered radiation intensity in counts per second. A log-log plot of the data in Figure 11 indicates the correlation

$$\% \text{ Cb} = \frac{I_p}{0.01 \ I_b^{2.1}}. \tag{5}$$

The correlations of peak intensity and background intensity indicated in equations (2), (3), (4), and (5) are easily established with 10 to 20 synthetic standards for an element with selected matrices to cover a range of background intensities. Contrasted to the type of calibration curves shown in Figure 3, the system of equations is much easier to establish and more convenient to apply. The use of linear equations for selected ranges of background intensity or the log-log equation for the over-all range of background intensity appears to offer the same general degree of accuracy as indicated in Table III for natural standards that have been chemically analyzed and synthetic standards that have been physically compounded.

In general, the data of Table III indicate agreement within $\pm 10\%$ of the amount of columbium present except for two of the chemical standards, and these are within 20%. The important consideration is that a curve established with synthetic standards yields equations that can be applied to other synthetic and natural samples with reasonable accuracy for the very simple experimental procedure used.

Table III. Columbium Concentration (% Cb).

Chemical assay (natural sample)	Calculated	
	Equation C (linear)	Equation E (log log)
0.24	0.24	0.25
0.26	0.27	0.26
0.29	0.30	0.30
0.31	0.31	0.29
0.42	0.48	0.49
0.43	0.44	0.46
0.46	0.46	0.48
0.72	0.82	0.85
Compounded value (synthetic)		
0.17	0.16	0.17
0.35	0.36	0.39
0.70	0.70	0.76
1.4	1.4	1.5
2.8	2.7	2.8
5.6	4.8	5.2

As calibration equations for several elements were compared, it was noted that curves for adjacent elements could be used interchangeably in many instances with reasonably accurate results. Extension of this study revealed that there is a general pattern for elements emitting K series X-rays in the region of 0.3 to 2.8 A, and that the exceptions to this general pattern are explainable. A similar pattern was also noted for elements exhibiting L series in this region. Both general patterns indicate an obvious dependence of K or L X-ray peak intensity on background intensity as well as concentration of the element emitting the X-ray.

Peak intensities for $K_{\alpha 1}$ X-rays for elements with atomic numbers between 21 and 57 and background intensities measured at the wavelengths of these X-rays are indicated in Figure 12. Up to atomic number 42 (molybdenum) there is a correlation of peak intensity and scattered background intensity, with the major exceptions of nickel and zinc. Elements No. 48 to 56 show decreasing peak intensities as the atomic number increases with

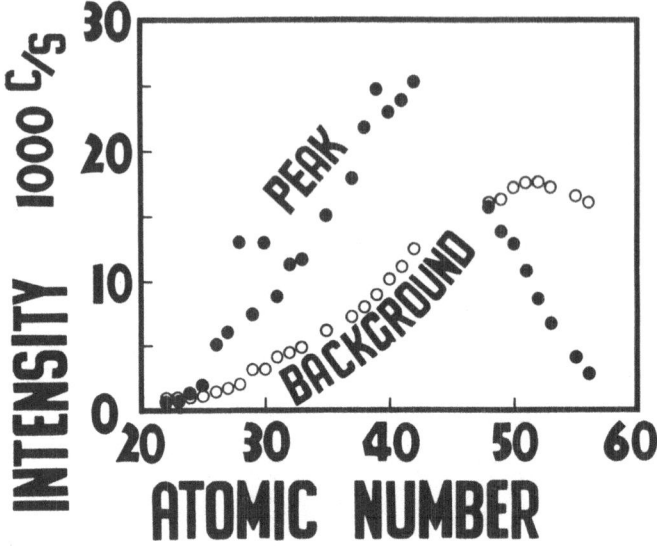

Fig. 12. Peak and background intensities for 1% standards.

a high and fairly constant background intensity. Considering the excitation potential of 50 kv, this departure from the general pattern by elements No. 48 to 56 is logical. Zinc and nickel peak intensities are above the general level for other adjacent elements, but this enhancement is explainable by noting the characteristic lines from the tungsten target of the X-ray tube that are in the vicinity of the zinc and nickel K-absorption edges.

The intensities indicated in Figure 12 were determined with synthetic standards containing 1% of the element in a silica matrix.

The dependence of peak intensity on background intensity is clearly shown in Figure 13 and the same exceptions to the general correlation are noted for the same reasons as mentioned previously. Due to the experimental procedure an alternate abscissa, "Atomic Number," ranging from 20 to 60 corresponding to the range of 0 to 2000 for "Background Intensity" will indicate association of elements with the plotted points relating peak and background intensities. An approximate linear correlation of peak and background intensities is noted by a straight line in Figure 13 and the equation of this line is as follows:

$$I_p = 25 I_b .\qquad(6)$$

Since equation (6) is based on 1% standards, it is possible to express concentration as a function of peak and background intensities:

$$\% \text{ elemental concentration} = \frac{I_p}{25 I_b},\qquad(7)$$

where I_p is the peak intensity in counts per second, and I_b is the scattered radiation intensity in counts per second.

One-percent standards were evaluated on the basis of equation (7) and the calculated concentration determined from experimental measurements of peak and background intensities of $K\alpha_1$ peaks was compared to the actual concentration for different elements. The resulting ratios of calculated to actual concentrations (concentration factors) are shown for the different elements in Figure 14. Concentration factors significantly greater than unity are indicated for iron, cobalt, nickel, and zinc. These

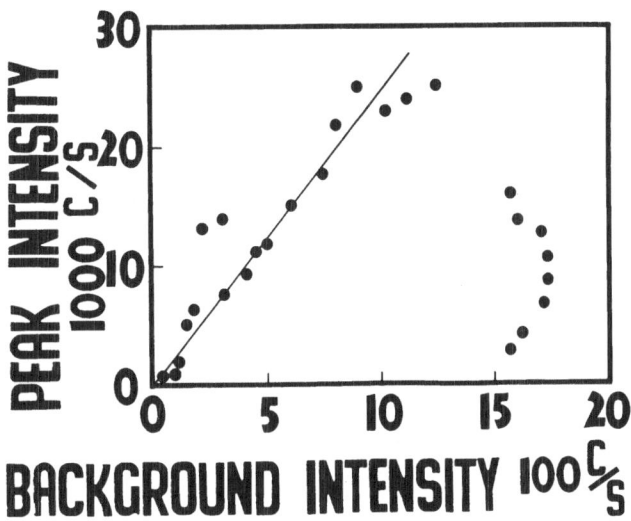

Figure 13. Peak vs background intensity. Alternate abscissa is "Atomic Number" with range of 20 to 60.

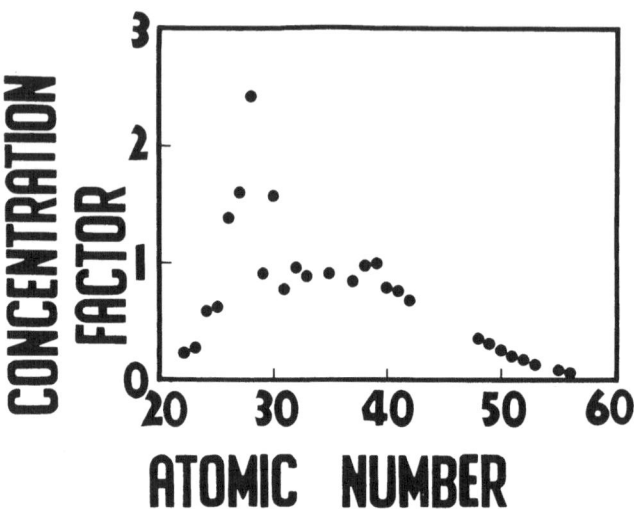

Figure 14. Concentration factors for elements No. 22 to 56. Based on K_{α_1} X-rays.

values are logical on the basis of enhancement effects due to tungsten radiation above the level of the average continuum. Elements No. 29 to 39 (except zinc) show concentration factors slightly less than unity. Titanium, vanadium, chromium, and manganese show factors ranging from 0.24 to 0.62 with concentrations calculated from equation (7) significantly less than actual concentrations. These results are similar to those noted in Table II when experimental background intensities in this wavelength region were less than the calculated intensity values. Absorption by the air path of the instrument is the logical explanation, especially when the increase in radiation is noted with the use of helium to displace air in the optical path of the spectrograph. Above atomic number 39 a steadily decreasing concentration factor is indicated, and consideration of the increasing excitation potential as the atomic number increases explains the trend in concentration factors as an effect of "less than optimum" excitation potential with the X-ray tube operated at 50 kv.

Table IV

	% Composition		% Deviation
	Chemical	X-ray	
TiO$_2$	0.49	0.31	37
	0.58	0.72	24
	1.29	1.2	7.0
	2.93	2.7	7.8
			Average 19
V$_2$O$_5$	0.45	0.53	18
	1.05	1.2	14
	1.35	1.6	19
	2.65	2.7	1.9
	3.73	3.6	1.9
			Average 11
Cu	0.100	0.11	1.0
	0.124	0.18	45
	0.144	0.14	2.8
	0.200	0.24	20
	0.224	0.22	1.8
	0.268	0.34	27
	0.320	0.38	19
	0.396	0.38	4.0
	0.552	0.51	7.6
	0.840	0.74	1.2
	0.948	0.95	0.2
			Average 12
MnO$_2$	4.17	3.6	14
	4.34	4.6	5.5
	9.21	9.7	5.3
	11.65	14.2	22
			Average 12
SnO$_2$	0.23	0.14	39
	0.96	0.64	33
	3.39	2.6	36
			Average 36

Modification of equation (7) to compensate for explainable deviations of individual elements from the generalized correlation of concentration as a function of peak and background intensities includes consideration of the individual concentration factors:

$$\% \text{ elemental concentration} = \frac{I_p}{25 f_c\, I_b}\,, \qquad (8)$$

where I_p is the peak intensity in counts per second, I_b is the scattered radiation intensity in counts per second, and f_c is the concentration factor. Several mineral samples were checked for concentrations of different elements with the aid of the generalized expression of equation (8). Results are compared to chemical values in Table IV. The average deviations are reasonable for semiquantitative analyses, and the maximum deviations are tolerable in consideration of the wide variation of sample types checked by the simple procedure of using equation (8) to determine concentration as a function of peak intensity, background intensity, and the concentration factor for individual elements. The intensity values are obtained by examination of the automatic chart recording, normally run as a basis for qualitative analysis of a mineral sample.

DISCUSSION AND CONCLUSIONS

The use of a multielement calibration system for semiquantitative analyses is advisable when a reasonable degree of accuracy can be obtained with very little more effort than that required for qualitative analysis of the sample. The system discussed is based on the use of loosely packed powder samples, with additional treatment of the sample required only in the case of high concentrations of an element in the mixture. This sample treatment is a simple dilution and physical mixing of a portion of the sample with silica.

The fundamental correlation of scattered radiation (background) intensity with the over-all mass absorption coefficient of the sample justifies to some extent the use of background intensity as a factor to aid in the correction for matrix absorption effects.

Calibration systems for individual elements may be established by the use of synthetic and natural standards to correlate concentration, peak intensity, and background intensity; and these systems can be expressed as families of curves relating concentration and peak intensity with parameters of background intensities, or by the expression of equations which indicate concentration as a function of peak and background intensities.

The multielement calibration system is based on a generalized expression of concentration as a function of peak intensity, background intensity, and a concentration factor. The concentration factor is used to compensate for the deviations of the individual elements from the general correlation of concentration as a function of peak and background intensities for elements with atomic numbers 22 to 56. The multielement system for these elements is established by observation of the K -series X-rays, but another system for higher-atomic-numbered elements can be established with the values for L series X-rays. Elements with atomic numbers less than 22 were not evaluated by the multielement system.

Although the multielement calibration system is based on very simple experimental procedures, attention to the following details is reflected in the range of accuracy obtained:

1. Consistency in sample preparation.
2. Reproducibility of operating conditions of the equipment from one workday to the next. The use of extensive reference standards is advisable with adjustment of the operating variables to obtain a constant value for the reference standard from day to day or even at periodic intervals during the day.
3. Precision of automatic chart recording with the optimum balance of recorder sensitivity, ratemeter and scaler stability, and goniometer scanning rate.
4. Consistency in chart interpretation to include intensity measurements of peak and background and notation of conditions affecting these intensities, such as adjacent elements interfering or causing enhancement of the peak intensities. Enhancement conditions should be studied independently for each element since the multielement calibration system requires some correction when enhancement of the peak intensity for the element is significant. A series of samples containing a range of concentrations of the element causing the enhancement can be used to establish empirical correction factors, or the net intensity at the absorption edge of the element exhibiting the enhanced peak intensity can be used in lieu of background intensity to calculate concentration.
5. Validity of calibration standard values. Proper techniques are necessary for the preparation of synthetic standards.

While the results presented here for the K series multielement calibration system were obtained with fixed operating conditions, the variation of collimation ratio, operating power of
X-ray tube, choice of X-ray tube (Mo vs W target tube), and other
systematic variables caused changes that were predictable within
the limits of the general system. The values of coefficients, ex-
ponents, and concentration factors are unique for a set of operating conditions and a particular instrument. The general equation
relating peak intensity, background intensity, concentration factor, and concentration of the element is established with known
standards containing elements with atomic numbers 29 to 39
(except zinc) which have concentration factors close to unity.
Concentration factors for other elements are determined by
measuring peak and background intensities with known standards
containing these elements, and comparing the experimentally
calculated concentrations from the general equation with an
assumed value of one for the concentration factor to the known
concentrations in the standards. The ratio of experimentally
calculated concentration to the actual concentration is the concentration factor for future use of the general equation for that
particular element. A similar procedure is used to establish
the general equation for the elements exhibiting L series X-rays
in the range of 0.3 to 2.8 A.

REFERENCES

[1] M. L. Salmon and J. P. Blackledge, Pittsburgh Conference on
Analytical Chemistry and Applied Spectroscopy, Pittsburgh,
Pennsylvania, February, 1955.
[2] M. L. Salmon and J. P. Blackledge, Pittsburgh Conference on
Analytical Chemistry and Applied Spectroscopy, Pittsburgh,
Pennsylvania, March, 1956.
[3] J. W. Kemp and G. Andermann, Pittsburgh Conference on Analytical Chemistry and Applied Spectroscopy, Pittsburgh, Pennsylvania, March, 1956.
[4] G. Andermann and J. W. Kemp, Anal. Chem., Vol. 30, 1958,
p. 1306.
[5] M. L. Salmon, Ninth Annual Symposium on Spectroscopy, Amer-
ican Association of Spectrographers, Chicago, Illinois, June,
1958.
[6] J. A. Victoreen, J. Appl. Phys., Vol. 20, 1949, p. 1141.
[7] M. L. Salmon and J. P. Blackledge, Norelco Reptr. Vol. 3, 1956,
p. 68.
[8] M. L. Salmon, Seventh Annual Conference on Industrial Applications of X-Ray Analysis, Denver, Colorado, August, 1958.

AN X-RAY FLUORESCENCE METHOD OF ANALYSIS OF MICROSAMPLES

B. G. Reisdorf

United States Steel Corporation
Applied Research Laboratory, Monroeville, Pennsylvania

ABSTRACT

A method is described for the analysis of microscopic samples that utilizes a commercial electron diffraction unit for the excitation of fluorescent X-rays in a specimen and photographic recording of the resulting spectra by a small fluorescence-analysis camera placed within the electron diffraction unit. Applications of this method to the analysis of precipitates extracted from steels are included in the discussion.

INTRODUCTION

Experienced metallographers commonly identify the phases observed in micrographs by their similarity in appearance to previously ascertained structures; but the interpretation of micrographs must always, in the first instance, be founded on supplemental information concerning the elemental composition of the observed constituents. At times, this information can be obtained by chemical analysis. However, chemical analyses often require the separation and collection of minor constituents prior to their analysis, and in many instances of metallurgical importance, the separation and collection techniques are unsatisfactory. Another, and more usual, method is to employ electron diffraction, especially for analysis of fine precipitate particles. The diffraction pattern is obtained from an extraction replica, which consists of a thin plastic or carbon film with the particles adhering to the film. The extraction replica is prepared by etching the metal surface with an etchant that leaves the particles of interest in relief, depositing a plastic or carbon film on the etched surface, and etching a second time through the film to free the particles from the metal and leave them adhering to the film. At times, the diffraction patterns obtained from replicas cannot be identified, and in these instances any information regarding the chemical composition of the particles is extremely helpful. The weight of the particles adhering to the extraction replica is usually too small for microchemical or spectrographic

163

analysis. Recently, Fisher[1] has successfully used the electron-probe microanalyzer to obtain an elemental analysis of the submicron-size particles adhering to extraction replicas.

The purpose of this paper is to describe an inexpensive technique that can be used for the analysis of extraction replicas and single particles of microscopic size. The technique consists of recording the X-ray spectra emitted from a specimen under electron bombardment, by means of a miniature photographic X-ray fluorescence-analysis camera that is inserted within the vacuum chamber of an electron microscope or diffraction unit. This photographic technique is inferior to the electron-probe microanalyzer in several respects and is suggested primarily for use by laboratories that have only an occasional need for such an analysis or that because of other considerations cannot justify the expense of an electron-probe microanalyzer.

APPARATUS

The basic equipment used in this experiment was an RCA Model EMD-2 electron diffraction unit. The optical system of the diffraction unit was unmodified with the exception of the substitution of a 0.025-in.-diameter condenser aperture for the standard 0.001-in.-diameter aperture. This change provides a large increase in intensity with little impairment of the quality of the electron diffraction patterns obtained when the apparatus is used for diffraction purposes. To minimize modifications of the diffraction unit, the fluorescence-analysis camera was made small enough to fit entirely within the vacuum chamber of the diffraction unit. The obvious choice for the location of the camera is within the specimen chamber of the diffraction unit (Figure 1). In this position, modifications of the unit are not required and the camera is close to the specimen where the X-ray intensity is highest.

Figure 2 is a photograph of the camera with its cover removed. A slit on the side of the camera nearest the specimen limits the angular divergence of the X-ray beam entering the camera. Immediately behind the slit there is a thin foil opaque to light, but transparent to most X-rays. The foil is composed of a 0.00025-in.-thick polyester film, vacuum-aluminized on each side and having a total weight of 1 mg/cm². The X-ray beam entering the camera is diffracted by a rotating lithium fluoride crystal and recorded on Kodak No-Screen X-ray film.

---

[1] Superscripts pertain to references at the end of the paper.

Figure 1. Specimen chamber of RCA model EMD-2 electron diffraction unit.

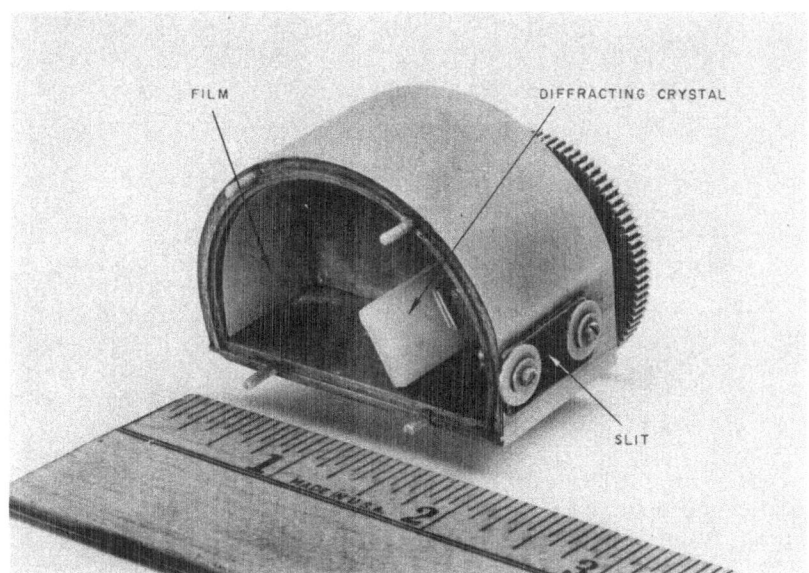

Figure 2. Miniature X-ray fluorescence-analysis camera.

Figure 3. X-ray fluorescence camera mounted on supporting fixture.

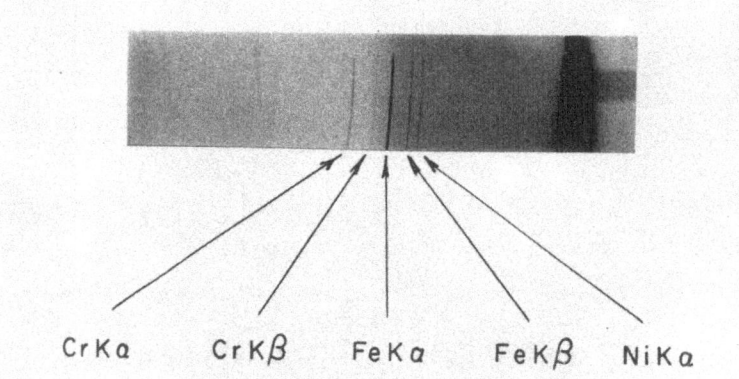

CrKα   CrKβ   FeKα   FeKβ   NiKα

Figure 4. Spectrogram from an 18 Cr−8 Ni stainless steel
specimen.

Means for supporting the camera within the diffraction unit
and for providing the necessary rotation of the crystal were
combined in one fixture which is attached to the diffraction unit
in place of the cover plate of the left accessory port of the specimen chamber. The fixture, with camera attached, is shown in
Figure 3. The central shaft of the fixture is rotated by a synchronous motor located outside of the vacuum chamber. The
rotation of this shaft is transmitted through a pair of gears to
the shaft on which the crystal is mounted. The combination of
motor speed and gearing is such that a scan of 180° (2θ) is completed in approximately 10 min.

RESULTS AND DISCUSSION

The radiation counters employed with conventional X-ray spectrographs are considerably more sensitive than film, so that higher X-ray intensities are necessary for their detection by film. The peak-to-background intensity ratio is also better with the counter method. This is because all of the film is continuously exposed to scattered radiation through an entire exposure, whereas with the counter method, the detector scans through the various angular positions and thus receives the scattered radiation at any one angular position for only a fraction of the total exposure time.

Because it was doubtful that there would be sufficient intensity to obtain a spectrogram from an extraction replica within a reasonable time by using photographic recording, the initial specimens examined were bulk samples where the intensity is a maximum. An exposure of a solid iron specimen resulted in a general blackening of the film, but no spectral lines were visible. The major portion of the darkening was caused by soft X-rays that could be filtered out by a 1-mil-thick aluminum foil. With the filter in place, iron K_α and iron K_β spectral lines of good intensity were obtained in a single 10-min scan. Examination of a solid specimen of an 18 Cr–8 Ni stainless steel yielded the K spectral lines of iron, chromium, and nickel in one scan. A spectrogram from the stainless steel specimen is shown in Figure 4.

The next step was to determine whether the particles adhering to an extraction replica emitted a sufficient quantity of X-radiation to be detected. Replicas are supported on a fine screen, commonly having 200 openings per inch. A magnified image of the area of specimen irradiated by the electron beam can be observed on the fluorescent viewing screen of the diffraction unit. Figure 5 shows three images obtained from a 200-mesh grid as the position of the focal point of the electron beam is brought successively closer to the plane of the grid. Figure 5C is the image obtained with the electron beam focused in the plane of the grid and aimed between the grid wires. From this figure it appears that the entire beam can pass between the wires of the 200-mesh grid so that an analysis should be possible without interfering effects from the elements in the grid wires. However, in addition to the highly intense focal spot, there is a general illumination of low intensity, probably caused by scattering of electrons from the walls of the electron optical system. Although the intensity of scattered electrons is low, they strike the entire $\frac{1}{8}$-in. grid screen, and during long ex-

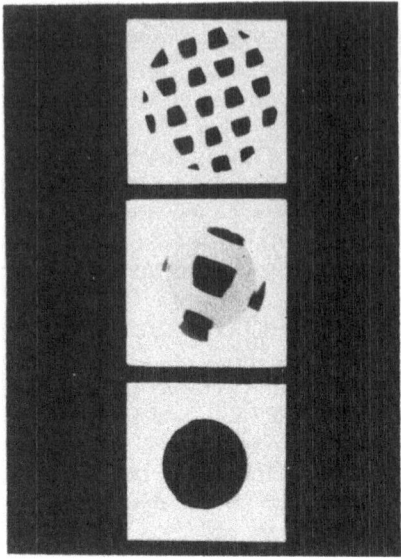

Figure 5. Effect of focusing on size
of irradiated area.

posures, interfering spectral lines from the elements in the grid
wires are obtained.

An extraction replica that contained particles that had previously been identified by selected area diffraction as TiC was
examined with the camera. The replica was prepared from a
titanium-bearing precipitation-hardening stainless steel. An
electron micrograph of the replica is presented in Figure 6.
After a 1-hr exposure, a very faint titanium K_α line was barely
visible in addition to the iron, chromium, and nickel lines from
the stainless steel grid. An edge of the replica was curled over
so that if that region were irradiated, several thicknesses of
replica would be presented to the electron beam. An electron
micrograph of the curled-over region is shown in Figure 7. In
this region, the density of particles is, of course, much greater.
A 2-hr exposure of the curled-over region resulted in a spectrogram on which the titanium K_α was clearly visible.

Several other replicas were prepared, and because it was
expected that they would contain iron or chromium compounds,
the replicas were supported on copper mesh. All replicas were
made of carbon because plastic replicas were torn by the high
electron beam intensity.

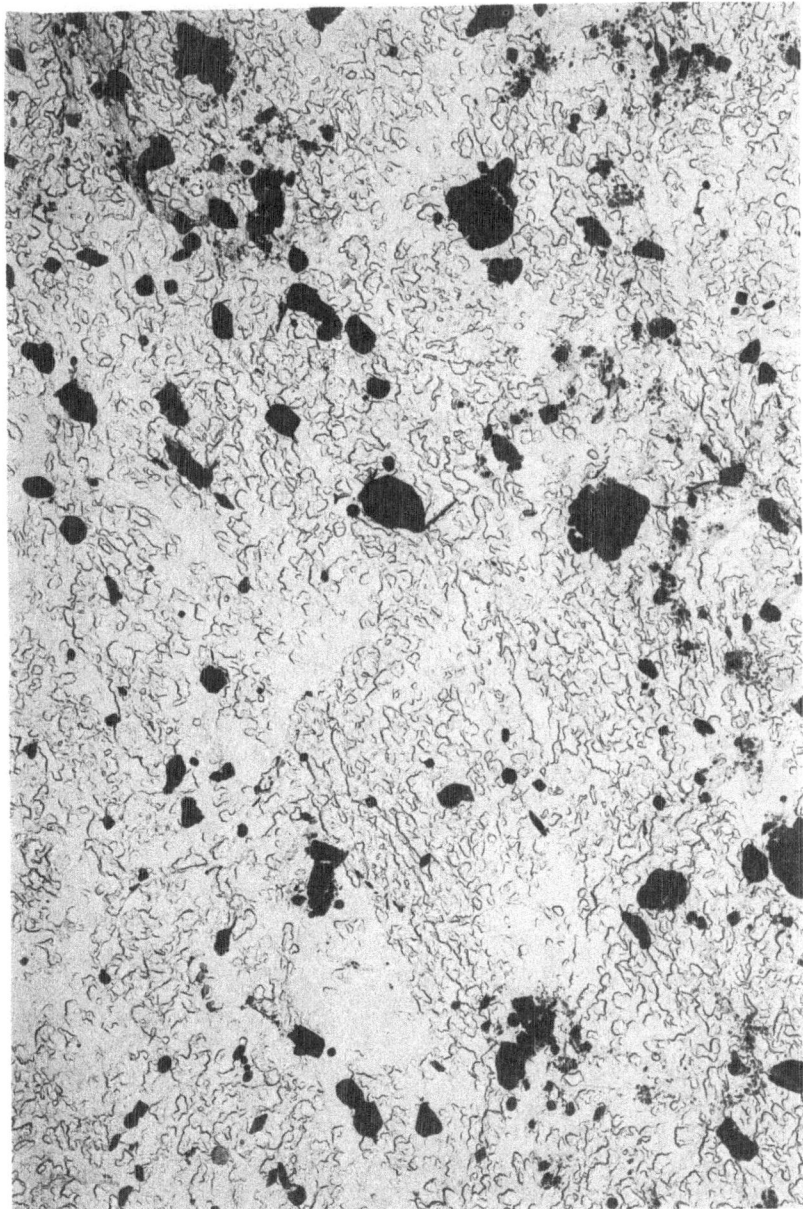

Figure 6. Electron micrograph of carbon extraction replica from a titanium-bearing stainless steel (× 7500).

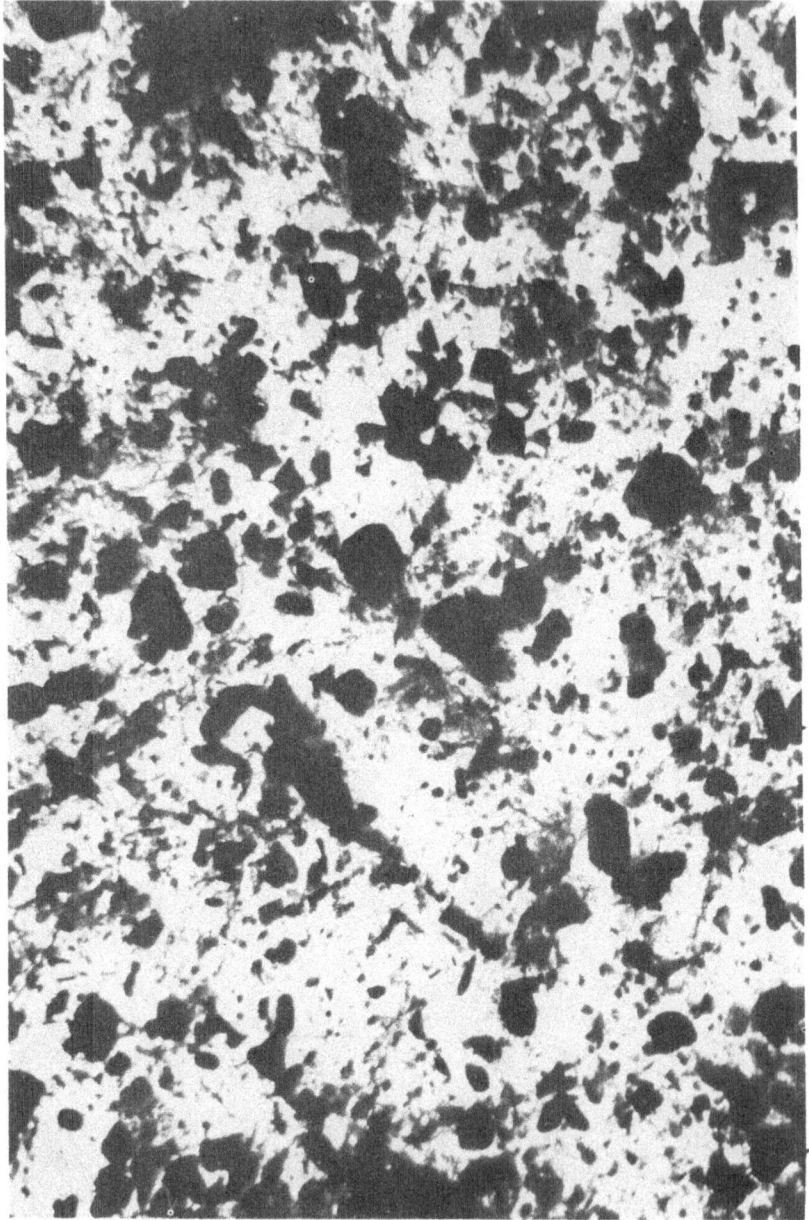

Figure 7. Electron micrograph of a folded portion of an extraction replica from a titanium-bearing stainless steel (× 7500).

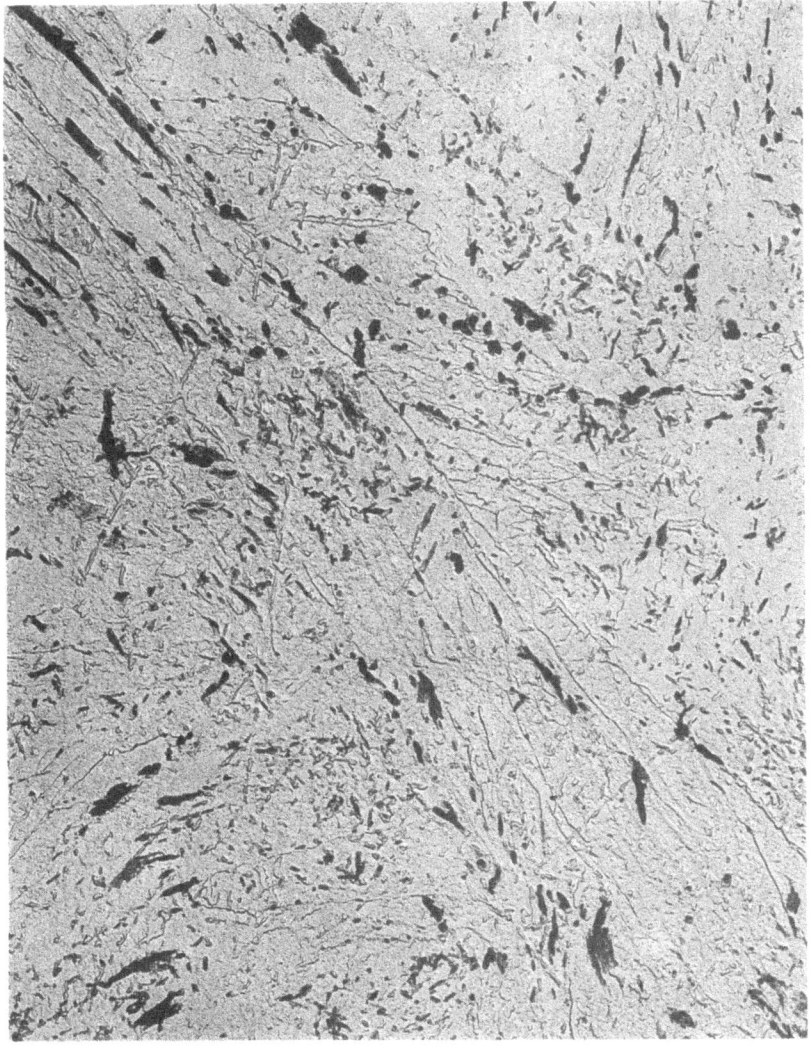

Figure 8. Electron micrograph of extraction replica from a low-alloy steel (× 15,000).

Figure 8 is an electron micrograph of an extraction replica prepared from a low-alloy steel. The particles shown in the micrograph were identified as Fe_3C by electron diffraction. A 50-min exposure of a curled-over region yielded the K spectrum of iron.

Figure 9. Electron micrograph of extraction replica from a low-alloy steel (×15,000).

Figure 9 is an electron micrograph prepared from another low-alloy steel. In addition to some small particles, the replica is primarily composed of large patches of cementite that have been etched free from the pearlite. As expected, the iron spectrum was obtained.

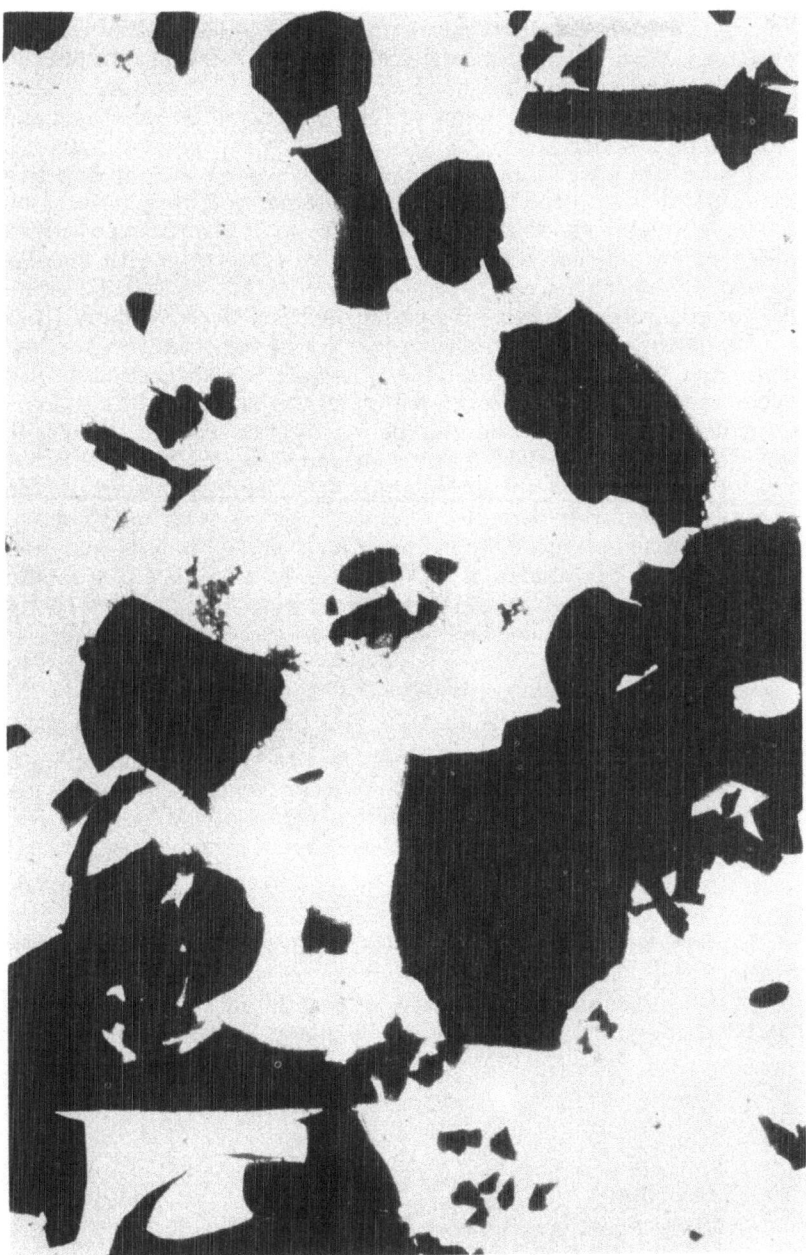

Figure 10. Electron micrograph of replica from a stainless steel sample (× 7500).

Figure 10 is an electron micrograph of an extraction replica prepared from a precipitation-hardening stainless steel. The particles were identified as $(Fe, Cr)_{23}C_6$ by diffraction. A 1-hr exposure of a curled region of the replica yielded the iron and chromium K spectra.

From the extraction replicas examined so far, it appears that this method does not detect the elements from particles on a single thickness of replica with a 1- to 2-hr exposure unless there is an unusually high particle density, but it will usually be successful if the replica is folded or curled, so that the number of particles irradiated by the electron beam is increased.

In addition to the analysis of extraction replicas, this method has been used for the analysis of small single particles. The specimen was prepared by transferring the particle with a sharpened needle, under a microscope, to a glass slide covered with wet Formvar. After the Formvar dried, a second film of Formvar was spread on the first and a grid screen placed in the Formvar over the particle. The film, along with the grid and particle, was stripped from the glass slide, and carbon was evaporated on both sides of the film. A 2-hr exposure of a single particle of 18 Cr—8 Ni stainless steel weighing about $3 \cdot 10^{-8}$ g yielded iron, chromium, and nickel lines.

SUMMARY

A miniature X-ray fluorescence-analysis camera has been designed for use within a commercial electron diffraction unit. The camera has been used for the qualitative elemental analysis of the particles adhering to extraction replicas and for the analysis of single particles weighing less than a microgram.

REFERENCES

[1] R. M. Fisher, "Precipitation of Carbides in Unalloyed and Low-Alloy Steels," paper at Symposium on Transformation and Precipitation Phenomena in Metals, Fourth International Congress in Electron Microscopy, Berlin, Germany, Sept., 1958.

X-RAY SPECTROGRAPHIC ANALYSIS OF MANGANESE NODULES

George M. Gordon, Donald J. McNely,
and John L. Mero

Department of Mineral Technology, University of California
Berkeley, California

ABSTRACT

Manganese nodules obtained from the ocean bottom are a vast potential source of minerals. These nodules, whose composition depend on their source, have an average composition of about 20% manganese, 14% iron, and 0.5% each of cobalt, nickel, and copper.

X-ray spectroscopy was investigated as a method for the analysis of the nodules, and analytical control during the process of extracting and recovering the minerals.

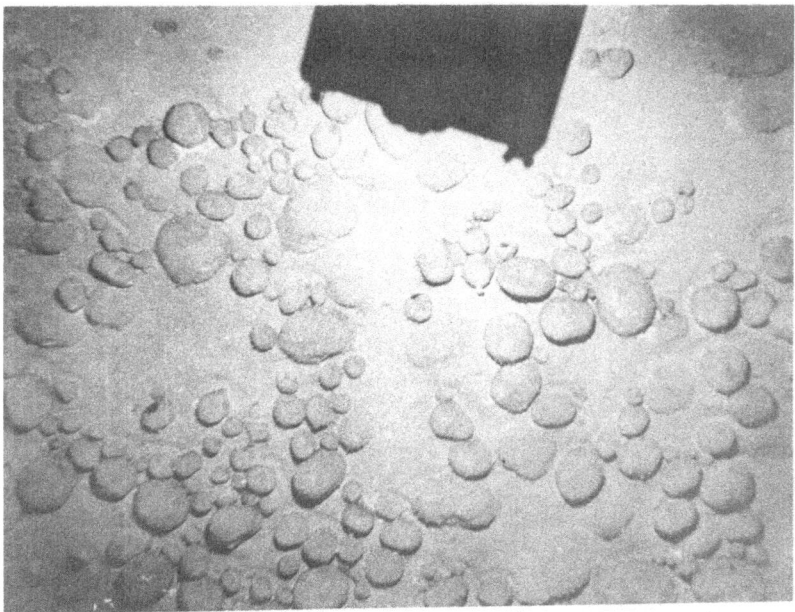

Figure 1. Underwater flash taken in 1948 in the Atlantic, location 30°-37' N lat., 59°-07' W long., depth 18,000 ft. Approximate area covered is 3 by 6 ft. Photo by D.M. Owen of Woods Hole Institute of Oceanography.

The basic borax fusion technique of Fernand Claisse (R. P. No. 327, Department of Mines, Quebec) was used. The influence of matrix elements was determined and corrections applied. The binary systems in borax were compared with the curves calculated from pure metals using the methods and calculations of Beattie and Brissey.

The results indicate that the fusion method is sufficiently rapid and accurate for use in the control analysis of manganese nodules.

* * *

Manganese nodules are lumps of impure complex manganese oxides obtained from the ocean bottom. Their size varies from that of a pea to very large lumps. A portion of one weighed over 250 lb. Preliminary estimates by Menard[1] and Mero[2] indicate that there are economic deposits of these minerals over a large area of the ocean floor. Underwater photographs show that in many areas the concentration of nodules is between 1 and 10 lb/ft^2. Figures 1 and 2 are underwater flash photographs of nodules showing a concentration of about 10 lb/ft^2. Figure 3 shows one type of dredge used to obtain the nodules, while Figure 4 shows a collection of nodules in the laboratory.

Figure 2. Underwater flash taken near Blake Plateau, depth 2700 ft. Photo by
D. M. Owen (1957).

---

[1]Superscripts pertain to references at the end of the paper.

Figure 3. Dredge aboard the "Horizon."

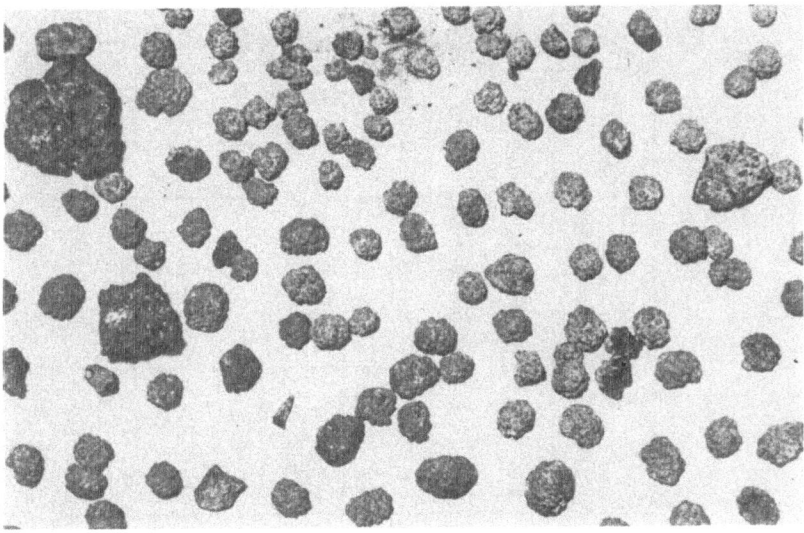

Figure 4. Collection of nodules; the shark's tooth in the lower center is about 1 in. long.

Table I. Chemical Analysis of a Selected Group of Nodules. Analysis Conducted by the U.S. Bureau of Mines, May, 1959

Sample	Location		Depth, ft	Mn	Fe	Cu	Ni	Co	SiO_2	Al_2O_3	CaO
	latitude	longitude									
P-1A	28-23N	126-57W	14,200	43.6	0.71	0.12	0.06	0.27	6.13	2.19	1.64
P-3	9-57N	137-47W	16,140	29.8	4.81	1.16	1.32	0.45	11.21	5.15	1.90
P-8	6-55S	83-34W	13,330	9.6	7.71	0.12	0.03	0.27	52.58	4.68	0.97
P-13	11-13.8S	89-35W	14,634	25.1	8.88	0.70	1.08	0.45	14.30	4.89	1.75
P-18	21-27N	126-43W	14,080	9.7	11.51	0.18	0.22	0.44	31.58	3.74	1.31
P-23	16-29S	145-33W	4,150	22.4	13.82	0.17	0.52	1.22	2.82	2.62	4.08
P-29	19-49N	121-43.5W	14,240	21.4	9.55	0.67	1.11	0.52	17.09	6.10	1.78
A-102	29-17N	57-23W	19,140	16.2	15.88	0.22	0.40	0.54	14.32	7.73	1.97
A-105	30-51N	78-27W	2,400	15.7	15.52	0.14	0.21	0.70	2.40	5.16	10.24

There has been much speculation on the origin of these nodules. They form principally where the sedimentation rate is very low. Goidberg,[3,4] Pettersson,[5] Clark,[6] and Dietz[7] have studied and contributed theories and information on the formation of the nodules. However, the mechanism of formation is still in question. The growth rate of nodules has been estimated at about 0.1 to 1μ per year. Low sedimentation rates and the motion of the ocean keep all sides exposed to the sea water and the nodules grow more or less spherical. There is evidence that if the nodules become covered with sediment the growth stops.

Nodules nucleate and grow around foreign material such as clay, a shark's tooth, or a whale's earbone. Many grow with an onion-like layer structure. We have not investigated the composition of the layers nor attempted to correlate the composition of the nodules with their location. There are strong indications that a correlation exists; however, data available are too limited to draw any valid conclusions.

Table I shows the analysis of a number of nodules. These results and others reported by Mero[2] indicate a composition range about as shown in Table II.

Table II.   Average Composition of Nodules as
Metallic Elements (%)

	Low	High	Average
Manganese	10	45	25
Iron	1	20	10
Copper	0.1	1.8	0.4
Cobalt	0.1	1.2	0.5
Nickel	0.03	1.4	0.5

As part of a program to extract the valuable minerals from these nodules, X-ray spectroscopy was investigated as a control method. Preliminary attempts at direct analysis of ground nodules were not satisfactory, as the results were not reproducible. The borax fusion technique of Claisse[8] was investigated and found to be quite satisfactory. About 2000 g of nodules were ground to 200 mesh; five portions were weighed and fused with borax, adding zinc as an internal standard. The reproducibility was excellent as the variation between samples was the same as replicate counts on the same sample.

The X-ray equipment used was a General Electric XRD-3 instrument with a bent mica crystal used in transmission, and argon-filled Geiger tube detectors. All determinations were made by setting the spectrometer with a sample of pure metal and recording the time required to register 16,384 counts.

The sensitivity of the Claisse method was quite low; therefore the amount of sample was increased from the recommended 100 to 500 mg. The final procedure adopted was as follows:

Sample	500 mg
Borax glass (reagent grade)	9.5 g
Internal standard zinc oxide	24.9 mg

The weighed mixture was placed in a platinum crucible and fused until clear, then cast into a chromel ring placed on an aluminum plate heated to about 400°C. After a few minutes, the glass disk was pushed onto a heated transite plate to cool.

Standards were synthesized from reagent grade oxides of manganese, iron, copper, cobalt, and nickel. These standards included single elements, the binary systems, and some approximating the average nodule composition.

Binary metallic alloys of iron—manganese, iron—nickel, iron—cobalt, nickel—copper, and nickel—cobalt were studied. Curves showing the K_α radiation vs composition were prepared for each of the binary systems. The method of Beattie and Brissey[9] was used to calculate the curve from the pure elements and one known alloy of about 50-50 composition. This calculation applied to a binary system becomes

$$A_{ab} = \frac{\%\ a\ \text{in known sample}}{\%\ b\ \text{in known sample}} \times \left(\frac{\text{Intensity } a \text{ in pure metal}}{\text{Intensity } a \text{ in known sample}} - 1 \right) .$$

When A_{ab} is calculated from the known alloy, then the relative intensity ratio of one element may be calculated for any other composition of the alloy from

$$\frac{I_a\ \text{unknown}}{I_a\ \text{pure metal}} = \frac{1}{A_{ab}\left(\dfrac{\%\ b\ \text{in the unknown} + 1}{\%\ a\ \text{in the unknown}} \right)} .$$

Figures 5 and 6 show the calculated binary curves for iron—manganese and iron—nickel. The deviation of these curves from a straight line indicates that the intensity of one element is enhanced or suppressed by the other. From these curves, one would predict that iron would have little or no effect on manganese, while it should greatly reduce the nickel intensity. Figures 7 and 8 show the results actually obtained from fused samples for these systems. The predicted influence is observed as the

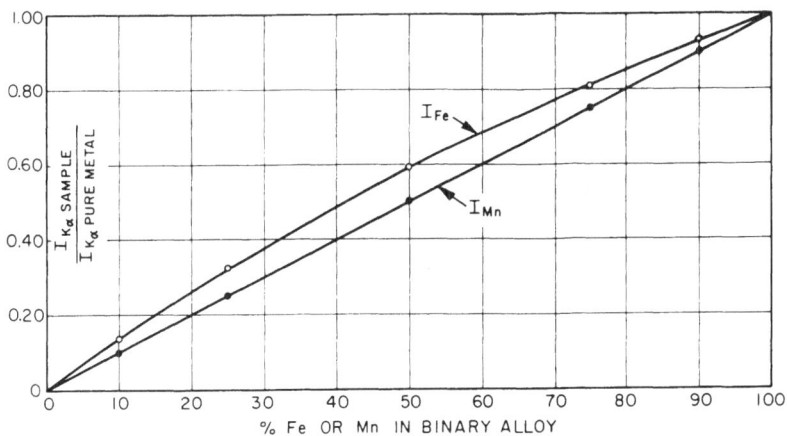

Figure 5. Iron—manganese binary system for metallic alloys.

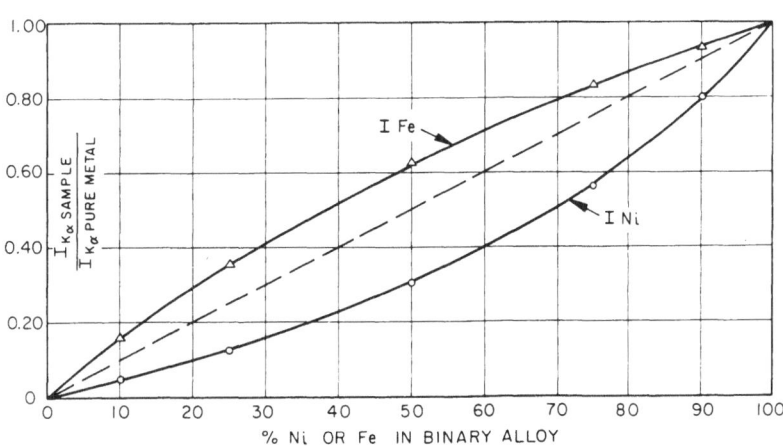

Figure 6. Iron—nickel binary system for metallic alloys.

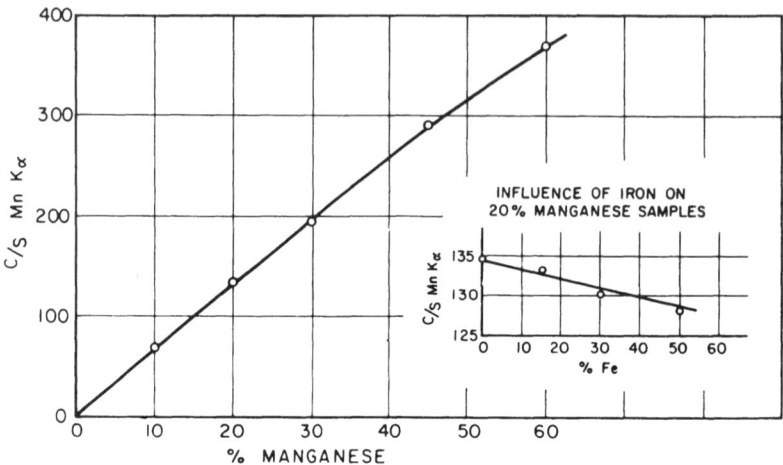

Figure 7. Analytical curve, manganese concentration vs intensity.

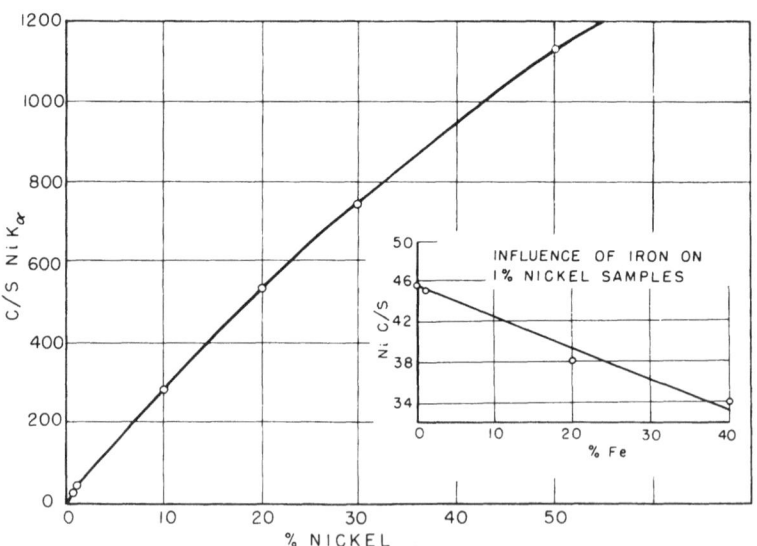

Figure 8. Analytical curve, nickel concentration vs intensity.

manganese intensity is reduced only 3%, while the nickel is reduced by more than 25% with the addition of 40% iron. Analytical curves, such as Figures 7 and 8, were prepared for each element from reagent grade oxide dissolved in borax.

For the analysis of nodules, a borax glass specimen was prepared as described, and the X-ray intensity determined for the K radiation of each element. (Note: recent work has shown that the internal standard could be eliminated by counting a standard specimen during each run.)

From the observed intensity, a corrected intensity was obtained by the use of a mathematical correction. The equations for this correction were calculated from the synthesized binary standards. The general form of the equations is illustrated by the correction formula for iron intensity:

$$\text{counts/sec } (Fe)_{corr} = \text{counts/sec } (Fe)_{obs} + \% \text{ Mn} \times (\% Fe)_{appar} \times 0.027.$$

In this example the % Mn was read directly from the manganese analytical curve and the apparent % Fe read from an iron analytical curve.

Empirical equations were calculated for each element. Fortunately, manganese was not seriously influenced by the other elements and could be read directly from the analytical curve.

Corrections applied only where the influence was an order of magnitude greater than the statistical accuracy. The following general influence of matrix elements was observed:

Manganese: no correction necessary for copper, nickel, or cobalt up to 20%. A slight correction was indicated for iron.

Iron: no correction necessary for copper, nickel, or cobalt up to 20%. A correction for manganese was applied.

Cobalt: no correction necessary for iron up to 40%, nickel and copper up to 10%. A correction for manganese was applied.

Nickel: the data on the influence of copper and cobalt are incomplete and indicate the cobalt oxide used to prepare the standards contained considerable nickel. Corrections for manganese and iron must be applied.

Copper: no corrections necessary for nickel up to 10%. Iron, manganese, and cobalt require corrections.

It is interesting to note that one could have predicted which matrix elements must be corrected for from the absorption coefficients of the elements.

CONCLUSION

Manganese nodules may be analyzed with sufficient accuracy for routine work using X-ray fluorescent techniques. These techniques will have value over normal chemical methods in that only a small quantity of sample is required and additional elements may be determined as instrumental techniques improve. By the use of more advanced instruments than were available for this investigation, both the range and accuracy may be increased.

At the present time none of the nodules have been analyzed both by chemical and X-ray methods. However, portions of the samples reported in Table I are available and will be used to obtain a better evaluation of the method, as applied to a large range of composition.

REFERENCES

[1] H. W. Menard and C. J. Shipek, Nature, Vol. 182, 1952, p. 1156.

[2] J. L. Mero, California Institute of Marine Resources, Preliminary Report, Jan., 1959.

[3] E. D. Goldberg, J. Geol., Vol. 62, 1954, p. 249.

[4] E. D. Goldberg and G. O. S. Arrhenius, Geochim. et Cosmochim. Acta, Vol. 13, 1952, p. 153.

[5] H. Pettersson and K. Goteborg, Vet. Somh. Hand. Sjatti Foljden, Ser. B, Vol. 2, No. 8, pp. 1-43.

[6] F. W. Clark, U.S. Geol. Survey Bull., No. 770, 1924.

[7] R. S. Dietz, Cal. J. Mines and Geol., Vol. 51, No. 3, 1955.

[8] F. Claisse, "Accurate X-Ray Fluorescent Analysis," R.P. 327, Dept. of Mines, Lab. Branch, Quebec, Canada.

[9] H. J. Beattie and R. M. Brissey, Anal. Chem., Vol. 26, 1954, p. 980.

INSTRUMENTATION FOR ELECTRON PROBE MICROANALYSIS

David B. Wittry*

Consultant, Applied Research Laboratories
Glendale, California

ABSTRACT

Since the development of the first practical electron probe microanalyzer by Castaing and Guinier in 1951, approximately 14 such instruments have been placed in operation throughout the world and at least 15 more are under construction. All of the existing instruments contain the following basic components: (1) An electron beam system, (2) a movable specimen stage, (3) a means for viewing the surface of the specimen, and (4) an X-ray analyzer. The integration of these components into a practical instrument requires certain compromises in the performance capability of the instrument. A critical evaluation of the various approaches to this problem is given and a new electron objective lens is described which results in fewer restrictions on the over-all performance. The basic design of an instrument employing this lens, which will be manufactured by the Applied Research Laboratories, is discussed.

INTRODUCTION

The electron probe microanalyzer is an instrument for local chemical analysis which makes use of a finely focused electron beam to excite characteristic X-radiation from a small volume at the surface of a solid specimen. The wavelength and intensity of the characteristic X-ray lines are then used to determine the elements which are present in the target and their relative mass concentrations. At the present time analyses can be performed for all elements of atomic number greater than 12. In typical cases, the spatial resolution of the method of analysis is about 2-3 μ, the sensitivity ranges from 1 part in 10^4 to 1 part in 10^3, and the relative accuracy is 1-2% if the concentration is greater than a few percent.

The idea of utilizing a focused electron beam for local X-ray emission spectrographic analysis was first conceived by Hillier in 1943. Hillier had been working on a method of local analysis

---

* Present address: University of Southern California, University Park, Los Angeles, California.

based on the energy losses of electrons transmitted by thin foils. In the course of this work, it occurred to him that a form of local analysis with somewhat poorer resolution could be achieved by utilizing the characteristic X-rays produced by a focused electron beam. Hillier's idea became the subject of a U.S. patent which was issued in 1947 but was not pursued to a practical stage.[1]

About 1949, the first operating electron probe microanalyzer was constructed by Raymond Castaing, while a student of Professor Andre Guinier, who independently conceived the principles underlying this method of analysis.[2] Castaing's first instrument was constructed at the Office Nationale d'Etude et Recherches Aeronautiques from an obsolete CSF electron microscope and utilized electrostatic focusing of the electron probe.[3] A second instrument was later developed by Castaing which utilized electromagnetic focusing.[4] This instrument is now being marketed by Cameca under license from O.N.E.R.A.

Shortly after the publication of Castaing's work, announcement was made of similar attempts being made by Borowskii in Russia.[5] Borowskii's microanalyzer employed magnetic focusing to obtain an electron focal spot of about 5 μ in diameter and was intended primarily for analysis of the heavier elements such as tungsten, molybdenum, and rhenium.[6]

In England, interest in microanalysis by means of electron probes led to development work on instruments at the Cavendish Laboratory,[7] Associated Electrical Industries,[8] Tube Investments,[9] the National Physical Laboratory,[10] and Imperial Chemical Industries, Ltd.[11]

In the United States, development of microanalyzers began at the California Institute of Technology,[12,13] and the Naval Research Laboratory.[14] This was followed by work at the U.S. Steel Company,[15] the Massachusetts Institute of Technology,[16] Battelle Memorial Institute,[17] the General Electric General Engineering Laboratory,[18,19] the Smithsonian Institution,[20] Westinghouse Research Laboratories,[21] Philips Electronics,[22] U.S. Geological Survey,[23] U.S. Bureau of Standards,[24] Advanced Metals Research,[25] and the Applied Research Laboratories.

Microanalyzers are also being developed at the Hitachi Central Research Laboratory in Japan,[26] and the Ontario Research Foundation in Canada.[27]

The number of instruments now in operation throughout the world totals at least 14, exclusive of any additional instruments which may be in use in Russia, and some 15 more are

---

[1]Superscripts pertain to references at the end of the paper.

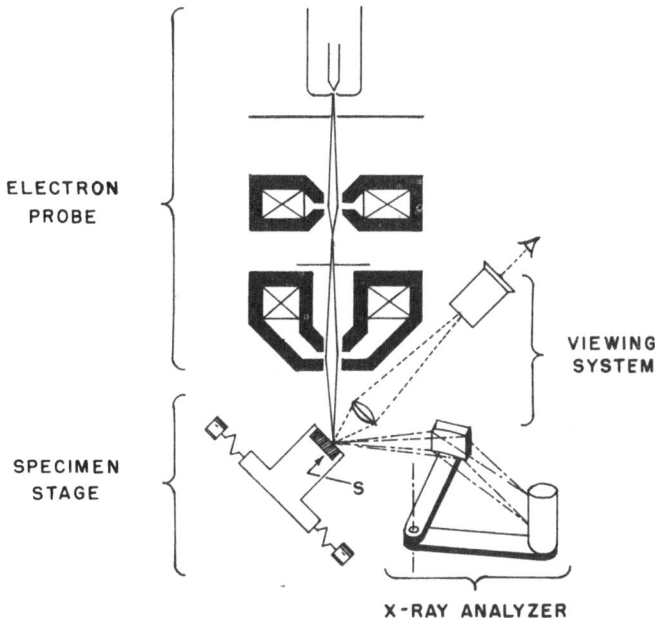

ELECTRON
PROBE

VIEWING
SYSTEM

SPECIMEN
STAGE

S

X-RAY ANALYZER

Figure 1. Schematic drawing of an electron probe microanalyzer showing the relationship of the electron probe, the specimen stage, the viewing system, and the X-ray analyzer to the specimen, S.

under construction at the present time. Of the instruments which are currently in operation the majority have been constructed by research laboratories for their own use. Sufficient experience has been accumulated through the use of these research-type instruments to permit evaluation of their most important limitations, and to arrive at the best compromise in the design of a general-purpose electron probe microanalyzer.

GENERAL CONSIDERATIONS

The basic components of the electron probe microanalyzer are: (1) an electron beam system or probe, (2) a moveable specimen stage, (3) a means for viewing the surface of the specimen, and (4) an X-ray analyzer. Figure 1 shows schematically the relative positions of these components. Since all of these components converge at the specimen position, compromises in the performance of each component are inevitable because of the limited space available around the specimen. The over-all design and such compromises as may be necessary should be decided on the basis of the following considerations, listed in order of their relative importance.

I. Stability

Accelerating voltage, electron probe current, probe size, and probe position.

II. Reliability
   a. Provision for monitoring target current.
   b. High emergence angle of X-rays to avoid surface effects and reduce self-absorption corrections.
   c. Minimum contamination rate of probe column and of specimen.
   d. Minimum interaction of focusing fields with specimen.

III. Quality of Electron Optics
   a. Magnification of $1/100$ or less.
   b. Proper aperturing through use of both field and aperture stops.
   c. Minimum spherical aberration.

IV. Quality of X-Ray Optics
   a. At least two spectrometers for coverage of all elements above atomic number 12.
   b. Maximum detection efficiency for characteristic radiation.
   c. Minimum detector dead time.

V. Quality of Viewing System
   a. Simultaneous viewing during analysis.
   b. Resolution of 1μ or better.

VI. Convenience of Specimen Handling
   a. Minimum downtime for sample change.
   b. Accommodation of standard-size metallurgical mounts.
   c. Provision for rotation so that scan may be accomplished in any direction.

In the light of these considerations, the limitations of most of the instruments currently in operation can be grouped in three categories: (1) limitations which resulted simply from economy or expediency in construction, such as poor stability, high contamination rate, and inadequate visual means; (2) limitations which resulted from a lack of understanding of the best means of accomplishing the desired performance, such as low brightness of the electron source, improper aperturing of the electron beam, poor choice of spectrometers, and faulty mechanical design; and (3) limitations which arose because of the compromises required in integrating the various components into a functioning instrument. For brevity, it is necessary to limit the present discussion to the latter category.

In many respects, the design of the final lens of the electron optical system can be regarded as predetermining the performance capability of the instrument as a whole. In the first place, the quality of this lens limits the minimum focal spot size which can be obtained with adequate current. In the second place, its size and shape limit the available space for the viewing system, the emergence angle of the X-rays, the smallest Bragg angles which can be obtained with a given focusing-type spectrometer, and in some cases, the maximum specimen size which can be accommodated.

THE OBJECTIVE ELECTRON LENS DESIGN

The requirement of minimum spherical aberration in the objective lens of the electron optical system can be met satisfactorily only by the use of magnetic focusing. This results from the fact that electrostatic lenses have excessive spherical aberration unless they are employed as immersion lenses. In the case of the magnetic objective lens, the spherical aberration increases in general with working distance. Thus, a minimum spherical aberration is not compatible with good quality in the viewing system, high emergence angle of the X-rays, or minimum interaction of the specimen and lens filled.

The various objective lenses which have been used in microanalysis are indicated in Figure 2. The configuration shown in Figure 2a minimizes the focal length of the lens and consequently its spherical aberration as well. However, the specimen is in a region of high magnetic field and this results in some inconvenience in studying ferromagnetic specimens since the position of the electron focal spot is shifted when the specimen is moved in the bore of the magnetic lens. There is also a severe limitation on the size of the specimen which can be accommodated since the advantage of a low spherical aberration is realized only if the size of the bore in the lens is not excessive. This type of lens is used in the microanalyzers constructed by Duncumb at Cambridge and the author at Caltech.

Figure 2b is the configuration of the objective lens used in the Cameca Microsonde,[4] in the instrument at MIT, and in several research-type instruments under construction. Its principal advantage is that there is adequate space above the specimen to place a light microscope objective of the refracting type coaxially with the electron beam. However, since the X-rays must pass

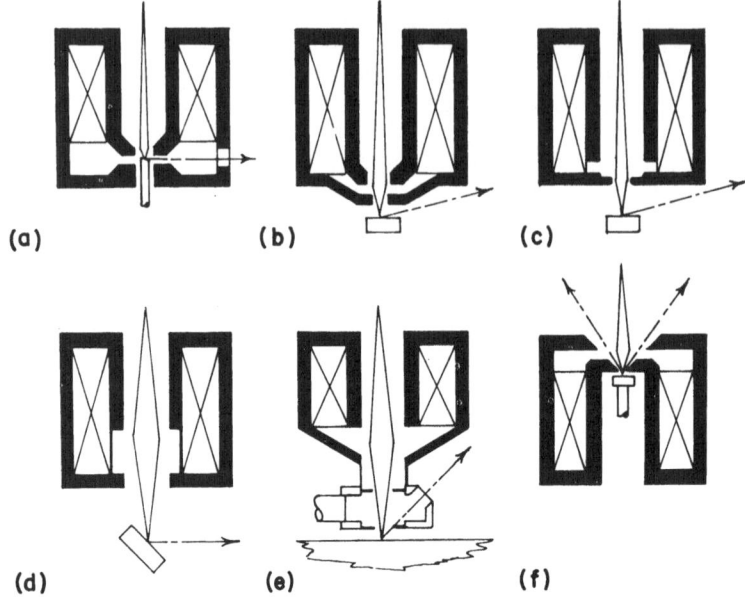

Figure 2. Objective electron lenses for electron probe microanalyzers.

to the spectrometer between the specimen and the lens, the emergence angle is necessarily small. In addition, the bore in the lower pole tip must be sufficiently large to accommodate the numerical aperture of the viewing objective, which leads to sufficient magnetic field at the surface of the specimen to cause some annoyance in analysis of ferromagnetic specimens.

The design of Mulvey, illustrated in Figure 2c, is intended to overcome the difficulties encountered in analysis of ferromagnetic specimens by maximizing the ratio of the working distance to the spherical aberration coefficient and by making the bore in the lower pole tip as small as possible.[8] The principal disadvantage is that examination of the specimen during analysis is impossible by optical means, and the X-ray emergence angle is small.

Figure 2d shows a novel approach to the problem of obtaining adequate working distance which was adopted by Fisher in modifying an RCA electron microscope for microanalysis.[15] The electron probe is formed first by an objective lens of short focal length (not shown) and this is then reimaged at the specimen by a repeater lens of approximately unity magnification. By inclining the specimen at 45° to the electron beam, it is possible to have an X-ray emergence angle of 45° and to utilize a simple optical viewing system. However, the repeater lens necessarily has a large spherical aberration which results not only from its long focal length, but also from the fact that it is used at nearly unity magnification. In addition, the high angle of incidence of the electron beam on the target is disadvantageous because of the distortion in the shape of the volume element excited and the possibility of enhanced scattering of electrons from the target.

The objective lens illustrated in Figure 2e represents an attempt to bring the electron probe out of the vacuum system. Probes of this type are being constructed by Schumacher at the Ontario Research Foundation for the study of volatile specimens[27] and by Lange and Riggs[20] of the Smithsonian Institution for the study of meteorites of very large lateral dimensions. * The extremely long focal length required in this lens and the consequent large spherical aberration combined with the scattering of the electron beam from the gas molecules restricts the resolution to such a degree that this approach is useful only for highly specialized applications.

The objective lens shown in Figure 2f was designed by the author for the instrument to be manufactured by the Applied Research Laboratories. It features normal incidence of the electron beam on the target with an emergence angle of the X-rays of 52.5°. There is adequate space either for the use of a reflecting objective of 0.4 numerical aperture or for the use of deflecting coils for image reproduction by the flying spot method using X-rays or secondary electrons. The lens coil has been made sufficiently large to accommodate specimens up to 1.25 in. in diameter. In addition, the magnetic field at the surface of the specimen is only 10 gauss for normal operating conditions.

THE ARL MICROANALYZER

The integration of the objective lens shown in Figure 2f into the over-all design of a microanalyzer is shown in Figure 3. This microanalyzer will be manufactured by the Applied Research Laboratories of Glendale, California. In this instrument the attempt was made to satisfy all of the design considerations mentioned in the preceding list.

It was decided that in the first commercial instruments of this type, the most satisfactory viewing means should provide

---

*For illustrative purposes all lenses in Figure 1 were drawn to represent magnetic focusing objectives. The probe being constructed by Lange and Riggs actually uses electrostatic focusing.

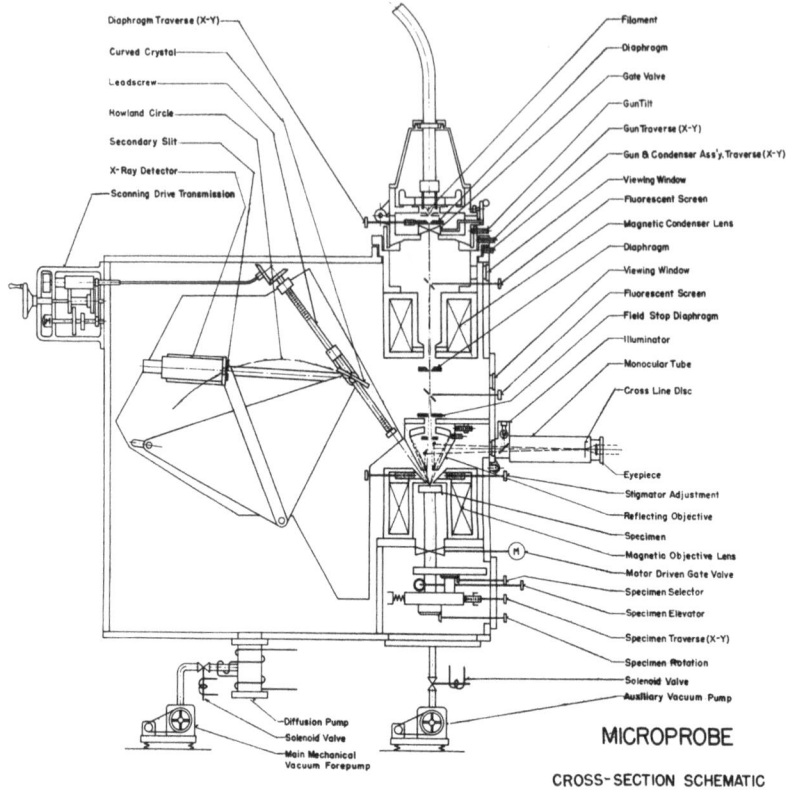

Figure 3. Schematic drawing of the Applied Research Laboratories' microanalyzer.

an image of the type that one is accustomed to seeing through conventional metallurgical microscopes. The viewing system therefore was designed to employ the elements of a Bausch and Lomb reflecting objective of 0.4 numerical aperture. It will be noted from the schematic drawing that the problem of bringing the image from objective to eyepiece is simplified by having the electron objective lens inverted. In other microanalyzers using reflecting objectives, as many as three plane mirrors are required in the optical path because of the space occupied by the objective lens coil. The entire optical viewing system is mounted on a door on the front of the instrument and can be removed as a unit for alignment purposes or for insertion of other viewing means.

The triode electron gun is a special design which permits concentration of the filament with respect to the aperture in the grid while the gun is in operation. This feature permits a rapid compensation of one of the principal causes of beam current drift, namely, filament warpage. In addition, a valve which functions automatically when the gun is opened enables the filament to be changed without raising the entire vacuum system to atmospheric pressure.

The filament of the electron gun is powered by dc from a transistorized, regulated power supply which uses a Zener diode as a reference. The use of dc for heating the filament is desirable in order to avoid magnetic modulation of the beam current.[28] Furthermore, the use of voltage regulation in this supply is believed to be superior to the feedback system used by Buschmann and Norton[19] as a means of insuring long-term stability of the beam current. The use of feedback to the filament power supply of a signal proportional to the collected target current is of questionable advantage when the electron gun is self-biased.

An airlock chamber is provided in order to minimize the time required for changing specimens. Up to eight specimens can be accommodated at one time on a selector disk, and when the airlock chamber is evacuated any one of these eight specimens can be raised into position for analysis. The stage assembly permits a translation of the specimen in two mutually perpendicular directions a distance of 0.4 in. and a rotation of the specimen about its center. This permits complete coverage of an area on the specimen's surface 0.8 in in diameter and enables the specimen to be scanned in any arbitrary direction over an area 0.4 mm in diameter.

The X-ray spectrometers are a modification of those currently in use on other ARL X-ray equipment. Three such spectrometers are used in the vacuum in the microanalyzer and are arranged around the probe column in such a way that the X-rays accepted by all spectrometers have the same effective emergence angle from the specimen surface. In addition, the use of a sine screw to control the setting of the spectrometers results in this emergence angle being constant for all wavelengths. The spectrometers are of the focusing type and use the following crystals: LiF with a focal circle radius of 11 in. for 0.36 to 1.76 A, SiO_2 with a focal circle radius of 11 in. for 0.60 to 2.92 A, and ADP with a focal circle radius of 4 in. for 2.66 to 10 A.

REFERENCES

[1] J. Hillier, "Electron Probe Analysis Employing X-ray Spectrography," U.S. Patent No. 2,418,029, Applied for: October, 1943, Issued: March, 1947.

[2] R. Castaing and A. Guinier, Proceedings of a conference on electron microscopy at Delft, 1949, Delft, 1950. Comptes rendus du premier congres international de microscopie electronique 1950. Editions de la Revue d'Optique, Paris, 1953.

[3] R. Castaing, "Application des Sondes Electroniques a Une Methode d'Analyse Ponctuelle Chimique et Cristallographique," These de Doctorat, Publication ONERA, No. 55, 1951.

[4] R. Castaing, "Recent Developments in X-ray Spectrographic Spot Analysis," Laboratories, 1956, p. 7.

[5] I. B. Borowskii, Symposium of problems in metallurgy, Academy of Sciences, USSR, 1953.

[6] I. B. Borowskii and N. O. Ilyan, "A New Method for Investigating the Chemical Composition of Microvolumes," Doklady Akad. Nauk, Vol. 106 (4) 1955.

[7] V. E. Cosslett and P. Duncumb, "Microanalysis by a Flying Spot X-ray Method," Nature, Vol. 177, 1956, p. 1172.

[8] T. Mulvey, "Improved Electron Optical System for X-ray Microanalyzer," Proceedings of a conference on electron microscopy at Reading, 1956. "An X-Ray Microanalyzer of Improved Design," Proceedings of the International Conference on Electron Microscopy, Berlin, September, 1958.

[9] P. Duncumb and D. A. Melford, "Design Considerations of an X-ray Scanning Microanalyzer Used Mainly for Metallurgical Applications," Proceedings of the Second International Symposium on X-ray Microscopy and Microanalysis, Stockholm, July, 1959. (To be published.)

[10] A. Franks, National Physical Laboratory, Teddington, England. (Private communication.)

[11] M. Bloom, "N" Building, Imperial Chemical Industries, Ltd., Billingham Co., Durham, England. (Private communication.)

[12] D. B. Wittry. J. W. M. DuMond, and Pol Duwez, "An Electron Microprobe for Local Analysis by Means of X-rays," Bull. Am. Phys. Soc.. Vol. 2, 1957, p. 213. (Abstract of paper presented at the Am. Phys. Soc. meeting, Washington, May, 1957.)

[13] D. B. Wittry, "An Electron Probe for Local Analysis by Means of X-rays," Ph.D. Thesis, California Institute of Technology, Pasadena, California, 1957.

[14] L. S. Birks and E. J. Brooks, "Electron Probe Microanalyzer," Rev. Sci. Instr., Vol. 28, 1957, p. 709.

[15] R. M. Fisher, "Working Distance in Electron Probe Instruments," J. Appl. Phys., Vol. 28, 1957, p. 1377. (Abstract of paper presented at the fifteenth annual meeting of the Electron Microscope Society of America, Boston, September, 1957.) "Modification, An Electron Microscope for Microprobe X-ray

Spectroscopy," Fifth Annual Symposium on Industrial Applications of X-ray Analysis, Denver, August, 1956 (unpublished).

[16] V. G. Macres and R. Ogilvie, Massachusetts Institute of Technology, Boston, Massachusetts (private communication).

[17] C. S. Schwartz and A. E. Austin, "Microbeam Analyzer at Battelle Memorial Institute," J. Appl. Phys., Vol. 28, 1957, p. 1368. (Abstract of paper presented at the fifteenth annual meeting of the Electron Microscope Society of America, Boston, September, 1957.)

[18] E. C. Buschmann, "A New Microemission X-ray Spectrograph—Design and Operation of the Direct Emission Curved Crystal Instrument," Proceedings of the Sixth Annual Conference on Industrial Applications of X-ray Analysis, University of Denver, 1957.

[19] E. C. Buschmann and J. F. Norton, "Microemission X-ray Spectrograph," General Engineering Laboratory, General Electric, Report No. 57GL201, June, 1957.

[20] A. Lang and F. B. Riggs, Smithsonian Institution, Astrophysical Observatory, 60 Garden Street, Cambridge 38, Massachusetts. (Private communication.)

[21] A. Taylor, Westinghouse Research Laboratories, Pittsburgh, Pennsylvania. (Private communication.)

[22] D. C. Miller, Philips Electronics, Inc., 750 Fulton Avenue, Mount Vernon, New York. (Private communication.)

[23] L. Adler, U.S. Department of the Interior, Geological Survey, Washington 25, D.C. (Private communication.)

[24] J. R. Cuthill, National Bureau of Standards, Washington 25, D.C. (Private communication.)

[25] R. Ogilvie, Advanced Metals Research Corporation, 625 McGrath Highway, Somerville 45, Massachusetts. (Private communication.)

[26] H. Watanabe, Hitachi Central Research Laboratories, Kokubunji, Tokyo, Japan. (Private communication.)

[27] B. W. Schumacher, Ontario Research Foundation, 43 Queens Park, Toronto, Ontario, Canada. (Private communication.)

[28] D. B. Wittry, "Two Improvements in Electron Sources for Electron Probes," Rev. Sci. Instr., Vol. 28, 1957, p. 58.

METALLURGICAL APPLICATIONS OF ELECTRON PROBE MICROANALYSIS

David B. Wittry *

California Institute of Technology, Pasadena, California

ABSTRACT

The electron probe microanalyzer has found numerous applications in physical metallurgy. Some recent problems which have been studied with the microanalyzer constructed by the author at the California Institute of Technology include (1) determination of the penetration of chromium in chromized iron, (2) determinations of the concentration of nickel in taenite bands occurring in a fragment of meteorite from Canyon Diablo, (3) identification of intermediate phases in the as-cast structure of an alloy of titanium and copper, (4) identification of phases in the titanium−silver binary system, (5) determination of concentration fluctuations in a gold−copper alloy resulting from slow solidification, and (6) determination of the diffusion profile of nickel in single crystal sapphire.

INTRODUCTION

One of the major advances in new analytical techniques during the last ten years was the development of electron probe microanalysis by Castaing and Guinier.[1] In this method of analysis, the characteristic X-radiation excited by a focused electron beam is utilized to obtain the concentration of elements present in a few cubic microns at the surface of a solid specimen. This method is far more practical than any other method of local analysis of comparable sensitivity and accuracy and, in addition, is nondestructive − a feature which is extremely useful in practical applications because it enables repetition of unusual or unexpected results which can often occur when one is dealing with micron-sized regions.

The spectrum of applications of the electron probe microanalyzer has ranged from purely practical problems involving the manufacture of such diverse products as photographic film, transistors, and corrosion-resistant coatings to matters primarily of scientific curiosity, such as meteorites from outer

---

*Present address: University of Southern California, University Park, Los Angeles, California.

[1] Superscripts pertain to references at the end of the paper.

space and sediments from the bottom of the ocean. However, the majority of applications up to the present have been to problems in physical metallurgy.

The potential of the method for problems in physical metallurgy was demonstrated by the early work by Castaing, which consisted of the analysis of a Cu–Zn diffusion couple, analysis of precipitates in Al–Cu and CuSb–Sn alloys, and a local crystallographic analysis (using the Kossel line technique) of aluminum and Al–Cu alloy films.[2] In the eight years since this pioneering work was done, the technique has been applied to numerous other problems involving diffusion in binary alloys, the analysis of precipitates, segregates, and inclusions, the identification of phases, and the determination of solid solubilities.[3-14]

The purpose of this paper is to describe some of the problems in these areas of physical metallurgy which have been studied with the microanalyzer constructed by the author at the California Institute of Technology.[15] This instrument, which was completed in the latter part of 1956, is shown in Figure 1. It utilizes a self-biased triode electron gun, a single magnetic immersion lens with a magnification of about $1/80$, and a vacuum X-ray spectrometer with a range of 1.1 to 3.5 A.

THE CONCENTRATION PROFILE OF CHROMIUM
IN CHROMIZED IRON

A specimen of electrolytic iron was impregnated with chromium by the process known as chromizing. In one form of this process the iron is maintained at a temperature from 1100-1300°C, and the chromium which is deposited at the surface of the iron by a replacement reaction diffuses into the iron to a depth which depends on the time and temperature. In this form of the chromizing process, the surface concentration is limited by the equilibrium of the replacement reaction and therefore will be somewhat less than 100%. It is desirable to determine the surface concentration of the chromium and the depth of penetration of the chromium into the iron.

After chromizing, the specimen was plated with 0.010 in. of hard nickel to prevent rounding of the edge when the section was polished. The nickel and iron concentrations were measured on a section of the specimen at varying distances from the surface. The results of this analysis are shown in Figure 2.

In the case of binary alloys of iron and chromium, the intensities of the iron K_α line and the nickel K_α line are not proportional to the mass concentration of iron and nickel, respectively.

Figure 1. The Caltech electron probe microanalyzer.

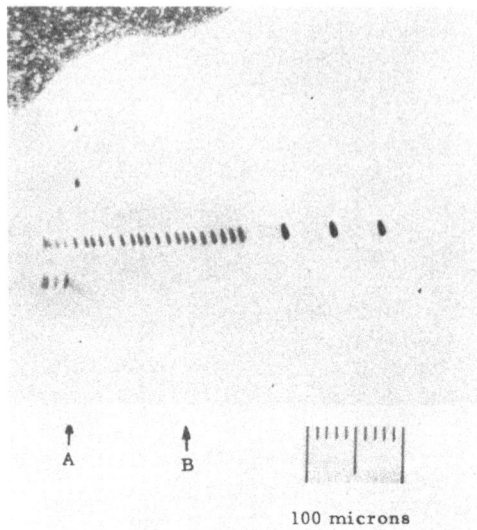

100 microns

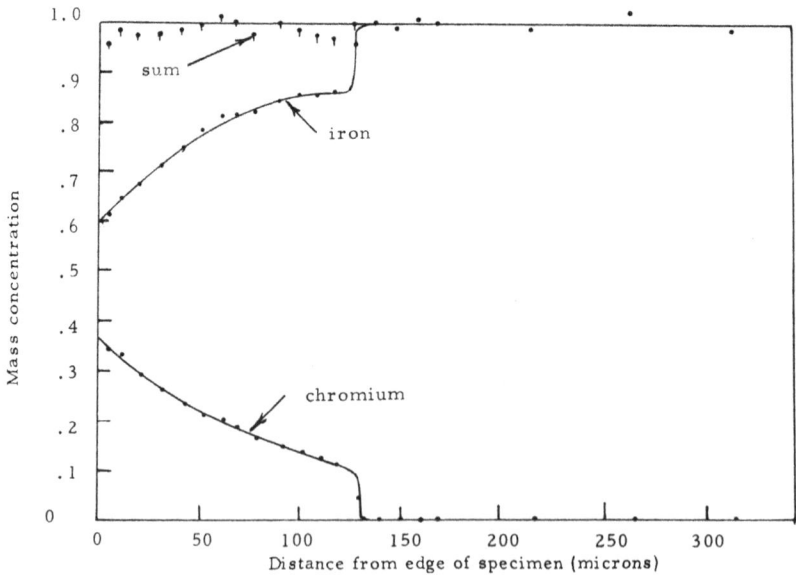

Figure 2. Concentration profile of chromium in chromized iron.

Instead, the iron K_α intensity is reduced by the strong absorption in chromium, and the chromium K_α intensity is enhanced by fluorescence. This is usually the case for specimens containing elements having atomic numbers differing by more than two. The corrections for fluorescence and self-absorption can be calculated, but since they are large, the accuracy of the final result which is plotted in Figure 2 is at most about 2%.

The experiment shows that the concentration of chromium at the edge of the specimen is about $37 \pm 2\%$. The chromium concentration decreases with depth in a manner which appears to obey Fick's law of diffusion to a depth of about 130 μ, where an abrupt decrease occurs. The abrupt decrease in chromium concentration corresponds to the concentration at which a phase change from α to γ occurs. If the diffusion constant of chromium in these two phases were nearly equal, the discontinuity would be at most the width of the two-phase field in the binary phase diagram. Therefore, it must be concluded that the diffusion rate of chromium in the γ phase is much less than that in the α phase. It is interesting to note that the position of the discontinuity in chromium concentration is visible (arrow B) in the micrograph. The surface of the specimen was polished with 1 μ diamond paste and was not subsequently etched.

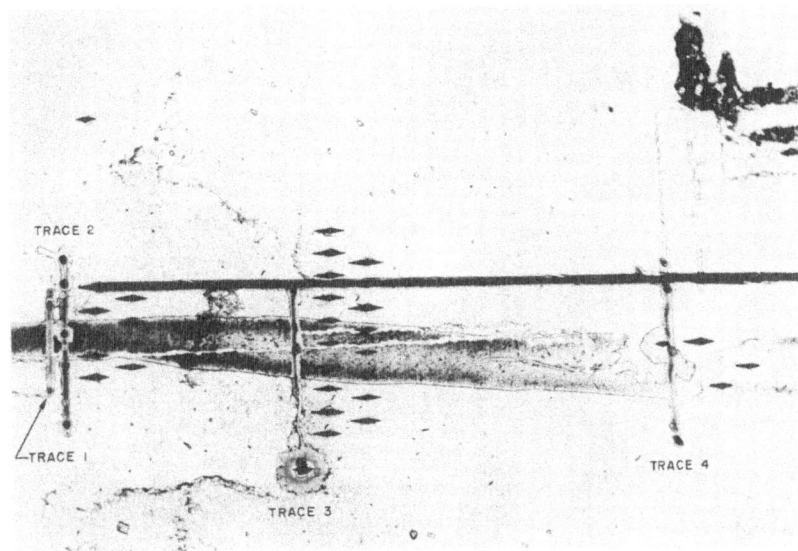

Figure 3. Micrograph of a portion of a taenite band in Canyon Diablo meteorite, fragment A.

ANALYSIS OF THE WIDMANSTÄTTEN STRUCTURE
IN METEORITIC IRON

In iron—nickel meteorites, the over-all nickel concentration is generally low except for narrow bands. Figure 3 shows a micrograph of a portion of such a band itself. The regions of low Ni concentration are called kamacite while those of high Ni concentration, which occur in the band itself, are called taenite. It is often the case that the interior of the bands appear after a picral etch to be a mixture of two phases which is called plessite. The structure of these bands have been extensively studied with the electron probe microanalyzer.[11,12,13] Most of the existing results confirm the conclusions which came from an independent analysis performed with the Caltech microanalyzer. In general, it is found that the edges of the band contain a high concentration of nickel, while the interior of the band contains a concentration intermediate between that of the matrix (kamacite) and the edge of the band (taenite).

Traces which were made with the microanalyzer can be seen by the contamination formed under the electron probe. The dark triangular marks are indentations formed by a Knoop micro-

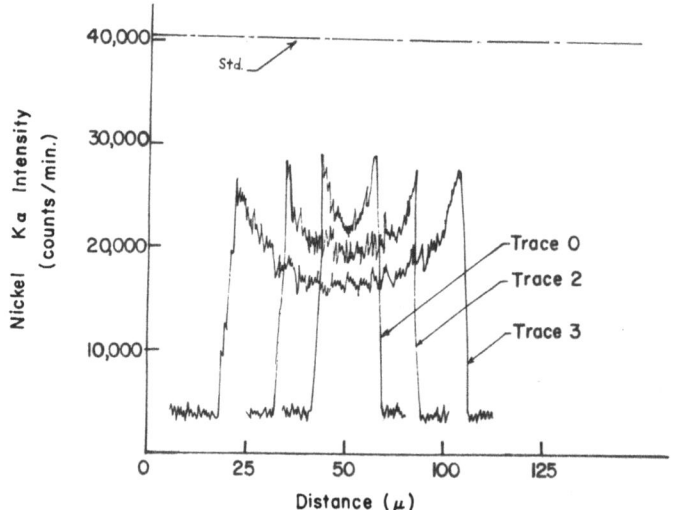

TAENITE BANDS IN CANYON DIABLO METEORITE
FRAGMENT A

Figure 4. Nickel K_α counting rate as a function of distance for three traverses over different portions of the band shown in Figure 3.

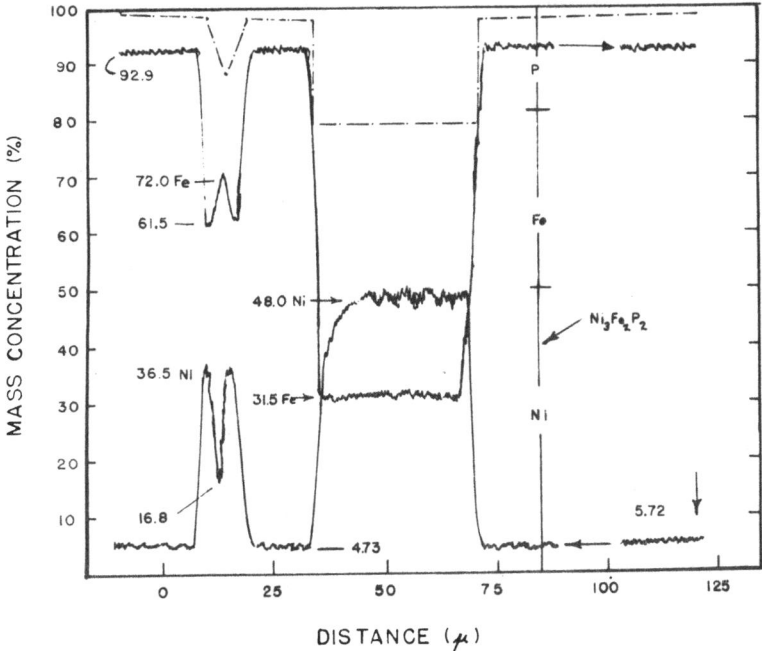

Figure 5. Microanalysis of a particle found in a discontinuous part of the taenite band shown in Figure 3.

hardness tester using a 10 g load. The horizontal white line on the band is not a true feature of the microstructure but results from a contamination deposit which has prevented the etch from attacking the underlying material. The fourth trace passes over a discontinuous part of the band and the microanalysis shows that the particle at this point is of a different nature than the band itself (see Figures 4 and 5).

The analysis which was performed with the Caltech microanalyzer was made on a narrow band in Canyon Diablo meteorite fragment A. Figure 4 shows the nickel K_α count rate as the specimen was traversed under the electron probe at points along the band where its width varied. The nickel concentration at the center of the band appears to increase as the width of the band decreases. As the probe moved from kamacite (low Ni) to taenite (high Ni), the counting rate rose to its maximum value in a distance of about 2.5-3 μ, which is approximately the width of the region excited by the electron probe. The intensity of a standard containing 50% nickel is shown as a dashed line in the figure. The maximum nickel concentration in the taenite appears to be about 38-40%.

This figure exhibits two features of interest which are significantly different from the result of Maringer et al. on the Grant meteorite. In the first place, Maringer's analysis, which was made on a band 450 μ wide, shows a nearly constant nickel concentration of 13% in the interior of the band up to 50 μ from the edge, whereas the present analysis shows a gradual change in nickel concentration across the band. The minimum nickel concentration depends on the width of the band and varies from about 25% when the width of the band is 75 μ to 15% when the width is only 15 μ. In the second place, the present analysis shows that the nickel concentration undoubtedly changes abruptly from kamacite to taenite since the distance over which the count rate goes from minimum to maximum is of the order of the width of the region excited by the electron probe, that is, about 2.5 μ.

Figure 5 shows an analysis of a particle which was found in the taenite band and which has been tentatively identified as Ni_3Fe_2P.

The Ni K_α intensity was recorded in one direction (note arrows) and then the stage reversed and the Fe K_α intensity recorded (note that on return the probe did not exactly retrace its original path). The numbers given are mass concentration obtained after correcting the Ni K_α intensity for self-absorption and correcting the Fe.K_α intensity for fluorescence excitation. The constant concentration of Ni and Fe across the particle clearly indicates that it is different from the band itself.

The assumption was made that the deficiency in the sum of the concentration of nickel and iron was due to phosphorus. Unfortunately, it was not possible to measure the concentration of the phosphorus since its characteristic radiation fell outside the range of the spectrometer.

PHASE IDENTIFICATION IN TITANIUM—COPPER
AND TITANIUM—SILVER ALLOYS

One of the important areas of application of the microanalyzer is the identification of phases formed in alloys after solidification from the melt. The positive identification of the phases formed can be very important in the process of interpreting the metallography to determine the phase diagram. It is also possible that one can get a more accurate picture of the changes which may occur in the equilibrium phase diagram as a result of nonequilibrium conditions.

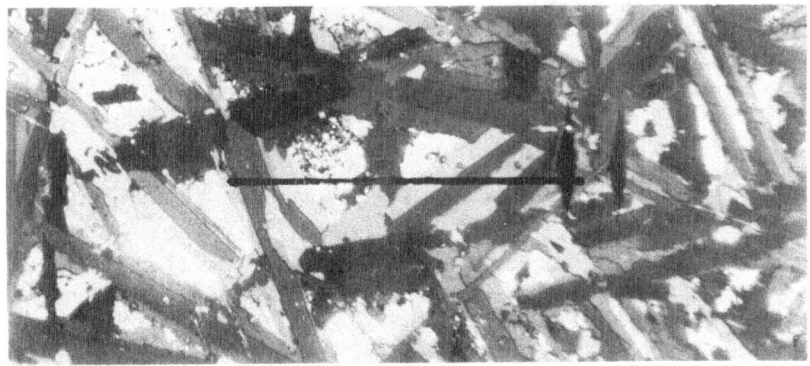

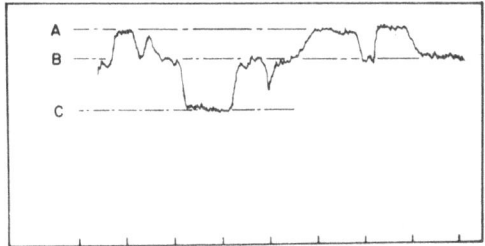

Figure 6. Microanalysis of a titanium–copper alloy.

One example of such an application is shown in Figure 6. This figure shows a micrograph of an alloy of titanium and copper which was cooled from the melt. The micrograph was taken with polarized light before analysis and the path of the electron probe indicated by an inked line. The graph shows the intensity of the titanium K_α line as a function of distance (1 div = $= 25\,\mu$). The three plateaus shown correspond to atomic concentrations of titanium as follows: A–50%, B–41%, C–26.6%. The first crystals to form (dark blue) were known to be the compound TiCu (50 at.% titanium) and this was therefore used as a standard for determination of the composition of the other phases.

The second phase to form under solidification appears light blue in the metallograph and is due to a peritectic reaction. After correction for the fluorescence excitation of the titanium by copper, it is found that the titanium concentration of this phase is approximately 41 at.%, which corresponds within experimental error to Ti_2C_3. The third plateau is the residual liquid, which freezes last and corresponds to the known regions. The titanium concentration in these regions was found to be 26.6 at. %.

Another composition intermediate between the peritectic wall and the residual liquid was observed on traces which were made on other parts of this same specimen. The titanium concentration in this phase appears to be a minimum of 37.4 at. %, with the concentration being somewhat higher at the boundary between this phase and the residual liquid.

A second problem involving the identification of phases with the microanalyzer is shown in Figure 7. In this experiment, a titanium rod was immersed in molten silver at 1000°C for two days. Almost all of the titanium went into solution in the silver, so that after cooling, the remaining part of the rod appeared as a central core surrounded by a wall. The rest of the specimen consisted of crystals which grew from the melt and were dispersed throughout the solidified residual liquid. The micrograph was taken after analysis and the contamination streak shows the path of the electron probe.

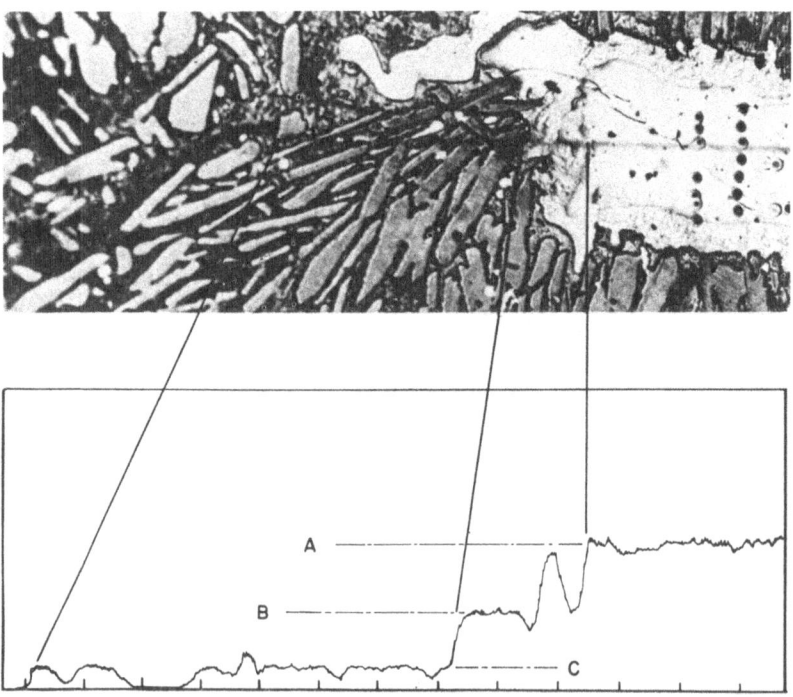

Figure 7. Microanalysis of a titanium—silver specimen.

The K_α intensity of titanium was recorded as a function of distance (1 div $= 25\,\mu$) and is shown in the graph of Figure 7. The intensity of Ti K_α from the central core, which is indicated by the level A, appears to be about 10% less than that from pure titanium. Unfortunately, however, it is not possible with the present instrumentation to record any of the characteristic radiation from silver. The plateau at level B corresponds to the wall, and it can be seen that the titanium concentration is constant except for the peak which can be seen from the micrograph to result from a projection of the core. The concentration of this plateau is somewhere between 79 and 85 at.%, the uncertainty being due to the uncertain concentration of the core. The constant concentration at this level cannot be understood in the light of present knowledge of the Ag–Ti phase diagram. The plateaus at level C correspond to the crystals dispersed throughout the solidified liquid. The titanium concentration in these crystals was determined from the microanalysis to be 50.12 at.%. This is in agreement with a crystallographic analysis performed on this phase by Dr. P. Pietrokowsky of Caltech.

Because of the lack of agreement of the high-Ti phases in this specimen with the existing phase diagram, a Ti–Ag diffusion couple formed at 960° was studied. The result of this analysis shows that the phase diagram given by Hansen[16] is qualitatively correct at this temperature. This diffusion couple may be described as follows, starting from the titanium side: the titanium concentration falls gradually over a distance of 250 μ and then drops abruptly to a level corresponding to Ti–Ag. The titanium concentration is constant at this new level for 60-70 μ and then drops sharply. Beyond this point, there is some indication of diffusion of titanium into the silver for a distance of about 15 μ.

CONCENTRATION FLUCTUATIONS IN SOLID-SOLUTION ALLOYS

The freezing of a liquid solution containing two metals which are completely miscible in the solid state usually is accompanied by some degree of segregation. Assuming that the liquid is nearly homogeneous before freezing occurs, the degree of segregation will be limited by diffusion and hence will be dependent on how rapidly the temperature is lowered from the temperature at which freezing first occurs to that at which solidification is complete.

If the solidification is sufficiently slow so that the system is nearly in thermodynamic equilibrium at all times, the concentration fluctuations can be predicted from the equilibrium phase diagram. The determination of the concentration fluctuations therefore provides a possible means of determining the liquidus and solidus lines.

It is more often the case, however, that the segregation on solidification is a source of annoyance in the preparation of homogeneous solid-solution alloys, and one would like to achieve a high cooling rate in order to reduce segregation. The cooling rate which is required to achieve an arbitrary degree of homogeneity in the solid solution is seldom known. The usual procedure is to cool as rapidly as the experimental arrangement permits and then to homogenize the solid alloy by diffusion at a temperature below the melting point.

In the preparation of a homogeneous alloy of Au–Cu containing 56 wt. % gold it was suspected from the behavior of the specimen under etching that the initial alloy was inhomogeneous. The alloy had been melted in vacuum in an alundum crucible of $\frac{1}{4}$ in. inner diameter and $\frac{1}{16}$ in. wall using induction from a 450-kc rf generator. The temperature was maintained just above the melting point of copper for about 5 min, after which the power was shut off and the specimen allowed to cool with normal radiative losses. Figure 8 shows a microanalysis which was made on sections at the top (A), center (B), and bottom (C) of the resulting ingot.

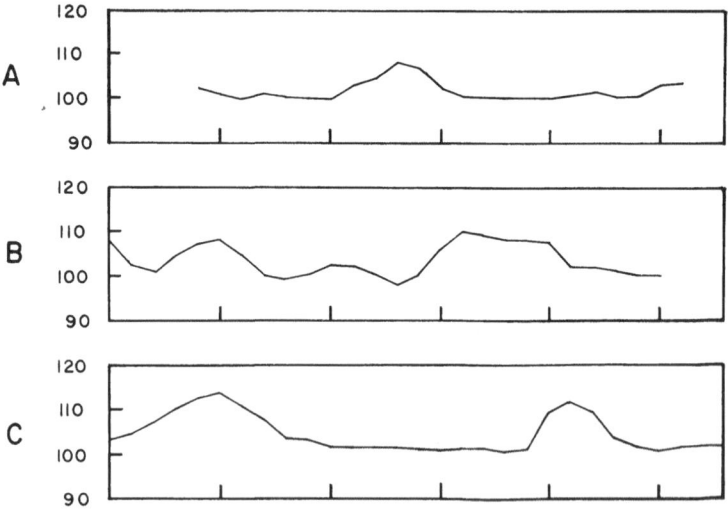

Figure 8. Concentration fluctuations in an as-cast alloy of gold and copper containing 56 wt.% of gold.

The vertical scale is thousands of counts on the Cu K_α line and each division on the horizontal scale is 50 μ. It can be seen that the average concentration is nearly the same at the top, center, and bottom of the ingot and that the copper concentration fluctuates over distances of 50-100 μ by as much as 10%. After cold rolling to about 50% reduction in area and heat treating overnight at 900°C, no concentration fluctuations could be observed with the microanalyzer.

THE DIFFUSION OF NICKEL IONS INTO
SINGLE CRYSTAL SAPPHIRE

Since most of the applications which have been made of the microanalyzer involve metallic specimens, it is desirable to illustrate its use in one case involving insulators. The problem of diffusion of transition metal oxides into single crystal Al_2O_3 was called to our attention by Dr. H. Palmour, III, of North Carolina State College. Dr. Palmour provided us with a specimen of oriented single crystal sapphire rod 0.090 in. in diameter, which had been treated in the presence of NiO powder for 2 hr at 1550°C.

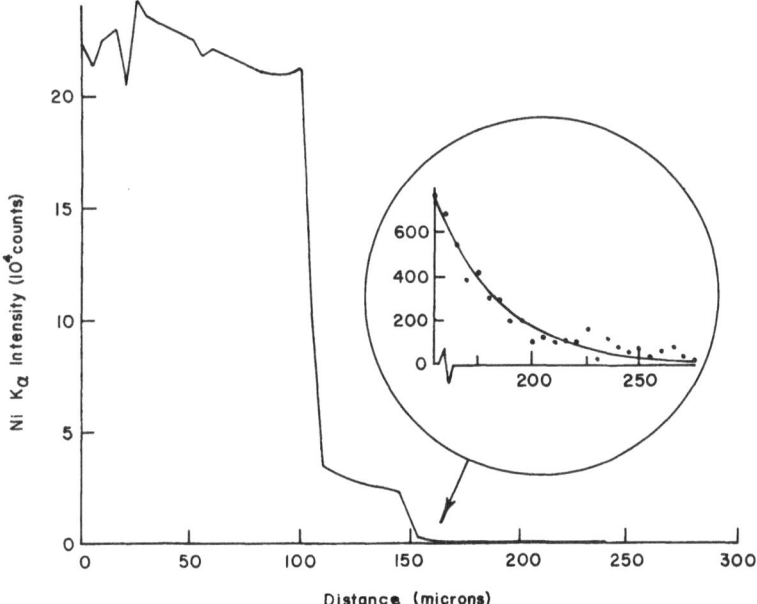

Figure 9. Intensity of the nickel K_α line as a function of distance in a specimen of single crystal sapphire treated in the presence of NiO for 2 hr at 1750°C.

After polishing the specimen and depositing an optically opaque layer of aluminum on its surface, a microanalysis was made. The initial results were promising and additional specimens were treated in our laboratory with NiO for 29 hr at 1350°C, and with Cr_2O_3 for 14 hr at 1350°C. A specimen was also treated in NiO for about 2 hr at 1750°C through the cooperation of Dr. P. Johnson of Atomics International.

The last of these specimens exhibited the most interesting results. Figure 9 shows the Ni K_α intensity as a function of distance for this specimen. The c axis of the sapphire was normal to the direction of diffusion. The intensity is very high for a distance of about 100 μ from the surface of the specimen. This is believed to correspond to the NiO $\cdot$ Al_2O_3 spinel and was observed on all of the specimens which had been treated in the presence of NiO. In addition, however, there is a region of substantially lower Ni concentration which extends over distances ranging from 10 to 50 μ, depending on the location in the specimen. This plateau was well confirmed by several measurements and in addition the regions in which it occurs on the specimen can be visibly distinguished from the others when observed through a metallurgical microscope.

Beyond the second plateau in the nickel K_α intensity there is a decay of the intensity as shown in the magnified view of Figure 9. This decay is similar to that observed on other specimens and was first attributed to the diffusion of the foreign ions into the sapphire. However, the magnitude of the intensity observed just inside the Al_2O_3 and the depth of decay can both be accounted for on the basis of fluorescence excitation of the region of high concentration by the bremsstrahlung produced under the electron probe. Experiments are now in progress to determine if any diffusion can be observed when the fluorescence excitation is eliminated by removing the spinel layer.

ACKNOWLEDGMENTS

This work was supported by the Office of Ordnance Research, U.S. Army. The specimen of chromized iron was furnished through the courtesy of Richard P. Seelig of the Chromalloy Corporation. The author would like to thank Dr. Walter Nichiporuk for suggesting the meteorite analysis and furnishing the specimens. Dr. Paul Pietrokowsky suggested the study of titanium—copper and titanium—aluminum alloys and furnished these specimens. The author is indebted to his colleagues, especially Dr. Nichiporuk, Dr. Pietrokowsky, and Professor Pol Duwez, for many helpful discussions on the interpretation of the results. Valuable technical assistance was rendered by Concetto Geremia, Stan Sajdera, and Frank Youngkin.

REFERENCES

[1] R. Castaing and A. Guinier, Proceedings of a conference on electron microscopy at Delft, 1949, Delft, 1950. Comptes rendus du premier congres international de microscopie electronique 1950, Editions de la Revue d'Optique, Paris, 1953.

[2] R. Castaing, "Application des Sondes Electroniques a Une Methode d'Analyse Ponctuelle Chimique et Cristallographique," These de Doctorat 1951. Publication ONERA, No. 55.

[3] R. Castaing, J. Philibert, and C. Crussard, "Electron Probe Microanalyzer and Its Application to Ferrous Metallurgy," J. of Metals, April, 1957, p. 389.

[4] J. Philibert and C. Crussard, "Applications of the Electron Probe Microanalyzer," J. of the Iron and Steel Inst., Vol. 183, 1956, p. 42.

[5] L. S. Birks and E. J. Brooks, "Applications of the Electron Probe Microanalyzer," Proceedings of the Sixth Annual Conference on Industrial Applications of X-ray Analysis, University of Denver, 1957.

[6] J. F. Norton, "Application of the Microemission X-ray Spectrograph — Comparison of the Analyses from Small Areas," Proceedings of the Sixth Annual Conference on Industrial Applications of X-ray Analysis, University of Denver, 1957.

[7] L. S. Birks and E. J. Brooks, "Applications of the Electron Probe," Sixteenth Annual Pittsburgh Diffraction Conference, Pittsburgh, November, 1958 (unpublished).

[8] V. Macres, J. T. Norton, and R. B. Ogilvie, "Interface Compositions in CuZn Diffusion Couples," Sixteenth Annual Pittsburgh Diffraction Conference, Pittsburgh, November, 1958 (unpublished).

[9] R. M. Fisher, "Electron Probe Microanalysis of Submicron Precipitates in Steels," J. Appl. Phys., Vol. 28, 1957, p. 1379; (abstract of paper presented at the fifteenth annual meeting of the Electron Microscope Society of America, Boston, September, 1957).

[10] C. S. Schwartz, A. E. Austin, and N. A. Richard, "Application of the Electron Probe Microanalyzer to Segregation and Diffusion," Sixteenth Annual Pittsburgh Diffraction Conference, Pittsburgh, November, 1958 (unpublished).

[11] R. K. Lewis and R. M. Fisher, "Electron Probe Microanalysis of Meteorites," Sixteenth Annual Pittsburgh Diffraction Conference, Pittsburgh, November, 1958 (unpublished).

[12] A.A. Yavnel, I.B. Borowskii, N.P. Ilyan, and I.O. Marchukova, "Determination of the Composition of the Phases of Meteoritic Iron by Means of the Local X-ray Spectroscopic Analysis," Meteorite Conference, Moscow, June 4, 1958.

[13] R. E. Maringer, N. A. Richard, and A. E. Austin, "Microbeam Analysis of Widmanstätten Structure in Meteoritic Iron,"Transactions of the Metallurgical Society of AIME, Vol. 215, 1959, p. 56.

[14] D. B. Wittry, J. W. M. DuMond, and Pol Duwez, "Use of the Electron Probe Microanalyzer in the Study of Semiconductor Alloys," Conference on Properties of Elemental and Compound Semiconductors, Boston, September, 1959 (Proceedings to be published by Interscience).

[15] D. B. Wittry, "An Electron Probe for Local Analysis by Means of X-rays," Ph.D. Thesis, California Institute of Technology, Pasadena, California, 1957.

[16] Max Hansen, The Constitution of Binary Alloys, McGraw-Hill, New York, 1958, p. 59.

AN INTRODUCTION TO TOTAL REFLECTION
X-RAY MICROSCOPY*

James F. McGee

Saint Louis University, Saint Louis, Missouri

ABSTRACT

The reflection X-ray microscope is a microscope in the conventional sense. X-rays are actually focused in the formation of magnified images of minute objects. Its principal component is a pair of concave glass mirrors which have been polished and figured with great care. In common with other optical devices, the reflection X-ray microscope suffers from various aberrations which limit its resolution. Until recently its narrow field and lack of contrast have made it impractical for the examination of histological or metallurgical specimens.

The X-ray microscope has the theoretical advantage of resolution when compared with the optical microscope. While it cannot compete with the ultimate resolution of the electron microscope, it has the advantage of being able to reveal the internal details of objects which are opaque to the electron microscope. Although there is no doubt that the X-ray microscope can contribute to our knowledge of structural details, perhaps its greatest potential lies in its analytical-chemistry aspects. Already a technique exists at unit magnification (contact microradiography) for weighing histological specimens whose cellular structures range from 10^{-12} to 10^{-14} g. The enlarged image provided by X-ray microscopy should greatly aid in a more extensive application of the technique.

The above advantages and possible areas of application are a source of continuous stimulation in the development of a reflection X-ray microscope which will operate close to its theoretical limit of resolution.

INTRODUCTION

For centuries man has been occupied with the problems of microscopy. In his attempts to form magnified images of minute objects he has successfully developed the light microscope and the electron microscope together with a host of techniques such as fluorescent microscopy, dark-field illumination, interference microscopy, and polarized-light microscopy.

*Work supported in part by the Research Corporation and the National Science Foundation.

213

A measure of the quality of a microscope is its resolution or ability to distinguish two points or lines. The smaller the separation for which they can be distinguished as such, the better the resolution. Since resolution depends directly on the wavelength of the light, it is little wonder that the first improvement was made by using light in the ultraviolet part of the spectrum.

While the next region of smaller wavelength is the X-ray spectrum, it has been ignored for years, possibly because the literature contains statements to the effect that X-rays cannot be focused. It is interesting to note that Roentgen in 1895 convinced himself that X-rays could not be concentrated by lenses. He, of course, was referring to focusing by refraction. It is now understood why the latter method is of little practical value. Since the refractive index of matter for X-rays is less than unity by a few parts in 10^5, the focal length of a refracting X-ray lens would be approximately 10^4 times its radius of curvature. Object and image distances would be reckoned in miles. While focusing of X-rays by refraction has been eliminated on physical grounds, there still remains the possibility of using reflecting components in an X-ray microscope. It may be recalled that Sir Isaac Newton developed the first reflecting light microscope in 1672. Between 1672 and 1850 both reflecting and refracting light-microscope objective lenses were developed. However, with the introduction of higher quality optical glass and the achromatic objective in 1824 the reflecting light microscope fell into oblivion until the early part of the twentieth century.

While history records the suggestions of F. Jentzsch[1] concerning the possibility of a total-reflection X-ray microscope, it was almost twenty years later that P. Kirkpatrick[2] and W. Ehrenberg,[3] independently of one another, demonstrated that the focusing of X-rays by total reflection is practical. Kirkpatrick's first microscope stimulated more interest throughout the world. Already, some ten years later, two international symposia have been held on the subject. The first was held at the Cavendish Laboratory, England, in 1956, and the second at the Karolinska Institute, Stockholm, in 1959.

GEOMETRICAL OPTICS

The fact that the index of refraction is less than unity makes possible the total reflection of X-rays in the less dense medium and the use of concave reflecting surfaces as focusing elements. It is convenient to define the angle of grazing incidence as the

---

[1] Superscripts pertain to references at the end of the paper.

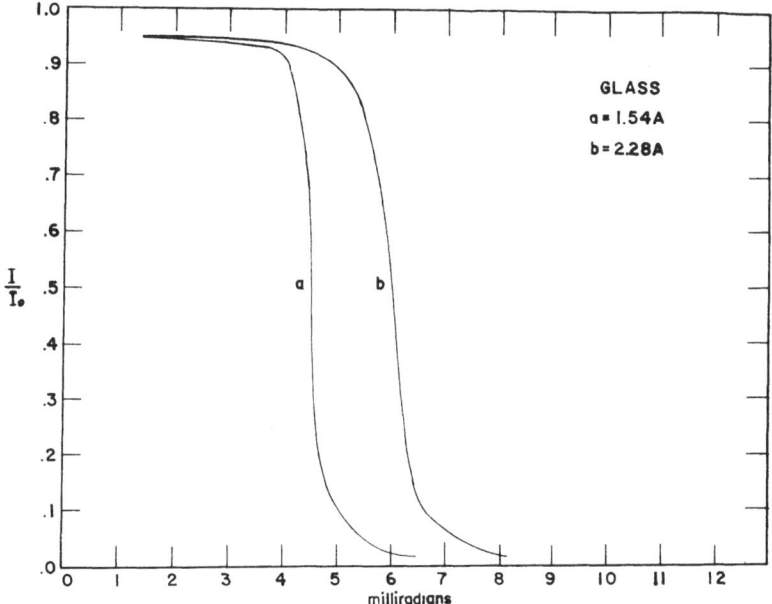

Figure 1. Reflection power of glass vs grazing angle of incidence.

complement of the conventional angle of incidence. The critical angle $\theta_c = \sqrt{2\delta}$, where $\delta = ne^2 z\lambda^2/2\pi mc^2$, is a well known result of the Lorentz theory. Thus, provided the X-rays strike the surface at a grazing angle less than θ_c, total reflection will take place. The reflecting power of glass is shown in Figure 1 as determined by Reiser.[4]

Figure 1, curve a shows that the critical angle (measured at 50% reflection power) is approximately 4.5 mrad. for $Cu K_\alpha$ radiation (1.54 A). For $Cr K_\alpha$ (2.28 A) the critical angle is approximately 6.0 mrad.

As shown by Kirkpatrick and Baez,[2] divergent rays from a point source which are incident on a cylindrical surface at an angle $\iota = \theta_c$ are brought to a line focus as shown in Figure 2a. The focal length f_m in the meridian plane is given by the equation $f_m = (R \sin i)/2$. Sagittal rays are focused at $f_s = R/(2 \sin i)$. For reflecting surfaces of small extent the object and image positions are given by

$$\frac{1}{p} + \frac{1}{q} = \frac{1}{f_m} = \frac{2}{R \sin i} , \qquad (1)$$

where p and q are the object and image distances, respectively, f_m the meridional focal length, R the radius of curvature of the concave reflector, and the angle of grazing incidence. A comparison of the two focal lengths f_m and f_s reveals the extreme astigmatic nature of a single concave surface as a focusing element. As seen in Figure 2a, a point object is focused into a line image which is perpendicular to the meridional plane. The extreme astigmatic performance may be removed according to Kirkpatrick and Baez by using two reflecting surfaces in tandem and at right angles, as seen in Figure 2b, or by orthogonally butting as in M. Montel's[5] arrangement. To a first approximation such a pair of crossed reflectors will produce real images of real objects.

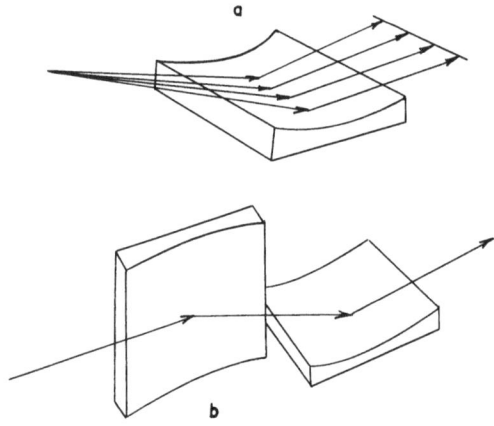

Figure 2. Astigmatic optics of a single mirror
at grazing angles of incidence.

In Figure 3 the pin-hole object is focused into a line image at h by the concave reflector H. The concave reflector V focuses the same object into the line image v. The combined action of both reflectors results in the point image hv of the pin-hole object. The image at d is due to rays that miss both mirrors.

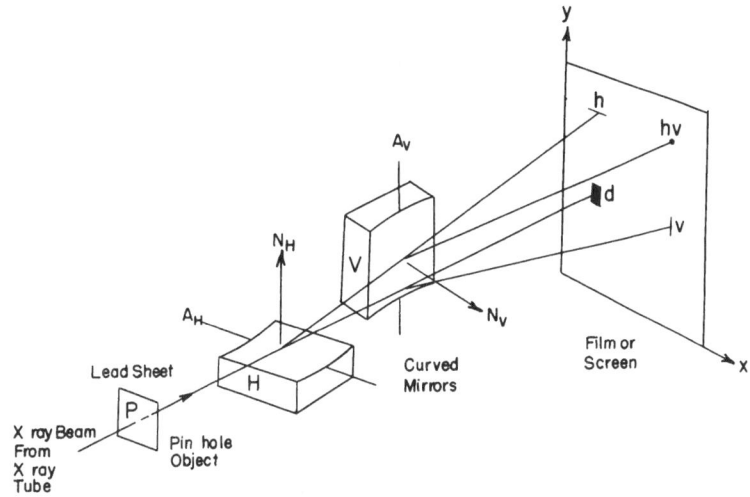

Figure 3. Two-dimensional image formation by crossed mirrors.

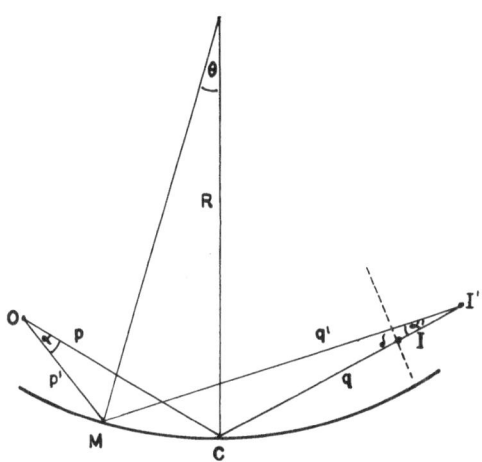

Figure 4. Spherical aberration.

ABERRATION THEORY

Spherical Aberration. If we confine our attention to the two-dimensional optics of a plane, the most prominent optical

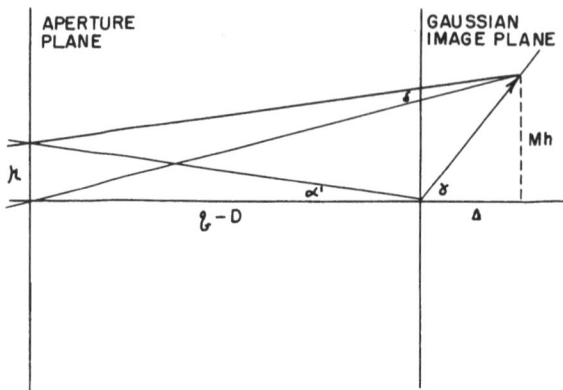

Figure 5. The obliquity aberration.

defect of a reflecting surface is the aperture defect known as spherical aberration. In Figure 4, a narrow bundle of rays from the point object O strikes a narrow segment of reflector at C, and is imaged at I according to equation (1). Another narrow bundle of rays from O strikes the reflector at M and is imaged at I' just beyond the Gaussian plane. The intercept δ is a measure of the transverse spherical aberration and for a small reflector segment $\overline{CM}$ is given by

$$\delta = \frac{3\,Ms^2}{2\,R} \tag{2}$$

when the magnification M is much greater than unity.

Obliquity Aberration. An object which is perpendicular to the principal ray in object space will produce an image which is quite oblique with the conjugate ray as shown in Figure 5.

Because of the obliquity aberration, the image plane is tilted at an angle $(90-\gamma)$ with respect to the Gaussian image plane. Obliquity produces an image defect of extent δ on a photographic film placed in the Gaussian image plane even when spherical aberration is absent.

While it is generally known that a correctly positioned aperture stop will improve the performance of many optical systems, it is only recently[6] that an exact expression has been given for the position of an aperture stop which will erect at least one image point into the Gaussian plane. It has been shown that an

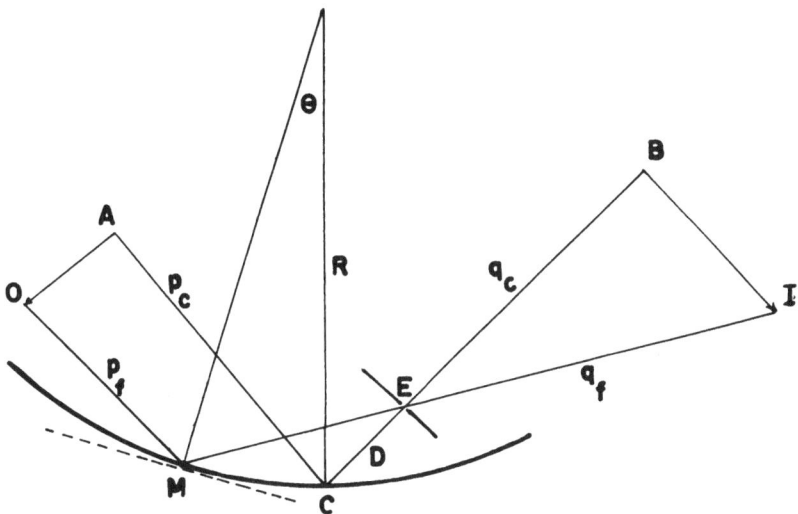

Figure 6. Correction of obliquity.

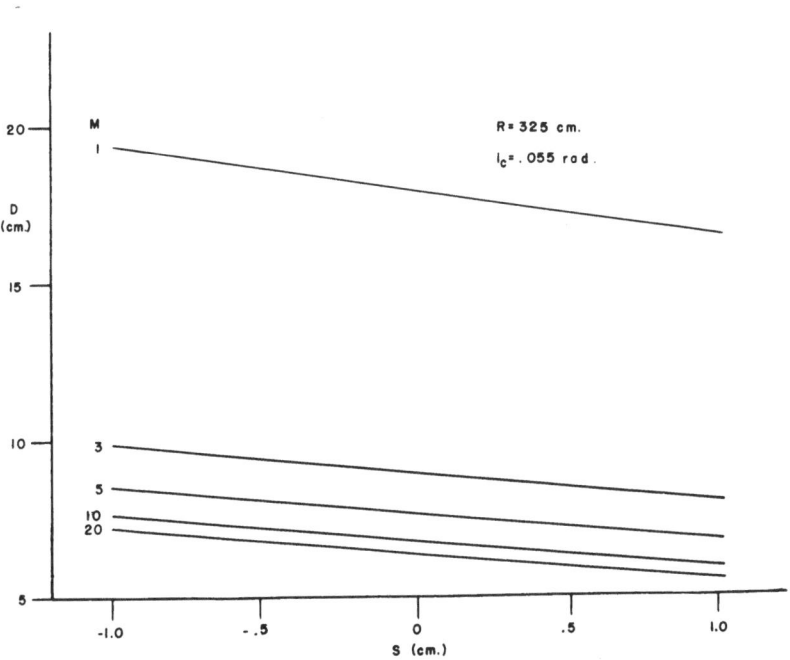

Figure 7. Infinitesimal aperture position.

infinitesimal aperture stop placed at E of Figure 6, which is a distance D from the center of the reflector C, will cause the image point I to lie in the Gaussian plane which is perpendicular to the conjugate principal ray q_c at B. The distance D is given by

$$D = \frac{4(1 + M)^2 \, f_c^2 + (3 - 5M)\,(1 + M)\,sf_c - 4Ms^2}{2(1 + M)\,(3M - 1)\,f_c + 4Ms}, \quad (3)$$

where M is magnification, $2s$ is mirror length, and $f_c = (R \sin i)/2$.

In Figure 7 the aperture position D is plotted versus the length $\pm s$ according to equation (3) for various values of the magnification M. The value of f_c was computed for a radius $R = 325$ cm and an angle $i_c = 0.055$ rad using the focal length expression from equation (1).

For infinite magnification equation (3) reduces to

$$D = \frac{R \sin i_c}{3} - \frac{5s}{6} ; \quad (4)$$

with $s = 0$ and i_c small, equation (4) yields Dyson's[7] result

$$D = \frac{R i_c}{3}, \quad (5)$$

A detailed ray-tracing analysis, made with the aid of an IBM computer, demonstrated that only one field point of the extended object in Figure 6 will fall in the Gaussian plane erected perpen-

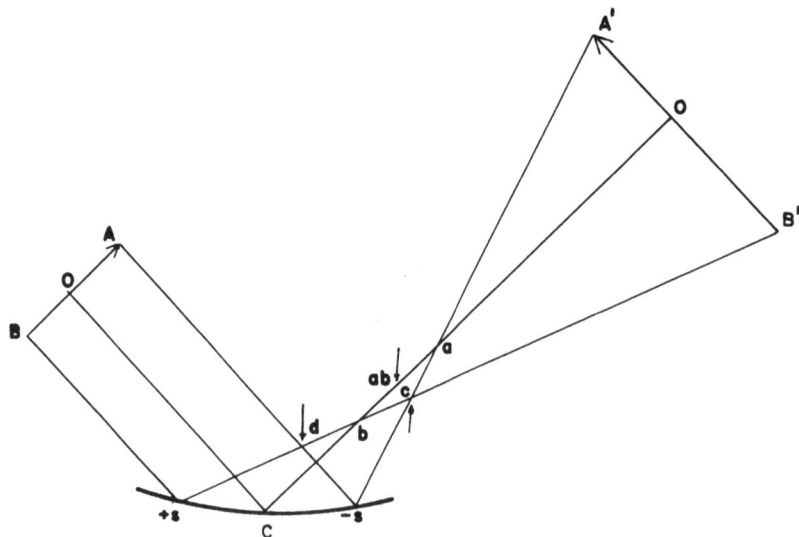

Figure 8. Limits of ray crossing points.

dicular to q_c at point B when all rays are made to go through the aperture E at a fixed distance D from the center of the reflector.

After some consideration of equation (3) and Figure 8, it may be seen that infinitesimal ray bundles, which originate at different object points along AB, must intersect the reflector at different values of s and cross the ray $\overline{CO'}$ at different points α and b, if they are to focus in the plane AOB. It would, thus, seem that an infinite number of narrow apertures would be required, if each and every point of the field is to be focused in the Gaussian plane and an erect image obtained. It is obvious that the placing of narrow apertures at a and b as well as at intermediate positions would not be satisfactory, because of the blockage of rays by the physical extension of each aperture. The ray diagram of Figure 8 suggests that perhaps half-apertures (single knife edges) located at the three cross-overs, a, b, and c would insure that the rays from field points A and B will cross $\overline{CO'}$ at a and b as given by equation (3). The latter solution is acceptable, until one traces the path of other infinitesimal ray bundles from other intermediate field points between A and B. They will intersect $\overline{CO'}$ at points between a and b and will consequently be cut off by the knife edges previously assumed in position at a and b, respectively To pass rays from intermediate field points, it becomes necessary to move the half-apertures at a and b toward one another until they finally coalesce into one at some intermediate point (ab) almost directly above the point c. It now becomes obvious that a single half-aperture at (ab) and a half-aperture at c constitute one finite aperture, as distinguished from the narrow (infinitesimal) apertures previously considered but tacitly assumed to be slightly finite in practice. Further examination of Figure 8 reveals that the new finite aperture $(ab)c$ is not of itself capable of directing the limiting rays to cross at a and b, respectively. If another half-aperture is placed at d the edges of the half-apertures c and d would then determine the ray $\overline{BB'}$ which should cross at the point b. In a similar manner the half-apertures c and d determine the direction and cross-over point a of the ray bundle $\overline{AA'}$ after reflection. It thus appears that a finite aperture $(ab)c$ and a half-aperture d will serve the same function as the originally required infinite number of narrow apertures.

RAY-TRACING ANALYSIS

Image Contours. To verify the success of equation (3) in the removal of obliquity and the consequent erection of the image of a extended object such as $\overline{AB}$ of Figure 8, a ray-tracing analysis was performed with the aid of an IBM-610 computer.[8] A reversed-ray analysis was used. After selecting R, the radius of the reflector, i_c, the grazing incidence angle of the principal reference ray $\overline{OC}$ in Figure 8, and a practical image distance q_c, the computation starts with the reverse ray which passes through an unknown point in the Gaussian image plane, but intersects the reflector at an assigned point s and intersects the principal reference ray $\overline{CO'}$ at the distance $D = D(M, s, f_c)$ as given by equation (3). This is sufficient to fix the reverse ray's direction in object space so that its intersection with the object plane AB can be computed. The distance from the latter intersection to the intersection s with the reflector is the object distance p_i which can be readily computed from analytical geometry. An application of equation (1) taking into account the new angle of grazing incidence determines the image distance which does not necessarily fall in the Gaussian image plane $A'B'$. The calculation was repeated for rays whose successive s values range from -1 to $+1$. The rectangular coordinates of the corresponding image points were obtained. The resulting two-dimensional image contours are shown in Figure 9.

The image contours were computed for three different cases. Case I was computed with $R = 850$ cm and $i_c = 0.015$ rad; Case II for $R = 325$ cm and $i_c = 0.015$ rad; and Case III for $R = 325$ cm and $i_c = 0.055$ rad.

It is to be noted that the rectangular coordinate system of Figure 9 has its origin at C of Figure 8, with its abscissa tangent to the reflector at C, and its ordinate directed along R.

The image contours of Figure 9 do not significantly depart from a straight line perpendicular to the principal reference ray which in Figure 9 would make a small angle with the abscissa. The deviation over a large portion of the contour in Case III is less than the thickness of the plane photographic emulsion on which the image would ordinarily be recorded. The maximum width of field corresponding to Case III of Figure 9 is 1288μ.

The image contours show that the obliquity aberration can be reduced to a negligible extent over a wide field provided the individual ray pencils can be made to cross at the intersection determined by equation (3).

Total Geometrical Transverse Aberration. The aperture-stop system proposed in Figure 8 will direct limiting pencils of rays to cross the principal reference ray at the correct point. Since the practical system operates with a finite but rela-

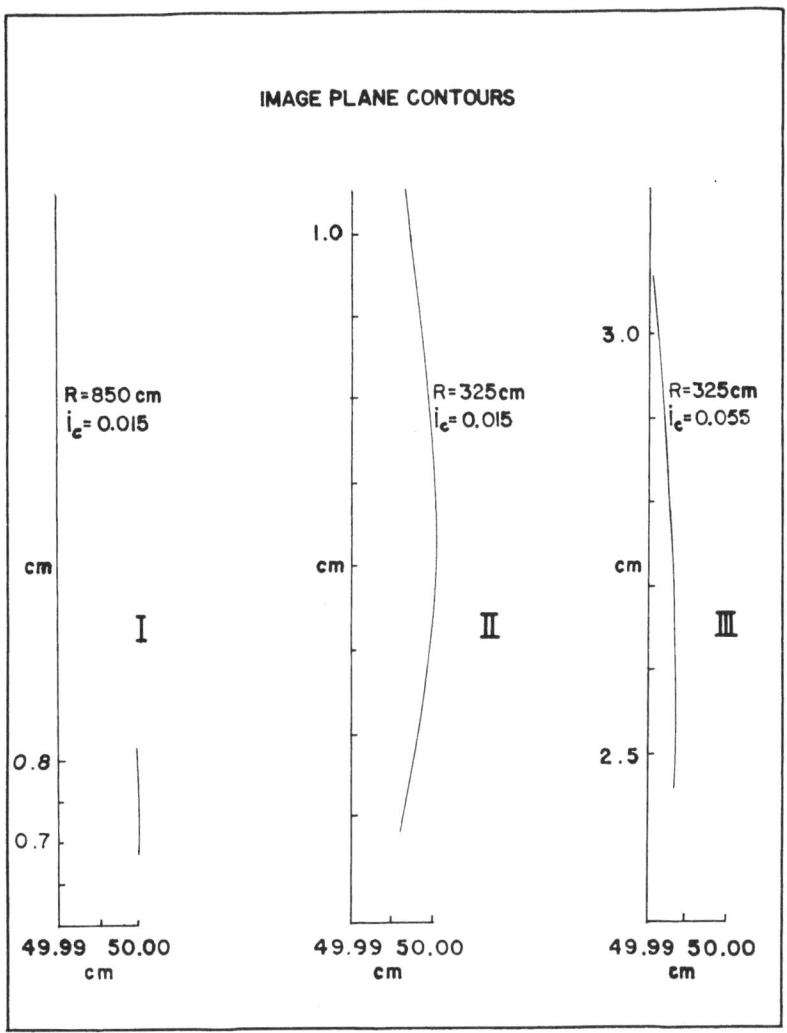

Figure 9. Two-dimensional image contours.

tively narrow bundle of rays from each point, it should be of interest to calculate the total transverse aberration produced by the proposed optics using the multiple apertures.

If each object point is considered the source of a divergent bundle of rays then an ideal aberrationless system would focus all the rays into a point image in the Gaussian plane. Actual systems do not direct the rays to intersect at a point in any one plane. Instead they will intersect a given plane over a length δ which is called the total aberration.

Assuming that Gaussian object and image planes have been erected, the path of any ray through the optical system is completely determined if h, the distance of the object point from the principal reference ray, and r, the coordinate of the ray's intersection with the aperture plane, are specified. In Gaussian optics the ray from a field point at h would intersect the Gaussian image plane at a distance Mh units from the reference ray where M is the magnification. An aberrant ray from the same field point intersects the Gaussian plane with an error which may be expressed as a power series in h and r as in equation (6);

$$
\begin{aligned}
\delta_t = a_{00} &+ a_{01}r + a_{02}r^2 + a_{03}r^3 + \ldots \\
&\ldots + a_{10}h + a_{11}hr + a_{12}hr^2 + \ldots \\
&\ldots + a_{20}h^2 + a_{21}h^2 r + \ldots \\
&\ldots + a_{30}h^3 + \ldots .
\end{aligned} \tag{6}
$$

The coefficients of each term are functions of the various system parameters such as object distance, magnification, angle of incidence i_c, and radius of curvature R. Since the optical system is unsymmetrical, the series may contain even as well as odd powers.

It is possible to interpret various terms of the power series as representing particular types of geometrical aberrations. When δ_t depends only on r^2 the coefficient a_{02} would be called the primary spherical aberration. Likewise, the coefficient a_{11} may be associated with the obliquity defect, while a_{21} and a_{12} are, respectively, associated with curvature of field and coma. The definition of the Gaussian plane requires that δ_t be zero when h and r are respectively zero; thus the coefficient a_{00} should be zero if the film and Gaussian plane also coincide.

A determination of the coefficients of equation (6) was made assuming that terms beyond the third order were negligible. The computation was made for a reflector with $R = 325$ cm and $i_c = 0.055$ rad. A standard method was used for triangulation of the matrix in (h, r) and the coefficients were improved in accuracy by a technique due to Crout.[9] A considerable loss in precision may result through rounding-off errors and failure to evaluate the coefficients in the proper order. The results are tabulated in Table I. The δ_t of equation (6) should be divided by the magnification when one calculates the field defect from the coefficients given in Table I.

Table I

$a_{01} = -0.0086267$	$a_{10} = 0.0010645$	$a_{20} = 3.9670$
$a_{02} = -1.8874$	$a_{11} = -0.53659$	$a_{21} = 22.387$
$a_{03} = -680.60$	$a_{12} = 56.179$	$a_{30} = 4.4605$

Intensity Distribution. In the present study a uniform angular distribution of rays emanating from a single source was traced through the optical system, and their intersections with Gaussian image plane determined. The linear density of intersection was expressed in arbitrary "intensity" units and plotted as a function of distance from the intersection of the chief ray (zero aperture) with the image plane. The process was repeated for another field point displaced $0.1\ \mu$ from the original point which was located a distance $h = 382\ \mu$ from the center of the field. The resulting "intensity" curve is shown in Figure 10. A total of 650 ray intersections were used for above calculations of the intensity distributions in image space.

On the basis of any reasonable resolution criterion it is clear that the geometrical resolution is less than $0.10\ \mu$ over a very large field. Previous pessimistic calculations have taken the maximum spread of the ray intersections as a measure of the geometrical resolution. In fact, Prince[10] equates the geometrical blur of Kirkpatrick and Baez, to the diffraction resolution to yield a minimum resolution of $0.34\ \mu$ at the uninteresting magnification of unity.

X-RAY WAVELENGTH CONSIDERATIONS

To secure suitable contrast in any type of X-ray microscope used for the examination of typical biological sections, such as those employed in histological and cytological studies, long-wavelength X-rays are needed. In general the wavelengths are greater than those which are commonly employed in medical radiography or in X-ray diffraction studies.

The choice of the most suitable wavelength for use with biological and other specimens involves many considerations pe-

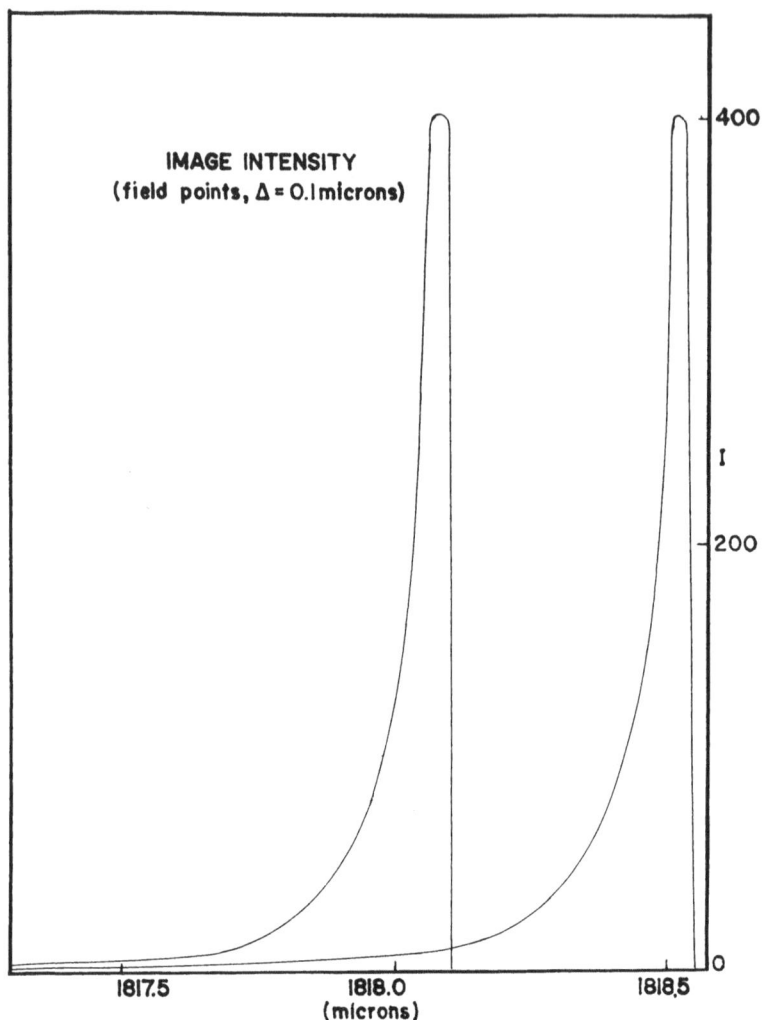

Figure 10. Distribution of image intensity.

culiar to the specific problem. Henke et al.[11] discuss these in great detail.

As a first approximation the specimen thickness and wavelength are chosen so that the transmitted intensity is reduced to $1/e$ of the incident intensity. For a biological section of average density the corresponding wavelength is in the region of 10 A. Unfortunately, suitable commercial X-ray tubes are not

Figure 11. Complete X-ray microscope system.

available for this and higher wavelengths. The development of an X-ray microscope invariably involves the additional development of a suitable X-ray source.

From an instrumental point of view the longer wavelengths are desirable because of the larger critical angle which in turn results in a larger aperture and increased speed.

Long-wavelength X-rays are further desirable because they lessen the requirements on the surface perfection of the reflectors. It has been shown by Jentzsch, that the surface roughness which can be tolerated should be equal to or less than $\lambda/8i$, assuming that $\frac{1}{4}$ wavelength of wave front distortion is acceptable. For a wavelength of 8.34 A and angles i varying between 10 and 50 mrad, the tolerance of the surface will lie between 100 and 20 A. The latter roughnesses lie in the experimentally determined range for optical flats by Koehler.[12] According to Ehrenberg[13] the criterion of $\lambda/8i$ is not small enough for good X-ray imaging. He suggests that the finest optical polishing leaves hills and valleys of about 1 mm width and about 10 A high. No one has yet explained the diffuseness of X-ray line images which Ehrenberg observed. Until his experiments are explained a shadow will remain over the possibility of ever attaining the theoretical resolution of reflection microscopy.

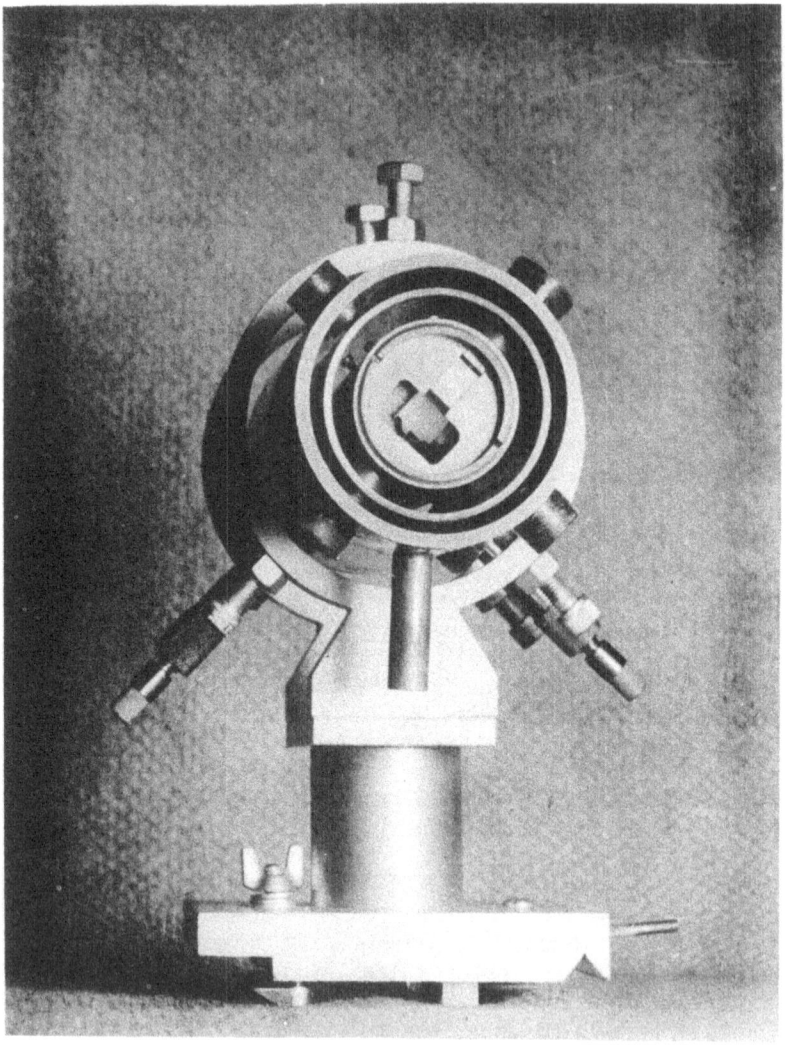

Figure 12. Complete mirror and stop assembly.

COMPLETE X-RAY REFLECTION MICROSCOPE

The complete long-wavelength microscope is shown in Figure 11. At the left end of the optical bench is a demountable aluminum-target X-ray tube containing a Mylar window. The target makes an angle of approximately 45° with the optic axis. A short length of flexible hose couples the X-ray tube window to the specimen holder mounted on the optical bench. The specimen holder can be moved in two mutually perpendicular directions at right angles to the X-ray beam. A larger-diameter flexible tube couples the specimen holder to the mirror and aperture-stop assembly, which is equipped with micrometers for adjusting the angles of the two mirrors independently. The adjustment of the stops is made through slots in the outer wall of the mirror and stop assembly. After adjustment the slots are covered with black plastic tape to keep the system gas-tight. The long metal tube is terminated by a photographic plate holder. The entire system is filled with helium to reduce the absorption of the aluminum K radiation. Auxiliary cooling and vacuum pump equipment for the X-ray tube may also be seen.

One of the two crossed mirrors is visible in the end view of the reflector and aperture-stop assembly shown in Figure 12. The two reflectors and their corresponding aperture stops are mounted in the central tube of Figure 12. The intermediate tube bears a micrometer which is located 10 cm from the axis of one reflector. This micrometer displaces the inner tube, causing the reflector to rotate about its axis with a least count of 0.0001 rad. The outermost tube bears a similar micrometer which deflects the intermediate tube, causing a rotation of the second reflector independent of the first.

If the innermost tube is removed and disassembled its parts will appear as in Figure 13. From left to right across the bottom are the two reflectors positioned at right angles to one another, a metal spacer, and finally two aperture stops, one for each reflector. The remaining half-aperture stops are contained in the reflector mounts. The ring at the upper left is simply a retaining ring for holding the components in the central tube shown at the top. Two adjustment screws are located on each full-aperture mount. One turn of the screw will displace the entire aperture 0.0177 in., while the second screw alongside but not visible opens or closes the aperture by 0.0014 in. for one complete turn. The adjustment screw on the half-aperture is the same as the coarse screw on the full apertures.

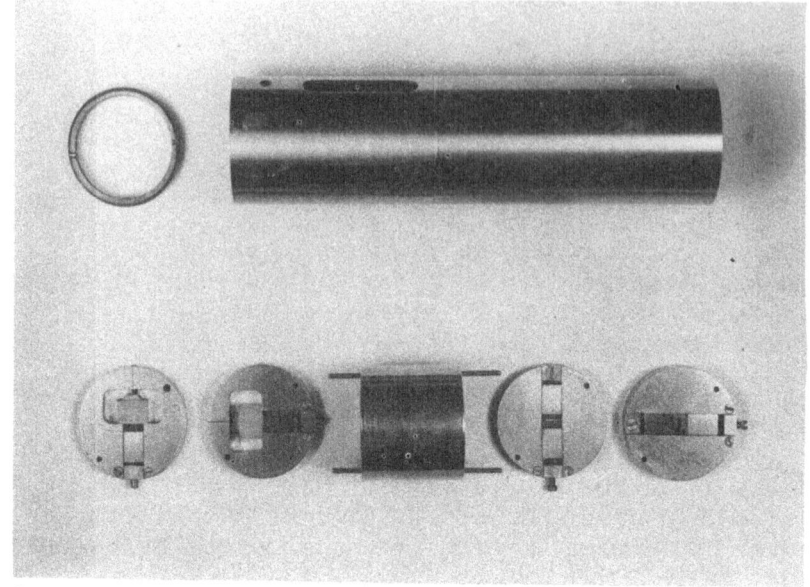

Figure 13. Individual mirror and aperture mounts.

Figure 14. X-ray micrograph of silver-mesh screen.

EXPERIMENTAL RESULTS

The present microscope, because of its limited resolution, is more or less confined to the examination of fairly large biological organisms such as plant and animal cells, red blood cells, and muscle fibrils whose gross aspects are of the order of a micron. The radiation emitted by the X-ray tube contains characteristic radiation of 8.34 A which is suitable for specimens approximately 5 μ thick. It should be noted that there is continuous X-radiation also incident on the specimen and the reflectors. The reflectors will of course only reflect the long-wavelength components of the continuous spectrum as determined by the angle at which the reflector is set.

In Figure 14 is seen a test picture of a silver-mesh screen of 1500 bars/in. The thickness of the bars is approximately 3 μ and the space between bars about 17 μ. The total magnification is 230 ×.

Figure 15 shows the X-ray microradiograph of a section of earthworm originally prepared for visible-light microscopy in a student biological laboratory. The specimen had been stained with haematoxylin and embedded in paraffin. The only additional preparation for X-ray microscopy consisted of removing the paraffin with xylene. The total magnification was also 230 ×. The visible graininess is due to the grain of the dental film used for the original exposure.

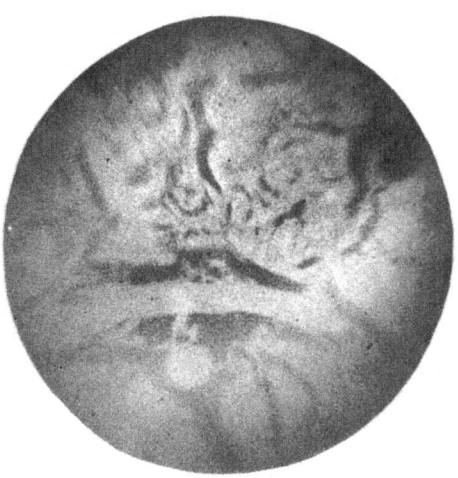

Figure 15. X-ray micrograph of earthworm gut.

The section of salamander kidney shown in Figure 16 was also 7 μ thick. It had previously been stained with osmic acid for routine light-microscope examination. The total magnification is again 230 ×.

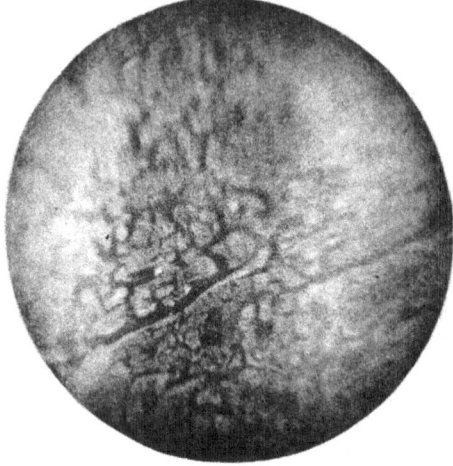

Figure 16. X-ray micrograph of salamander kidney.

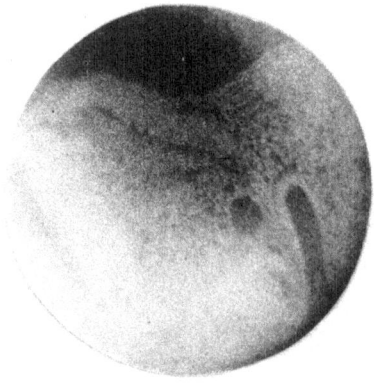

Figure 17. X-ray micrograph of embryo chick.

In Figure 17 the specimen was a 7-μ section of embryo chick which had been previously stained with eosin. The total magnification is 230 ×. The X-ray magnification of each of the above was 6.85 ×.

REFERENCES

[1] F. Jentzsch, Physik Z., Vol. 30, 1929, p. 268.

[2] P. Kirkpatrick and A. V. Baez, J. Opt. Soc. Am., Vol. 38, 1948, p. 766.

[3] W. Ehrenberg, Nature, Vol. 160, 1947, p. 330; J. Opt. Soc. Am., Vol. 39, 1949, p. 741; J. Opt. Soc. Am., Vol. 39, 1949, p. 746.

[4] L. M. Reiser, Jr., J. Opt. Soc. Am., Vol. 47, 1957, p. 987.

[5] M. Montel, Rev. Optique, Vol. 32, 1953, p. 585; Optica Acta, Vol. 1, 1954, p. 117; C. R. Acad. Sci. Paris, Vol. 239, 1954, p. 39.

[6] J. F. McGee, "A Long-Wavelength X-ray Reflection Microscope," X-Ray Microscopy and Microradiography, edited by Cosslett et al., Academic Press, 1957.

[7] Dyson. J. Proc. Phys. Soc. (London) B Vol. 65, 1952, p. 580.

[8] J. F. McGee and J. W. Milton, Proceedings of the Second International Symposium on X-Ray Microscopy and X-Ray Microanalysis, Stockholm, 1959 (in press).

[9] P. D. Crout, Trans. Am. Inst. Elec. Engrs., 1941, 60:1235-1240.

[10] E. Prince, J. Appl. Phys., Vol. 21, 1950, p. 698.

[11] B. L. Henke, B. Lundberg, and Engstrom, X-Ray Microscopy and Microradiography, edited by Cosslett et al., Academic Press, 1957.

[12] W. F. Koehler, J. Opt. Soc. Am., Vol. 43, 1953, p. 743; W. F. Koehler and W. C. White, J. Opt. Soc. Am., Vol. 45, 1955, p. 1011.

[13] W. Ehrenberg, J. Opt. Soc. Am., Vol. 39, 1949, p. 746.

DIFFERENTIATION OF SEVERAL RELATED CERAMIC BODIES BY X-RAY DIFFRACTION

Louis H. La Forge, Jr.

Sylvania Electric Products, Inc., Mountain View, California

ABSTRACT

For some vacuum tube work, it was desired to establish a reliable and moderately rapid means of differentiation among several related ceramic bodies, mostly among the high aluminas. Matured composition from a crystallographic standpoint was of primary interest; source and batch constituents were sought only as additional information for interpreting results on the matured ware.

Since Al_2O_3 represented the major crystal phase in most bodies, the indicators of difference were the minor crystal phases. A number of X-ray diffraction techniques were tried, as were machines of different manufacture. Camera studies with powdered samples were generally ineffective. Useful data were obtained using the Geiger tube—goniometer setup with experimentally determined settings for sensitivity, circuit time constant, sweep rate, beam and receiving slit widths, etc. Fixed-position peak-height counting was much slower, not practically better for peak-height determination, and introduced chances of missing peaks unless they were first determined by scanning and recording, thus further lengthening the analysis time considerably.

Satisfactory results have been obtained on a number of the bodies of interest. Generally, those bodies which are not differentiated by the X-ray technique outlined are not markedly differentiated in the vacuum tube application.

INTRODUCTION

During the course of an investigation of ceramic bodies in relation to their suitability for metallizing and sealing as parts of vacuum tube envelope structures, it became necessary to investigate the properties of the ceramics themselves. Although it seemed apparent that bodies in the high-alumina (i.e., 85% or more alumina) family were of prime importance because of strength and low dielectric loss factor, a number of other bodies held interest because of other desirable parameters.

Experience had shown that the metallizing process itself required tighter controls in some areas than were then included.

Unfortunately, it was not clear what the tighter controls needed to be, since it had not been possible to reproduce results and establish limits with controlled experiments. It was therefore suspected that the ceramics themselves might vary in ways which were significant in the results of the metallizing process. Real improvement of the metallizing processes might well be affected by the detailed nature of the ceramics and the utilization of new or different materials might be dependent upon inadequate knowledge. An additional factor, though of somewhat secondary importance, was the desire to be able to identify materials which might have become mixed in stockrooms or in handling during plant operations. In all cases, it was very desirable that such checking or investigation be nondestructive and rapid.

There are, of course, a number of well-known techniques for determining some of the properties of interest, but no one process is capable of establishing all of the necessary data. Slumping-temperature test, wet chemical analysis, light-spectrograph analysis, standardized metallizing test, microscopic examination of sections — both by reflected and transmitted light—and density and exqansion measurements are some of the more important possible procedures. Several of them are destructive, and only the light-spectrograph technique is reasonably fast—not rapid.

Because of these considerations, X-ray analysis seemed an excellent solution to the problem if the necessary information could be so obtained. In addition to meeting the above-mentioned requirements, it offered the possibility of being independent of operator techniques, of being accurate, and requiring only normal calibration procedures. Interpretation of the results is probably no more complicated than the interpretation of the results from the other testing techniques once the significant factors are well established. It should be possible to prepare samples in sufficiently standardized form that results could be reproducible from one sample to another equivalent sample.

CHOICE OF TECHNIQUE

Of the several well-known and readily available X-ray techniques, the three most likely to be adaptable to this operation are the Debye-Scherrer, the fluorescence spectroanalysis, and the G-M goniometer—recorder methods. In the first category, the exposures are frequently for periods of many hours and the powdered sample requires both long preparation and destruction of the sample. The fluorescence analysis could be useful in identifying bodies which had lost their "name plates," but would give very little information on the state of crystallinity, the phases present, and other possibly pertinent factors.

The G-M counter technique really has two basically different procedures, but the automatic-scan strip-chart recording system offers greater advantage in terms of speed and of interpretation than the step-scan counting-rate record method. Except for limitations on the size of sample which can be handled in the sample holder, there is no basic requirement for destruction of the sample. Therefore, it was decided to put the major effort on the automatic-scan procedure, with a few records made by the film technique in order to establish the validity of the original assumptions.

DEFINITION OF THE PROBLEM

As was mentioned earlier, it was suspected that there would be some characteristics of the ceramic bodies which were of major significance in processing the ceramic parts for the metal-ceramic seals. There was no real knowledge of what these factors were, but inferences from results in laboratory and factory processing pointed to variations in composition and firing. It was decided to use samples of several kinds of ceramic bodies which had proven suitable when "properly" processed and more samples of the same bodies which could not be processed satisfactorily with standard procedures for comparison.

The problem, therefore, was to determine whether X-ray diffraction techniques could establish differences in characteristics between supposedly similar samples which behaved differently in metallizing, and then to establish criteria by which future inspections or investigations could be guided.

EXPERIMENTAL PROCEDURE

Since there are X-ray diffraction equipments of different manufacture, each with its range of optional features, and within the available options, a variety of adjustments and conditions, it became necessary to establish that some technique and combination of the variables could yield useful data. In addition to this, it was also desirable, but not fundamental, to determine whether the same kind of information could be obtained with different instruments.

Available to the author were three machines, the G.E. XRD-3, the Norelco with vertical G-M goniometer, and the G.E. XRD-1. This last machine was used only for a few runs to check the Debye-Scherrer procedure on samples which were also run on the other machines. Although admittedly there was much less effort applied to the Debye-Scherrer technique, it was easily shown that because of sample destruction, exposure of many

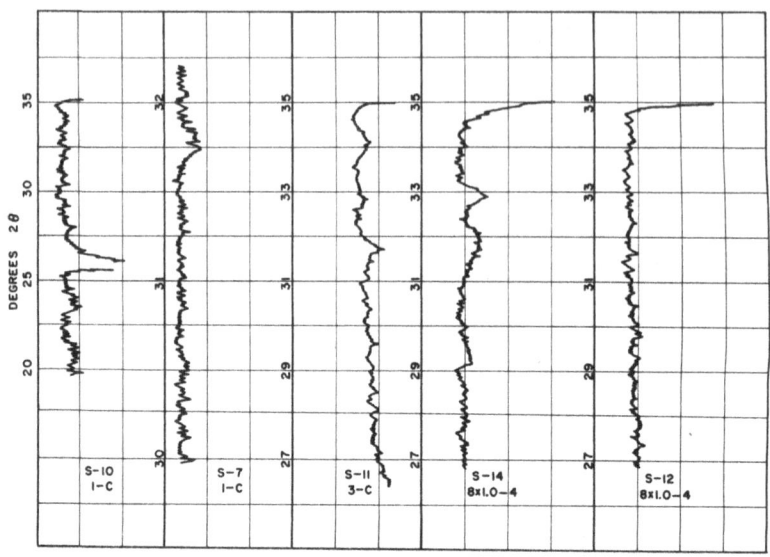

Figure 1. Alumina samples. Various machines and settings.

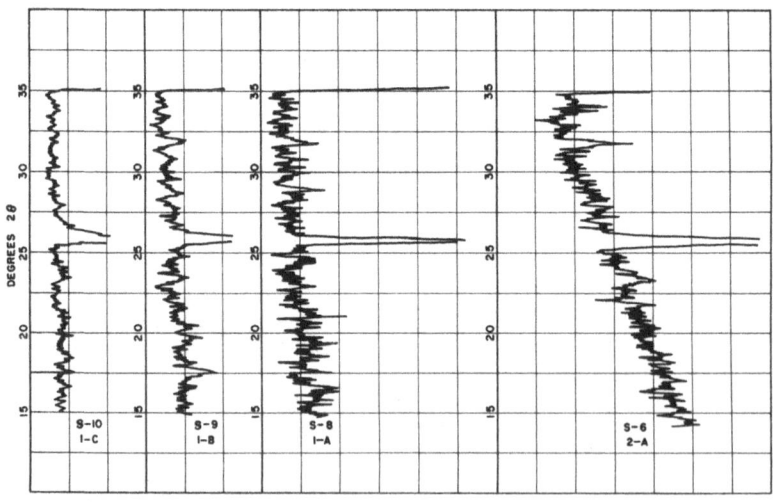

Figure 2. A-10 alumina sample. Sample and machine settings constant
except sensitivity and time constant, beam, and definition.

Operating Conditions for Experimental Data

Sample number	Type of ceramic	Machine	Target	Filter	Voltage (kv)	Current (ma)	Beam slit (deg)	Receiving coll.	Receiving slit (deg)	Sweep rate (deg/min)	Sensitivity setting	Time const. setting	Chart speed (in./deg 2θ)
S-6	A-10 Powder as rcd.	GE	Cu	Ni	50	16	1	MR	0.1	2	2	A	0.2
S-7	A-10 Powder as rcd.	GE	Cu	Ni	50	16	0.4	MR	0.05	0.2	1	C	2
S-8	A-10 Powder as rcd.	GE	Cu	Ni	50	16	0.4	MR	0.05	2	1	A	0.2
S-9	A-10 Powder as rcd.	GE	Cu	Ni	50	16	0.4	MR	0.05	2	1	B	0.2
S-10	A-10 Powder as rcd.	GE	Cu	Ni	50	16	0.4	MR	0.05	2	1	C	0.2
S-11	A-10 Powder as rcd.	GE	Cu	Ni	50	16	1	MR	0.05	2	3	C	0.5
S-12	A-14 Powder as rcd.	N	Cu	Ni	35	20	1	0.006	1	1	8×1.0	4	0.5
S-14	A-3 Powder as rcd.	N	Cu	Ni	35	20	1	0.006	1	1	8×1.0	4	0.5
5	93% Al$_2$O$_3$ body	GE	Cu	Ni	50	16	1	MR	0.05	0.8	3	C	0.5
6	93% Al$_2$O$_3$ body	GE	Cu	Ni	50	16	1	MR	0.05	0.8	3	C	0.5
10	89% Al$_2$O$_3$ body	GE	Cu	Ni	50	16	1	MR	0.05	0.8	3	C	0.5
16	89% Al$_2$O$_3$ body	N	Cu	Ni	35	20	1	0.006	1	1	8×1.0	4	0.5
29	93% Al$_2$O$_3$ body	N	Cu	Ni	35	20	1	0.006	1	1	8×1.0	4	0.5
42	93% Al$_2$O$_3$ body	N	Cu	Ni	35	20	1	0.006	1	1	8×1.0	4	0.5
43	93% Al$_2$O$_3$ body	N	Cu	Ni	35	20	1	0.006	1	1	8×1.0	4	0.5
55	92% Al$_2$O$_3$ body	GE	Cu	Ni	50	16	1	MR	0.05	0.8	3	C	0.5
56	92% Al$_2$O$_3$ body	GE	Cu	Ni	50	16	1	MR	0.05	0.8	2	C	0.5
58	93% Al$_2$O$_3$ body	GE	Cu	Ni	50	16	1	MR	0.05	0.8	3	C	0.5
62	94% Al$_2$O$_3$ body	GE	Cu	Ni	50	16	1	MR	0.05	0.8	3	C	0.5
63	94% Al$_2$O$_3$ body	GE	Co	Fe	50	6	1	MR	0.05	0.8	3	C	0.5
64	94% Al$_2$O$_3$ body	GE	Co	Fe	50	6	3	MR	0.02	2	2	B	0.2
66	94% Al$_2$O$_3$ body	N	Cu	Ni	35	20	1	0.006	1	0.5	8×1.0	4	0.5
70	94% Al$_2$O$_3$ body	N	Cu	Ni	35	20	1	0.006	1	0.5	8×1.0	4	0.5
85	97% Al$_2$O$_3$ body	GE	Cu	Ni	50	16	1	MR	0.05	0.8	3	C	0.5
86	97% Al$_2$O$_3$ body	N	Cu	Ni	35	20	1	0.006	1	1	8×1.0	4	0.5
87	85% Al$_2$O$_3$ body	N	Cu	Ni	35	20	1	0.006	1	1	8×1.0	4	0.5
88	94% Al$_2$O$_3$ body	GE	Cu	Ni	50	16	1	MR	0.05	0.8	3	C	0.5
89	99% Al$_2$O$_3$ body	N	Cu	Ni	35	20	1	0.006	1	1	8×1.0	4	0.5
90	95% Al$_2$O$_3$ body	N	Cu	Ni	35	20	1	0.006	1	1	8×1.0	4	0.5
92	89% Al$_2$O$_3$ body	N	Cu	Ni	35	20	1	0.006	1	1	8×1.0	4	0.5
93	89% Al$_2$O$_3$ body	N	Cu	Ni	35	20	1	0.006	1	1	8×0.6	4	0.5

hours, and difficulty in identifying crystal phases present in very small amounts, the ultimate utility of the method was at best questionable.

Several samples of nominally pure alumina grains were used as "standards" in setting up the equipment. The table shows the machine settings for the various recordings. As can be seen, several runs were made on these aluminas, portions of which runs are shown in Figure 1. The first three traces are of the same sample of A-10 run on the G.E. machine. (In the following, unless otherwise noted, both machines were run with copper tubes and nickel filters. The G.E. was set at 50 kv and 16 ma, and the Norelco at 35 kv and 20 ma.) Variations were made in beam size, sweep rate, detector slits, etc., to obtain these curves.

The remaining two curves, S-14 and S-12, were run on the Norelco machine and are, respectively, A-3 and A-14 alumina. The A-3 and A-14 samples are included in the figure with the A-10 samples to show the differences in the supposedly equivalent raw materials which go into the ceramic bodies.

It would be very difficult to establish the presence of a peak anywhere in S-10 except in the region of 25.6°. The irregularities in the curve might indicate something near 21°, 24°, and 28°. S-11 (the same material) shows that there is very little, if any, peak at 28°, while there are significant peaks at 31.7°, 32.8°, and 34.1° that are not obvious in S-10. The S-7 trace substantiates the 31.7° peak but, of course, this short selected section does not cover the remaining peaks which actually do show on the full

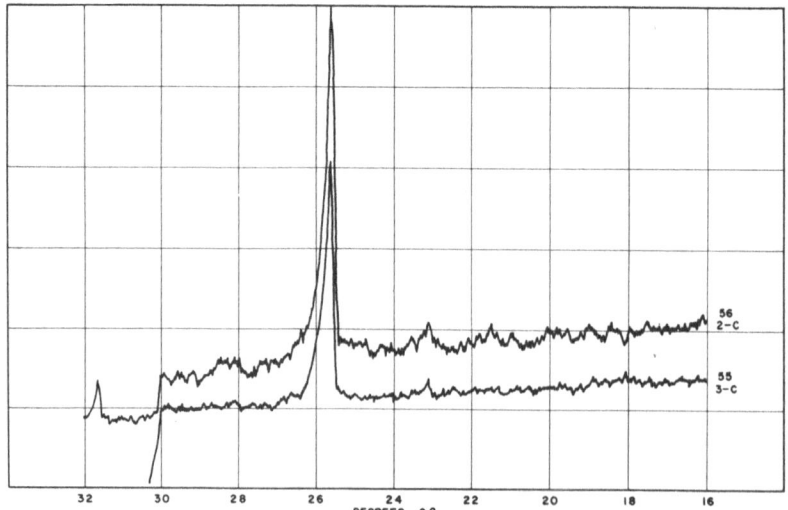

Figure 3. 92% alumina body.

S-7 curve. The conclusion from this is that the S-11 conditions are those which give us the most useful data at the highest speed.

Curve S-14 is similar to S-11 in a qualitative way, although it is definitely different in peak height and shape at 31.7° and 32.8°. There are also differences at 34.1° and 29.2°.

S-12 shows none of this detail, only indicating the beginning of the rise for the large peak above 35°. For purposes of later discussion, the difference among A-10, A-3, and A-14 should be kept in mind.

The curves shown in Figure 2 are all of the same A-10 alumina sample. For quick reference, S-10, which was shown in Figure 1, is repeated here. The differences among these curves are in sensitivity, time constant, beam size, and definition. Examination of these shows that, although there is a very large amount of detail in the curves run with the shorter time constants, the significance is extremely doubtful. Of the minor factors, the peak at 32° seems to be consistent in S-9, S-8, and S-6, with a possible indication in S-10. As we saw in Figure 1, S-11 indicates a peak at 31.7°. S-9 shows a good possibility at 17.5°, but this is not supported by the other curves.

Figure 3 shows the results of a similar test run on a matured, high-alumina ceramic body on the G.E. machine.

Curve 55 was recorded at 3-C, sensitivity and time constant, and 56 at 2-C, other conditions the same. After reaching the 30° point on curve 55, the beam was interrupted and the goniometer returned to the start of the sweep. The sensitivity switch was

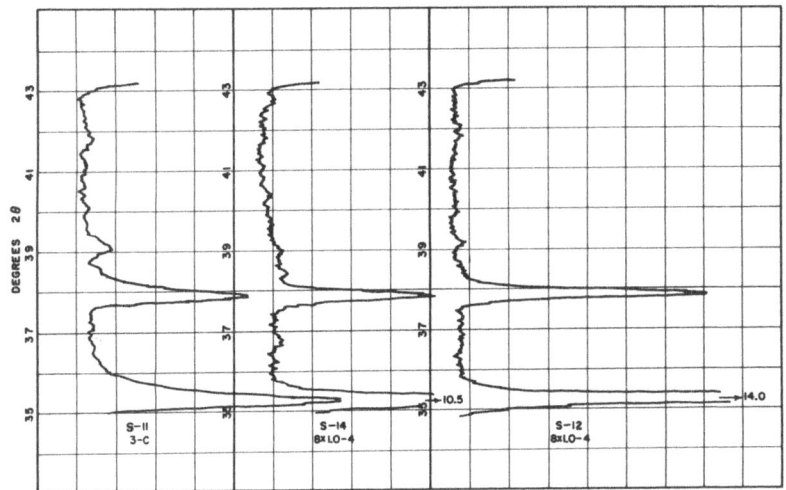

Figure 4. Alumina samples. Various machines and settings.

then changed to the higher range and the sweep redone, again to the 30° value of 2θ, at which place the sensitivity switch was returned to the original position so that the remaining record continued on as though it were part of curve 55. As can be seen, the continuation bridges onto the interrupted first curve very smoothly, showing good reproducibility with switch changing. However, the fine grain information in curve 56 is much more apparent than real since very little is substantiated by curve 55. Of the several small peaks indicated in trace 56, only the one at 23.1° seems to be repeated in curve 55.

In order to establish the interchangeability of the data from one machine to the other on the same sample, data were taken also on the Norelco unit on A-10 samples. No significant difference was observed in the records from the two machines.

Figure 4 again has data on the three alumina samples, A-10, A-3, and A-14, run with what were finally selected as the optimum conditions from the two machines. One item of possible significance which must be mentioned here in discussing the equivalence of the two X-ray machines is the fact that the G.E. unit records with logarithmic pen travel, while the Norelco uses a linear scale presentation. In most instances under discussion, the exact peak height is not significant in the conclusions; however, there are some exceptions to this in discussing the state of maturity of a ceramic body, as will be noted later.

The curves of Figure 4 are different parts of the runs illustrated previously under the corresponding S-11, S-14, and S-12; in fact, the three right-hand traces in Figure 1 continue directly onto these three traces, the breaking point being at 35°. Again the same observations can be made on the three samples: The A-10 material seems to have definite peaks at a much larger number of places along the curve than can be found either with the A-3 or the A-14. This condition is consistent throughout the remaining portions of the runs which are not shown in the figures.

Figure 5 shows the effect of changing the sensitivity on the Norelco machine. Presented here are portions of two runs on the same 89% alumina body. As was the case in practically all of the other runs, the machine was allowed to record for a short time (with the beam on and the counter working but without goniometer travel) to show the random noise which appeared in the potentiometer circuit. This is labeled "jitter" in the two curves. Although it is not presented in the other figures, it was used for the same purpose in all cases, i.e., to estimate the range and shape of the tracing without sweep.

As will be noticed, in the case of the higher sensitivity setting, i.e., with the multiplier 0.6, the full excursion of the jitter is approximately 3.7 divisions of the chart. The jitter at the 1.0

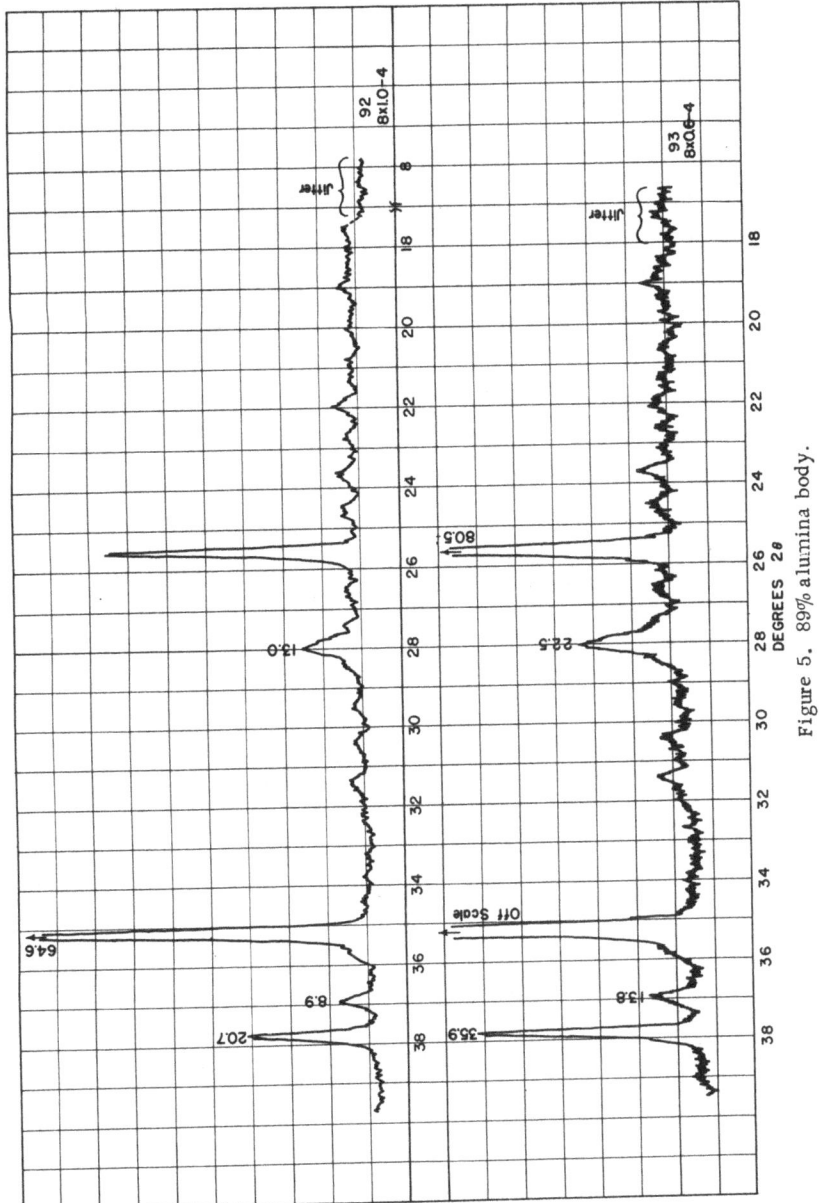

Figure 5. 89% alumina body.

multiplier setting is less than half this, being approximately 1.6 divisions. The ratio is about 2.3. A check of the peak-height ratios establishes the basis for using the lower sensitivities, since the peaks do not reach heights which have this ratio and, consequently, small peaks are not as easily observable in the presence of the noise. For instance, the peak at 37.8° in curve 93 shows a ratio of only 1.7 in relation to that in curve 92. The peak at 25.6° gives a ratio of approximately 1.6, again considerably lower than the jitter ratio would indicate. Other peaks are similar. If we draw in a base line through the estimated zero position, that is, at the background level, we still find that the ratios are off and the identification of the small peaks is still doubtful. The peak at 29.5°, for instance — although not large in curve 92 — is definitely lost in curve 93.

Figure 6 shows the change in the appearance of the spectrum when a cobalt target is used in place of the copper target. In this instance, curve 64 was run in the G.E. machine with a cobalt tube operating at 50 kv and 6 ma. Curve 62 on the same 94% alumina body shows the record with a copper tube under the standard conditions. Curve 63 is still another run with the cobalt tube and the same sample, but with machine settings the same as in curve 62, except for X-ray tube current.

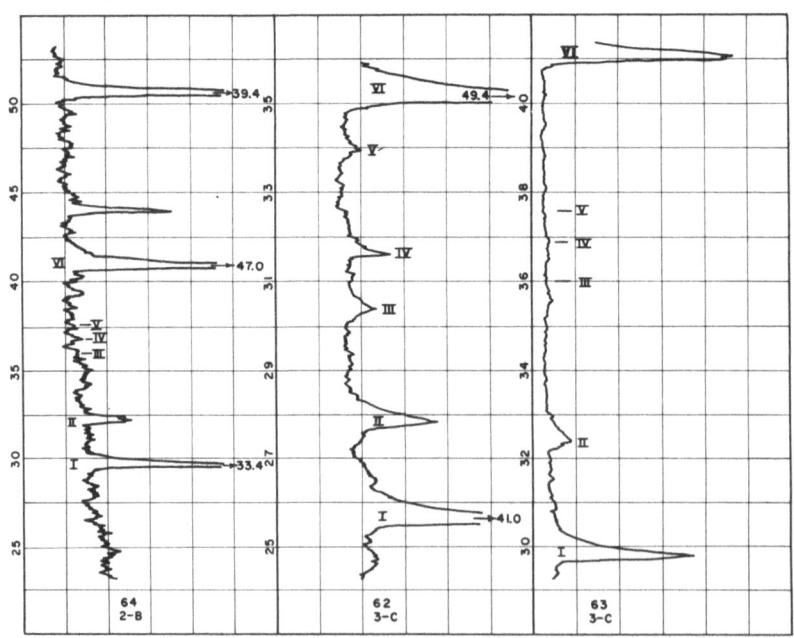

Figure 6. 94% alumina body. Cu vs Co tube.

Several peaks are identified on curve 62 as I, II, III, IV, V, and VI. The corresponding positions of these peaks with the cobalt radiation are shown on curves 63 and 64. As can be noted, only the larger peaks in the copper recording are retained in the cobalt runs, I, II, and VI being definite, though of considerably less height, while those corresponding to III, IV, and V are lost in the background. This is true whether the machine is running with the larger beam and higher sensitivity of curve 64 or the standard conditions of curve 63. The translation of the peaks, of course, is determined by the characteristic radiation (in this case, the K_α value) for both targets. The displacement of one vs the other is given by the formula

$$\sin\theta_{Co} = \frac{\lambda_{Co}}{\lambda_{Cu}}\sin\theta_{Cu} \cong 1.16\sin\theta_{Cu}$$

for a given d spacing.

The peaks as recorded with the cobalt tube are therefore somewhat farther apart than as recorded with the copper tube. It was felt that there might be some advantage in resolution of questionable peaks if the recording were so stretched out; however, because of the loss of sensitivity, there was a net loss in useful information.

Figures 7 and 8 need to be discussed together. Figure 7 shows three high-alumina bodies and A-14 alumina powder. Figure 8 shows again three high-alumina bodies and A-10 powder. It will be noted that curves 66, 86, and 89 differ very little from the alumina curve, S-12. The peaks recorded correspond to those determined by many workers in the field and can be considered typical of α-alumina. Some minor peaks do appear in the tracings in some instances that are not in others. Note that for S-12 and 66 there is a small peak at 39°, while 86 and 89 do not show this peak.

Figure 8 shows the same sort of information but with a number of minor peaks which are not well established in the literature; in fact, some of these peaks are not mentioned at all in the standard alumina references. It must be kept in mind also that the figures show only parts of the total spectra but are indicative of the full range of 2θ. The peaks shown in Figure 8 at 23.1°, 31.7°, and 34.1° are missing or extremely doubtful in the curves of Figure 7. The peak at 39° shows quite strongly in all of the curves of Figure 8. It should be remembered that curves 86 and 85 are for two different samples of a particular, very high-alumina body.

Figure 9 shows recordings for three different bodies in the 85 to 89% alumina range. One of these is the same body as was shown in the left-hand curve (10) in Figure 8. Again there are a number of minor nonalumina peaks which vary in intensity

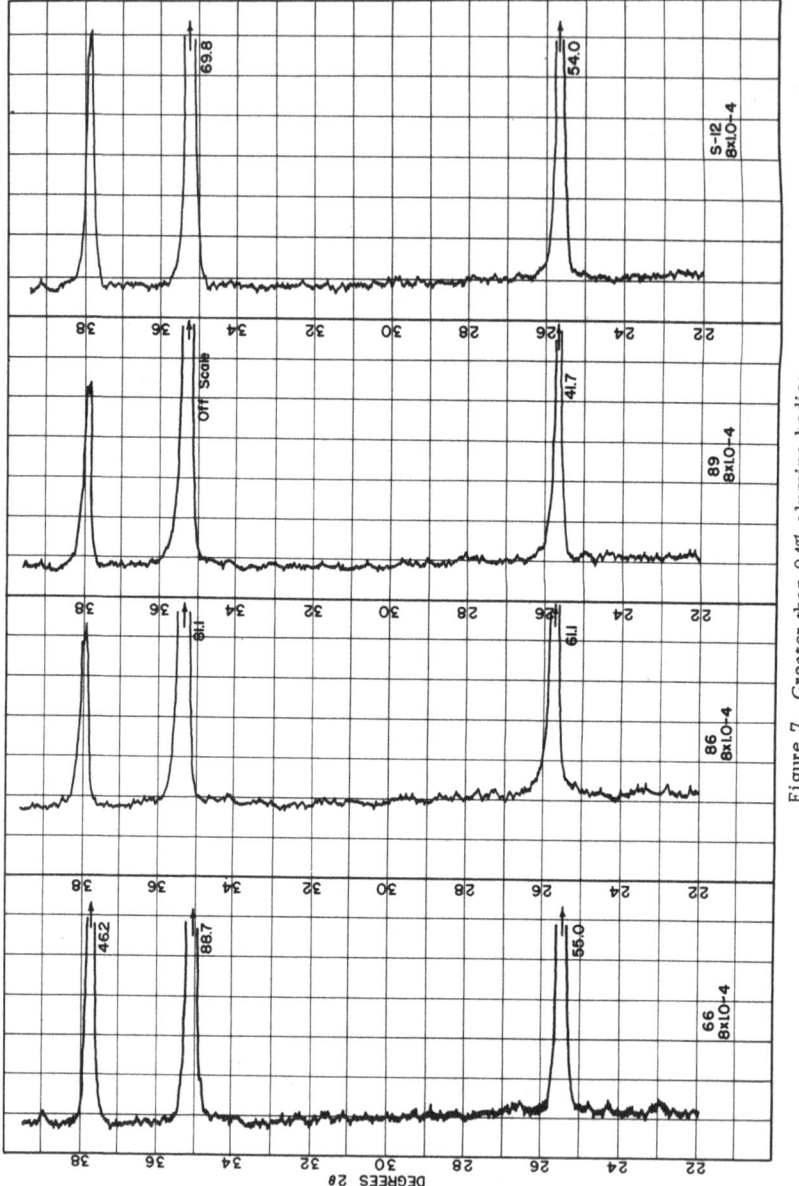

Figure 7. Greater than 94% alumina bodies.

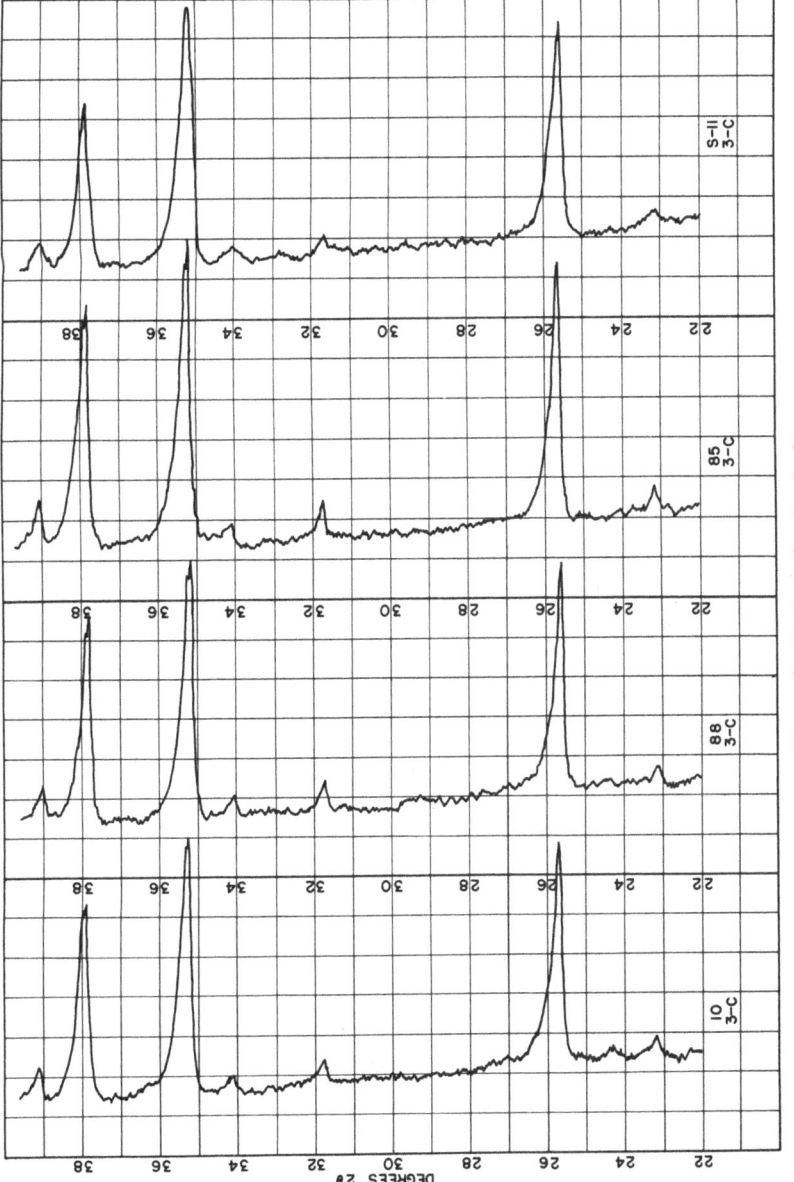

Figure 8. Various alumina bodies.

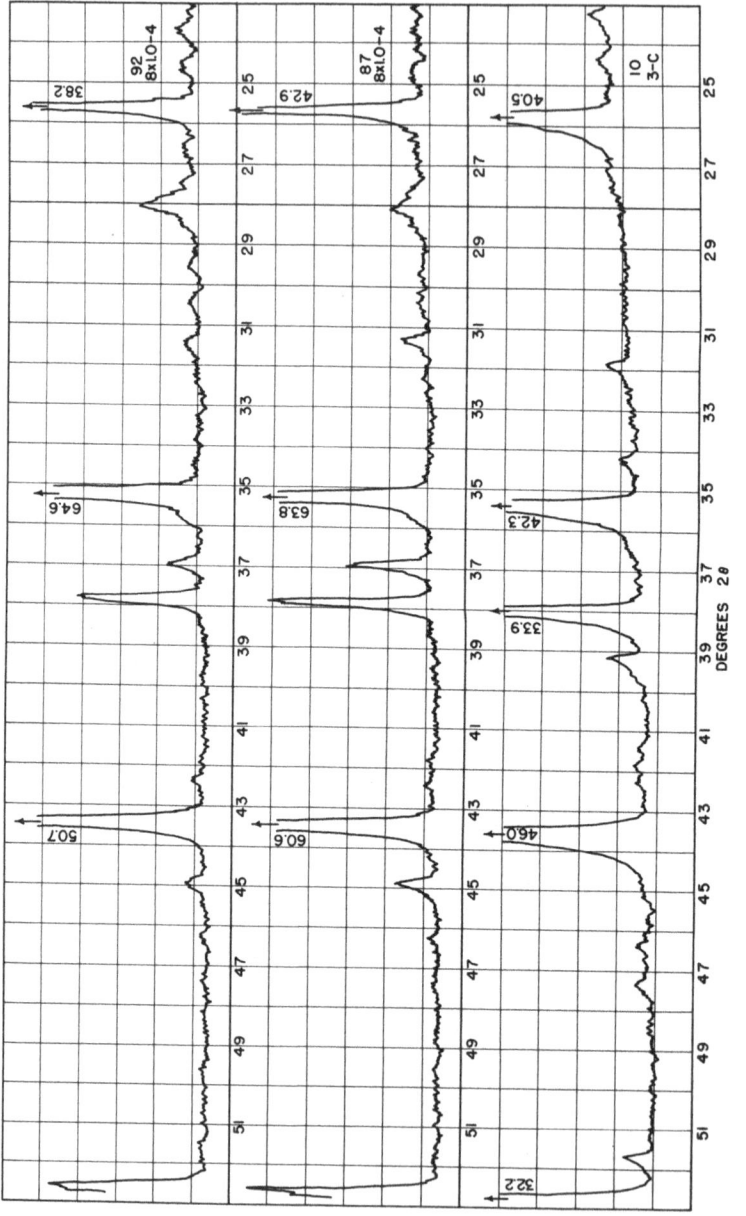

Figure 9. 85-89% alumina bodies.

from fairly strong to completely lacking from one body to the other. A very interesting peak is that at 37°, completely lacking in curve 10, strong in 87, and something in between in 92. The peak at 44.9° goes through the same sequence at a lower level. However, in order to prevent our jumping to the conclusion that these bodies vary in state of crystallization, we can look at the peaks at 28.1°, 31.3°, and 31.8°, which are not in the same sequences from body to body, and the peaks at 34.3° and 51.7° seem to be present only in curve 10.

In Figure 10, a group of alumina bodies in the 91 to 93% category are shown in the four curves to the left, with the lower-alumina body of curve 10 in the preceding figure being repeated on the right-hand side for reference. Again the presence and absence of several of the peaks can be very clearly shown. The peak at 31.7°, for instance, probably is the same as that 31.8° in curve 10. (There is in many of these cases a slight displacement of the peaks, depending upon the body and its condition. Even when the machines are carefully calibrated against the silicon standard — which was done at intervals in the course of this work — this displacement can be established and is not due only to recorder pen position error.)

It might be anticipated that bodies having smaller percentages of alumina and, therefore, possibilities of developing more and different crystal phases, would tend to show a larger number of nonalumina peaks than those having a higher percentage of alumina. Obviously, this is not necessarily so.

Figure 11 shows records on two different bodies, each with two different firing histories. These bodies are both in the same range of aluminas as those in the preceding figure; in fact, the right-hand curve of this figure is the same as the left-hand curve (6) in Figure 10. The body shown in curve 42 gives nearly the characteristic alumina record (cf. S-12 in Figure 7). There are some minor differences, however, such as the peak on the upper shoulder of the major alumina peak at 37.8°. Curve 43 shows considerably greater crystallization near 22° and from 26° through 29°, with several other minor peaks at higher angles.

The same sort of results shows on curves 5 and 6, but with differences in detail. Two rather prominent additional peaks are developed at 23.7° and 24.6°, with curve 43 showing only a weak indication at 24.6°. Curve 5, the body corresponding to that of curve 6, shows nothing at either of these two points, a peak at 24.2°, and both show peaks at 23.2° and 31.7°. There seems to be a fairly strong peak at 36.9° in curve 6, and this is absent in both 5 and 42, with some indication in 43. Both 5 and 6 show strong peaks at 39°. The differences between 42 and 43 and 5 and 6 are in the firing treatment.

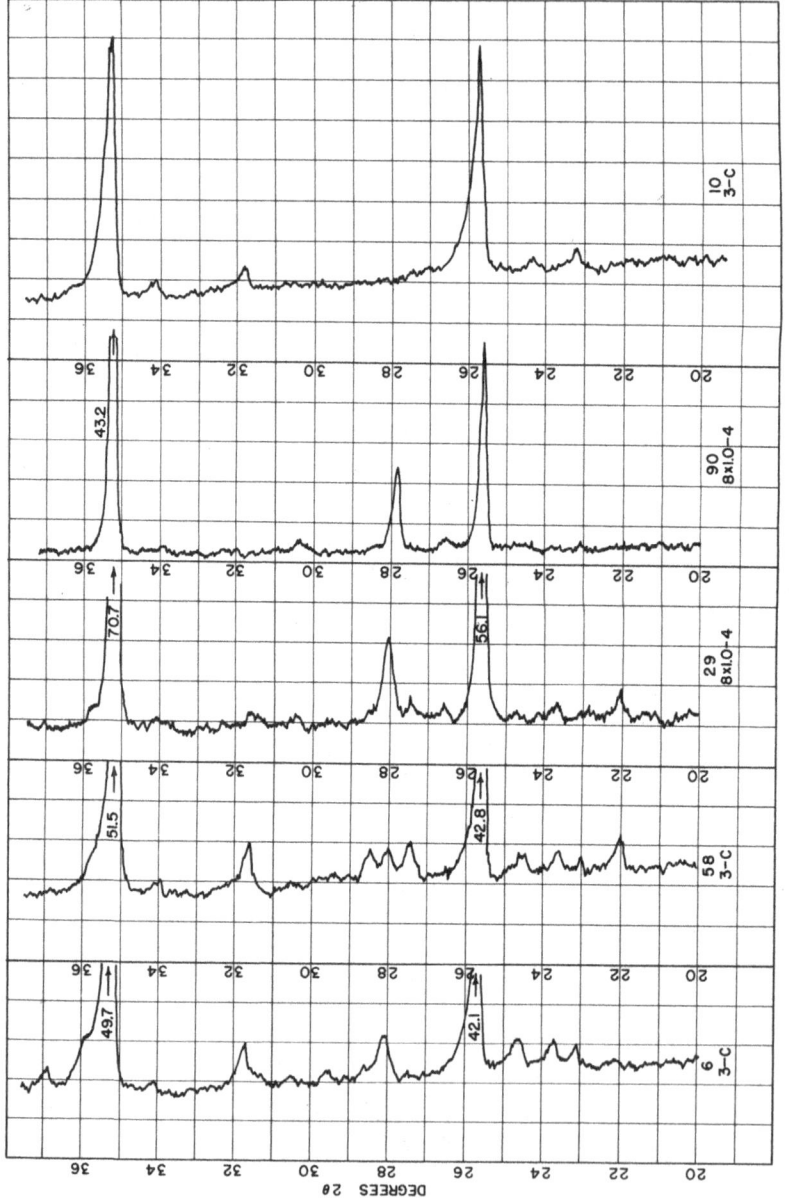

Figure 10. 91-93% alumina bodies.

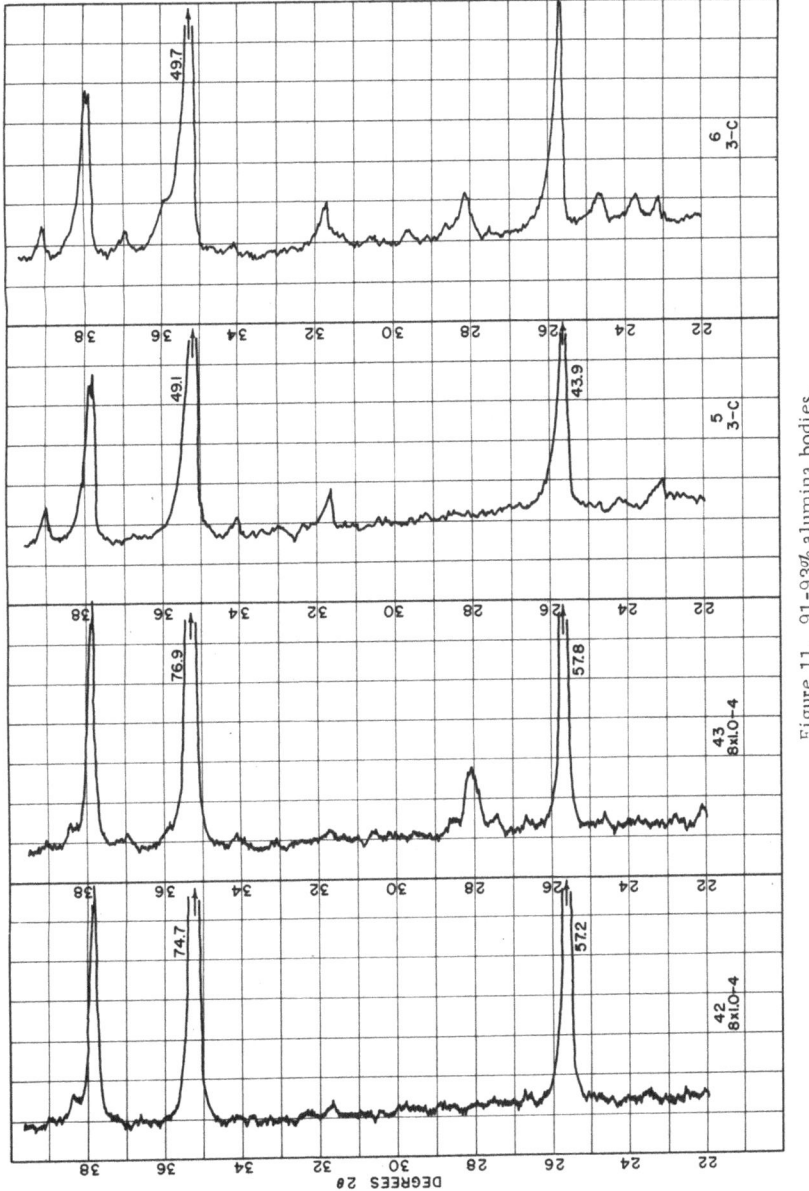

Figure 11. 91-93% alumina bodies.

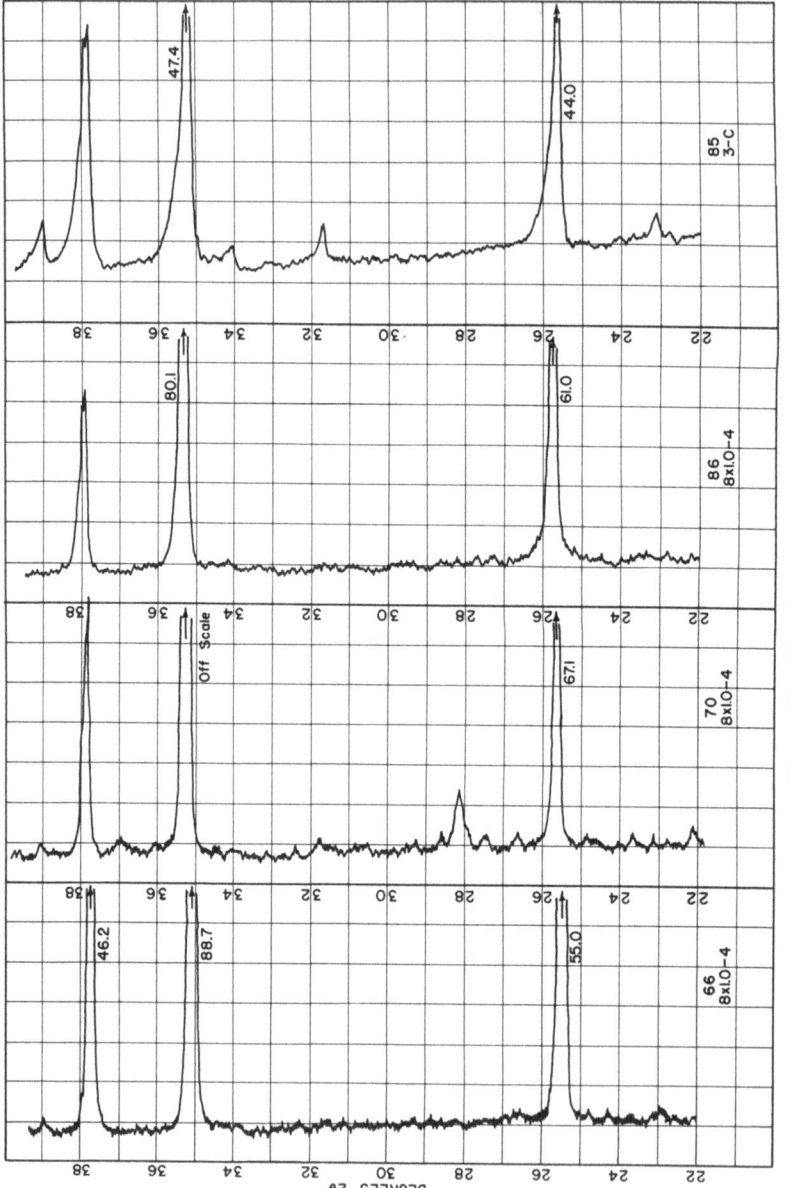

Figure 12. 94 and 97% alumina bodies.

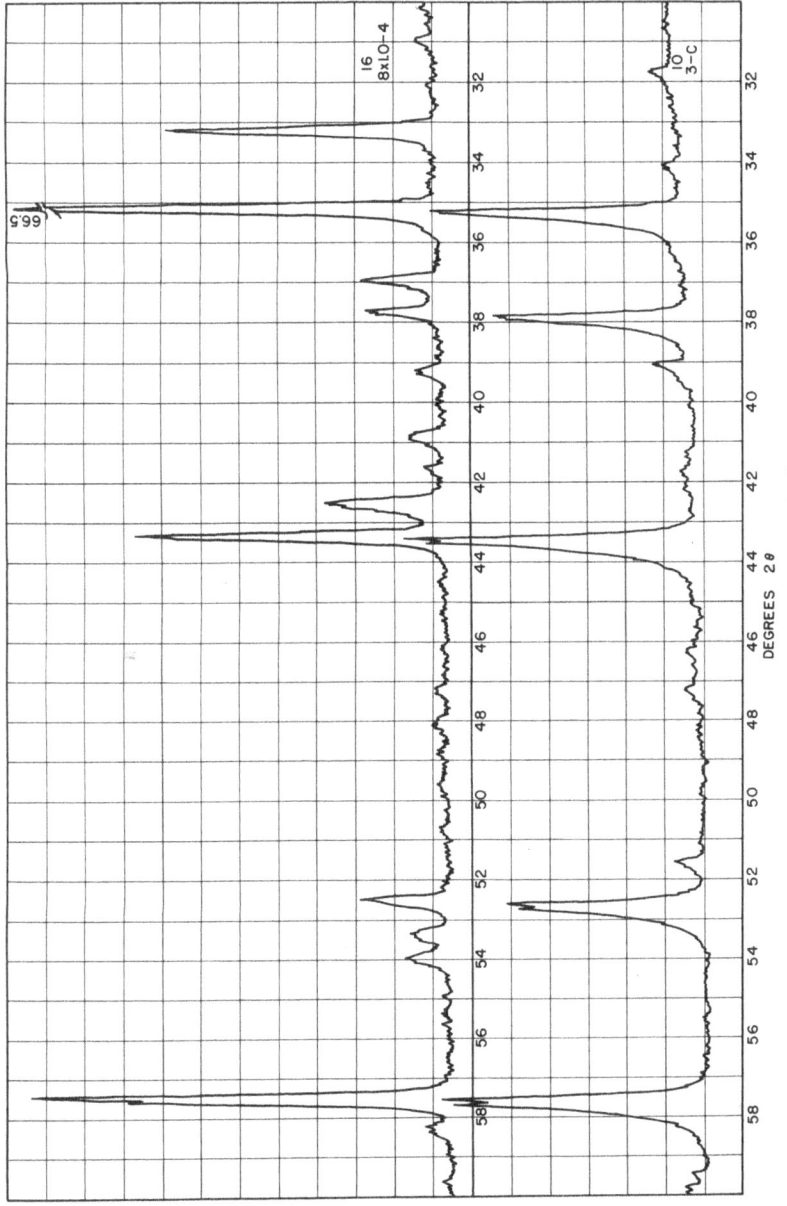

Figure 13. 89% alumina body.

Figure 12 is another illustration of the same thing. These are higher-alumina bodies than those in Figure 11, but the same variation in crystallinity is obvious. Curve 85 is a repetition of that shown in Figure 8. It is interesting to note here that the 39° peak is almost completely missing in curve 86 but is quite strong in 85, being definite but minor in 66 and 70. Significant here also is that the material analyzed in curves 85 and 86 is among the very high aluminas, i.e., in the 97% range, while that in curves 5 and 6 is several percentage points lower in alumina.

Figure 13 shows an 89% alumina body, as normally fired, in curve 10, and after a special firing treatment in curve 16. The peak at 37.8°, which is one of the very characteristic alumina peaks, is reduced drastically in magnitude in curve 16, and all of the other alumina peaks are significantly reduced. The peak at 52.6°, for instance, is much depressed in the upper curve. Some other peaks have disappeared completely and many new ones have appeared.

CONCLUSIONS

It seems to be well established that detailed differences in the composition and state of crystallinity of similar ceramic bodies can be determined by X-ray diffraction techniques, using procedures which are quick and reliable. The difficulty of interpretation is perhaps great if it is desired to understand the phenomena involved. This is a lengthy and complex procedure and not directly significant in this discussion. However, this does not seem to be necessary in order to determine whether a particular ceramic body is adaptable to the metallizing process, as significant differences between "good" and "poor" samples can be determined on an empirical basis.

It is apparently true that bodies which give equivalent results in metallizing do not necessarily have crystal phases in common. One thing does seem quite clear, however. The details of composition of the various bodies must be taken into account when interpreting the curves.

There have been many theories developed and supported with evidence to indicate the mechanism of metallizing of ceramic bodies. This is particularly so in the case of the so-called moly-manganese procedure, as contrasted with the active-metal method. The major advantage of the system outlined in this paper is its ability to yield useful data without dependence upon final verification of a theory of the metallizing mechanism. It is felt that some of the information accumulated in this investigation sheds light on that problem, but that discussion must be left for another occasion.

INTEGRATED X-RAY DIFFRACTION INTENSITIES FROM SINGLE CRYSTALS

Henry Chessin

United States Steel Corporation Research Center
Monroeville, Pennsylvania

ABSTRACT

The advantages and accuracy of counter equipment for the measurement of the intensity of scattered X-rays from small single crystals are discussed and illustrated. Several precautions necessary to obtain reliable intensity measurements are discussed. The equi-inclination method of inspecting all accessible reflections is treated in detail. The practicality of the stationary crystal method for determining integrated intensities, in principle the most accurate and rapid method, is demonstrated. Suggestions for improving the accuracy and speed of collecting data are made.

INTRODUCTION

The accurate determination of single crystal X-ray intensities is an important aspect of present-day crystallographic investigations. This report discusses the problem of collecting a large amount of data, including those on weak reflections at large Bragg angles, with a maximum statistical accuracy and economy of time for structural determinations; these lead to a detailed representation of the crystal structure by electron density maps or other presentations from Fourier syntheses. Our task is to set forth experimental and analytical precautions that must be taken, and to demonstrate what degree of accuracy is obtainable.

Previous workers have shown the feasibility of using counters to detect X-ray intensities. As early as 1914, the Braggs[1] used an ionization chamber to measure intensities. Lonsdale,[2] Cochran,[3] Birks and Wing,[4] and Jeffrey, Pringle, and Townsend[5] have made careful studies of accurate intensity determinations. Other workers such as Evans,[6] McLachlan,[7] and Furnas and Harker[8] have combined these studies with improvements in apparatus and special goniometers to be used for single crystal investi-

---

Note: This paper was presented at the Seventh Annual Conference. Approval for publication was received too late for inclusion in the Proceedings of that conference.

[1] Superscripts pertain to references at the end of the paper.

gations. Mention should be made of the extensive work done by W. Parrish and co-workers on the characteristics of various counters as applied to X-ray diffraction.

GEOMETRIC CONSIDERATIONS

Any instrumentation used must satisfy the requirements outlined as follows:

   a. It must provide for the recording of the greatest possible number of reflections accessible with any given radiation.
   b. It must record intensities with the greatest possible accuracy for a very wide intensity range, i.e., from 100 to 100,000 counts/sec.
   c. A minimum of computation and time should be required for the proper orientation of the crystal and of the counter tube.
   d. The motions required should permit the optimum simplicity in both design and operation.

The geometry of the reciprocal lattice rotating about any chosen axis and the reflection optics associated with the assembly of reciprocal lattice points passing through the sphere of reflection is thoroughly covered by Buerger.[10] We adopt, in what follows, the notation and experimental details covered in that book.

The considerations of convenience and time required to record a significantly large number of spectra dictate, to a great extent, the procedure used to collect the data. Of the many possible combinations of crystal and counter motions, our problem is to select the one that is the most convenient, simplest, and quickest. One procedure would be to take the reciprocal lattice points in radial lines, through the origin, each such line representing successive orders of reflection from a set of lattice planes. This method has the advantage that it permits the counter and crystal to rotate at a fixed two-to-one ratio, but it is not the most convenient or rapid because of the large number of separate lines to be scanned. A second procedure, the normal-beam method, is not satisfactory for the upper layers because it does not fulfill the conditions set forth in (a) above.

A third useful solution of this problem is the equi-inclination method.[10] Imagine the crystal rotating about, say, the *b* axis, thus restricting all further motions to one degree of freedom. The Bragg condition imposes one constraint on the motion of the crystal; the above imposes a second. If we confine our attention to the zero layer, then as the crystal rotates all the zero-layer reciprocal lattice points cut the sphere of reflection on its equa-

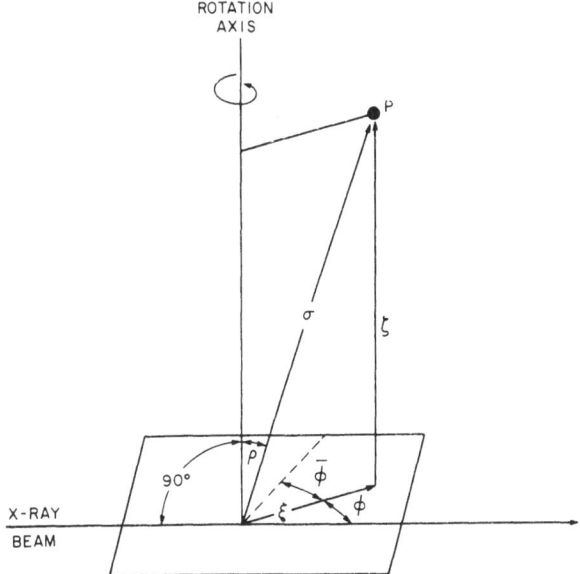

Figure 1. Location of reciprocal lattice point, P, in
cylindrical coordinates.

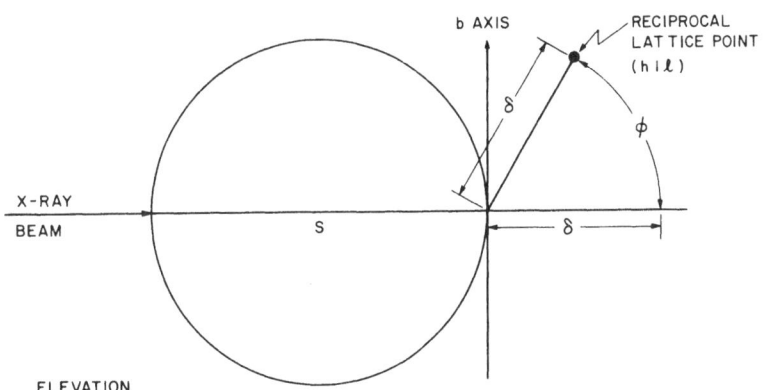

Figure 2. Bringing line $(h1l)$ into position for scanning by equatorial
method.

tor and the corresponding reflections all lie in the equatorial
plane. The location of a reciprocal lattice point in cylindrical
coordinates is shown in Figure 1. Figure 2 illustrates the geometry of bringing a point $(h1l)$ into the sphere of reflection. Fig-

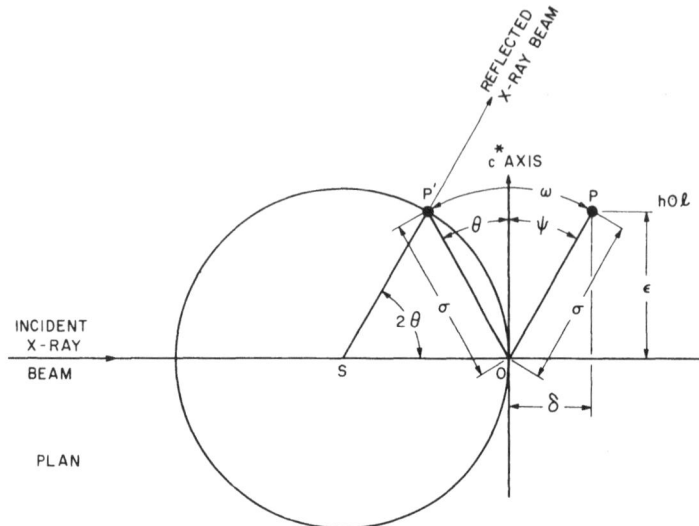

Figure 3. Measurement of $h0l$ reflection by equatorial method.

ure 3 shows the geometry of scanning the reciprocal lattice in the equatorial plane. Rotation of the crystal through the angle ω brings the reciprocal lattice point P onto the sphere of reflection at P'. In this way, we obtain the advantage of the counter tube moving in one circle, which for convenience is taken in the horizontal plane. Using the equi-inclination method involves only a one-circle crystal motion for each entire reciprocal lattice level, but the counter tube axis must travel in cones that have their apices at the crystal, apex angles being different for each level. With the b axis of the crystal vertical, the zero level is first scanned in the manner just described. For upper levels, the crystal and counter rotate about the b axis, which is inclined at a different angle for each level. Initially, the settings of the crystal and counter are determined once for an entire reciprocal lattice in the following way. For convenience, we take the $c*$ axis horizontal and perpendicular to the X-ray beam (see Figure 3). The crystal is rotated about the horizontal $c*$ axis through an angle ν such that

$$2 \sin \nu = \zeta,$$

where
$$\zeta = kb*$$

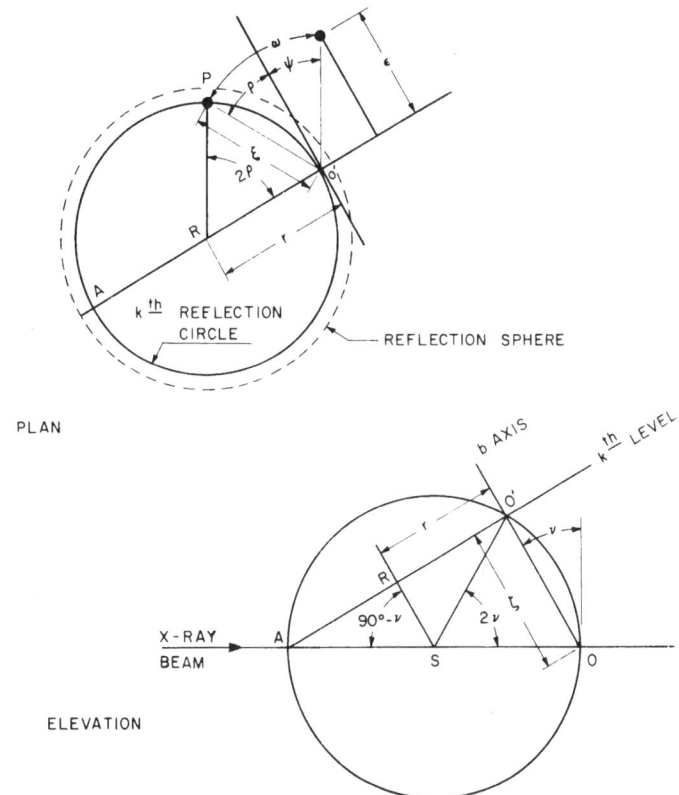

PLAN

ELEVATION

Figure 4. Geometry of *hkl* reflection. Equi-inclination method.

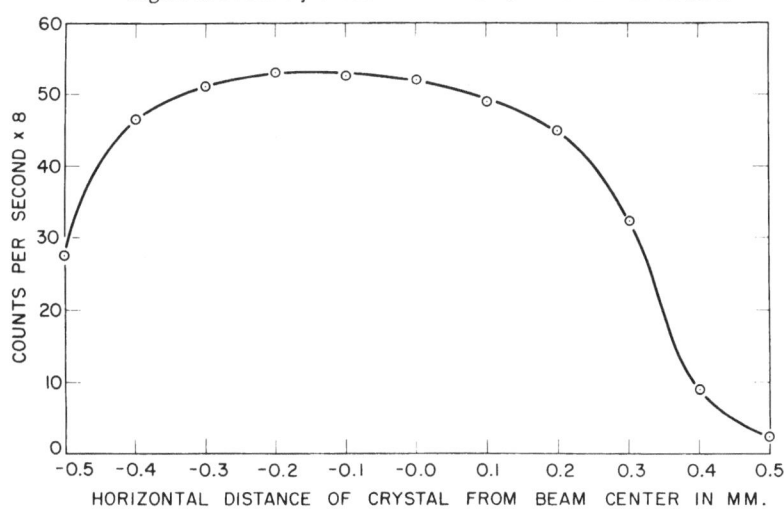

Figure 5.

is the perpendicular height of the kth level above the reciprocal lattice origin. As a result of this rotation, the kth level, which is inclined at an angle ν to the horizontal, intersects the sphere of reflection in a circle of radius

$$r = \cos \nu \, ,$$

which is the reflecting circle for this level (see Figure 4). The intersection of the b axis with the kth level plane lies on the re- flecting circle at the point O', in the vertical plane through SO, so that the radius SO' makes an angle of 2ν with SO. The counter is also tilted out of the horizontal plane so as to lie parallel to SO', and with the counter-tube axis always directed toward the crystal. The crystal is now in position for scanning of all the reciprocal lattice points of the kth level. The procedure is now similar to that for the zero layer. A given reciprocal lattice point, hkl, at a distance ξ from O' requires a crystal rotation of $\omega = \psi + \rho$ and a counter rotation of 2ρ in the plane of the reflecting circle, where ψ is the angle between the reciprocal lattice vector and the tangent to the reflecting circle at O', and ξ is defined by the equation

$$2r \sin \rho = \xi \, .$$

The major advantage of the equi-inclination method is the fact that the angle ν is fixed for each reciprocal lattice level. This advantage can be utilized to the fullest if the crystal is rotated about an axis such that the smallest dimension of the reciprocal unit is in the equatorial plane, or, what amounts to the same thing, the largest number of reciprocal lattice points are in a given layer. In this way, the counter and crystal are reset a minimum number of times. A further advantage becomes apparent when absorption corrections are to be made. If the crys- tal is cylindrical, it is possible to derive satisfactory correction factors that depend only on the relative orientations of the in- cident and reflected beams and the cylinder axis. This matter will be discussed later.

POSITION AND ABSORPTION OF CRYSTAL

In any method of surveying crystal reflections, it is essential that the position of the crystal remains unaltered with respect to the primary X-ray beam during all motions of the goniometer if the crystal is smaller than the beam. The variations in inten- sity along the cross section of the beam are shown in Figures 5 and 6. For this experiment a small, approximately cubic LiF crystal, about 70 μ on edge, was mounted on the end of a glass fiber and centered in the beam as observed by the cross hairs of a telescope while a scintillation counter having a wide open window was placed in the position to accept a medium weak reflection. The collimator was a circular aperture system. Figure 5 shows the results of moving the crystal along the horizontal diameter of the beam in steps of 0.1 mm, counts being taken at each 0.1-mm interval for 3 min. The same experiment is repeated in Figure 6, only the crystal is moved along the vertical diameter. In both cases, the collimator and associated goniometer were deliberately offset about 0.5 mm horizontally and vertically from the optimum position to illustrate the importance of aligning the apparatus with as much care as possible. The horizontal variation could be avoided by using a cylindrical crystal where possible, and some workers have used a square beam-defining collimator to minimize errors introduced by small vertical motions of the cylindrical crystal and consequent changes in effective volume irradiated.

In our case. we were limited to a very small spherical crystal and it was essential that the crystal remain stationary at all orientations. It was found that the settings of the goniometer head changed about 15 min of arc and about 0.1 mm of translation over the period of a week. Once the crystal was aligned with maximum precision, a drop of Duco cement was placed on the arc and translational ways, and no detectable change in crystal orientation was found in six months.

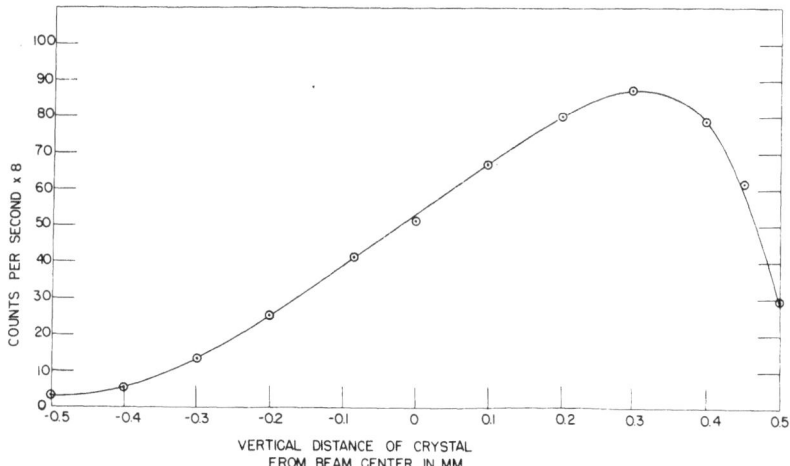

Figure 6.

It is advantageous to devote considerable effort to obtaining a small sphere or cylinder of the crystal. The absorption corrections for these two shapes have been listed in tabular form by Bond.[9] For crystals of irregular shapes, the absorption corrections are of such complexity that time devoted to obtaining a sphere or cylinder may be considered well spent. It may be possible, however, to use a small enough crystal (50 μ or so) provided a check is made to ascertain that absorption errors are negligible by comparing intensities of symmetry-related reflections. In this connection, where absorption is significant, there is some question whether the average or the largest intensity of the symmetry-equivalent reflections should be taken as the best value. If extinction is negligible, the largest intensity indicates the least absorption error.

CONVERSION OF PEAK HEIGHTS TO INTEGRATED INTENSITIES

The K_α doublet must be resolved if peak heights are to be used as a measure of integrated intensity.* A full discussion of this problem is beyond the scope of this report. With increasing Bragg angle, the separation of the doublet increases, and hence the contribution of the K_{α_2} component to the peak height diminishes accordingly. For certain optical conditions and dispersion, it may be possible to calculate or measure the separation of the K_α doublet and so resolve the K_{α_1} from the observed reflection profile by the usual methods described in the literature, such as that due to DuMond and Kirkpatrick.[11] However, in many experiments these conditions are not satisfied. Both Gerlach[12] and Richtmyer[13] observed the apparent doublet separation to be one-half of that calculated from Bragg's law. Both were able to determine the necessary corrections to their results after considerable theoretical and experimental work.

It is instructive to compare the strip-chart curves for two very different optical systems for the same single crystal reflections of the same crystal, as shown in Figure 7. The lower trace is the reflection profile recorded with a goniometer made by modifying a Weissenberg mount and the other trace is made with the diffractometer. The divergence of the beam for the first is about three times that for the second. The calculated and

---

* We shall only indicate the procedure which may be used to correct for the doublet separation. A discussion of the advantages of using peak heights as a measure of integrated intensities is found in Cochran.[3]

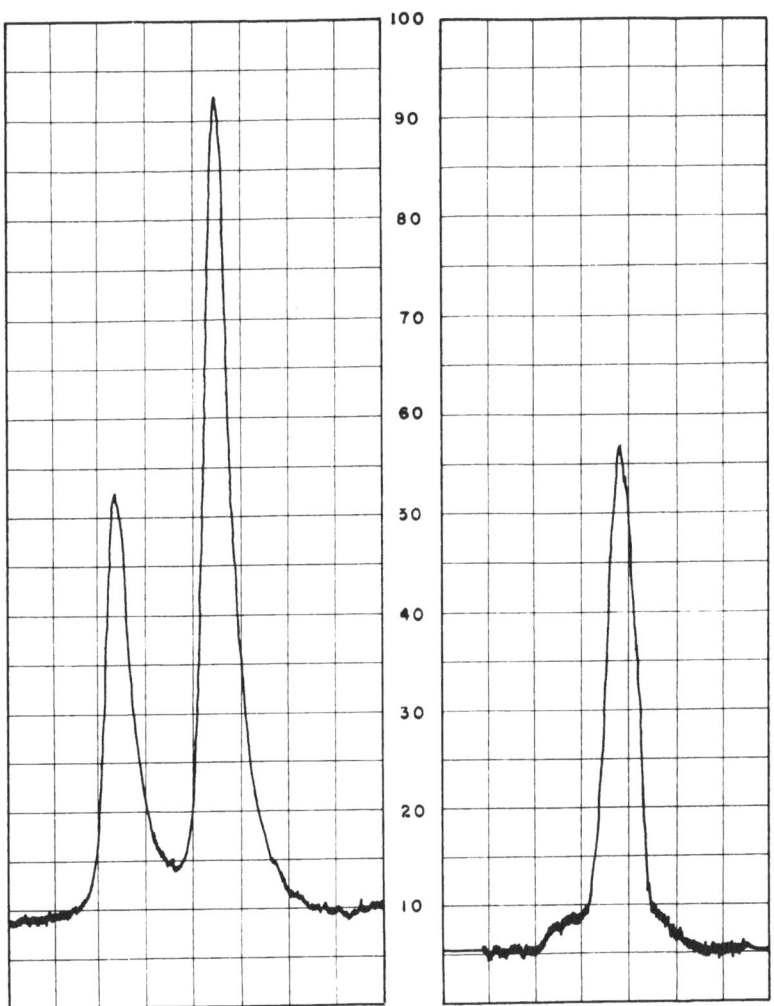

Figure 7. Comparison of line profiles for two different optical systems.

measured doublet separation agree exactly for the diffractometer trace, but this is hardly so for our experimental arrangement.

We chose, for convenience, a graphical method for correlating peak heights with integrated intensities. Our procedure in this case was as follows: we plotted the ratio of area of profile to peak height vs 2θ (Figure 8) for enough reflections in each layer to provide a smooth curve; then we converted all other peak-height measurements to the area of the profile and applied the usual corrections for the polarization and Lorentz factors (Cochran[14]). By making the correction in this way, it is not necessary to consider the proper use of the Lorentz factor and the omission of a velocity factor for the peak-height measurement.

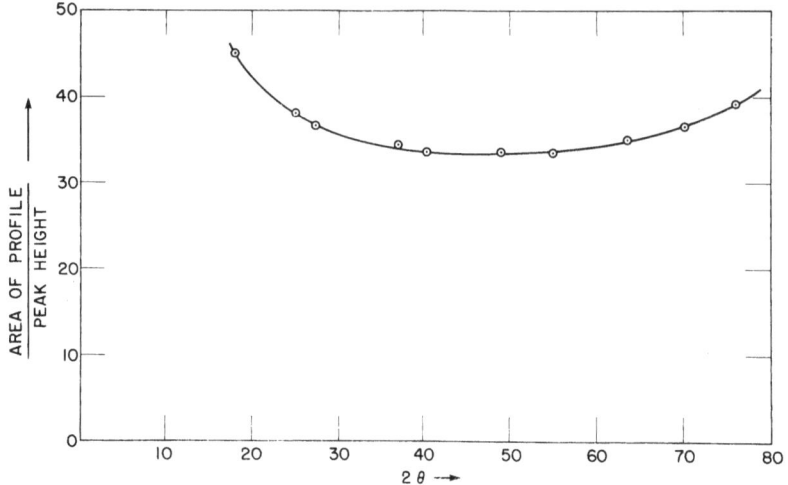

Figure 8. Conversion of peak heights to integrated intensities.

REPRODUCIBILITY OF INTENSITY MEASUREMENTS

The reflection intensities from the equivalent $\{211\}$ planes of an approximately spherical crystal of α-iron* 0.15 mm in diameter were measured accurately using Zr-filtered Mo radiation and a scintillation counter as a detector. The scintillation counter was held stationary at the 2θ position, and the crystal rotated through the reflection angles at an angular speed of 0.5 deg/min. A stereogram of the peak heights as a measure of the integrated intensities is shown in Figure 9. The peak heights can be measured, with care, with a reproducibility of 2.10%, but it must be emphasized that this is not a measure of the agreement between observed and calculated X-ray data.

---

* The single crystal of α-iron was annealed and furnace-cooled for a length of time such that no change in the breadth of the line profile was observed on further annealing.

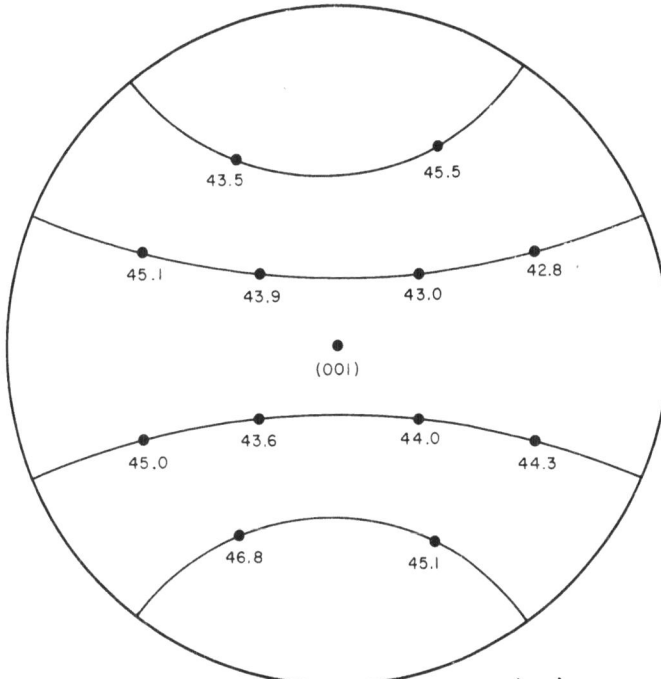

Figure 9. Stereogram of intensities from equivalent $\{211\}$ planes of α-iron. Zr-filtered Mo radiation. Spherical crystal 0.15 mm radius.

Average deviation from the mean $= \dfrac{\Sigma\,(\Delta I\,)}{\Sigma\,I} = 2.10\%$

SUGGESTED IMPROVEMENT IN PEAK-HEIGHT DETERMINATION

A further saving in time and increase in accuracy may be accomplished by applying the balanced filter technique to the measurement of peak heights. Figure 10 shows diagrammatically a reciprocal lattice point cutting through the sphere of reflection and thus satisfying Bragg's law. We usually measure the background intensity beyond the tails of the peaks as shown in the top and bottom sketches, and assume it is the same at the peak position as shown in the center sketch. If the background could be measured while the crystal is in the maximum reflecting position, the advantage would be twofold: (a) the time spent in collecting the data at both sides of the peak would be eliminated, and (b) the background of intensity, excepting that included in the pass band, could be evaluated when the crystal is at the peak position.

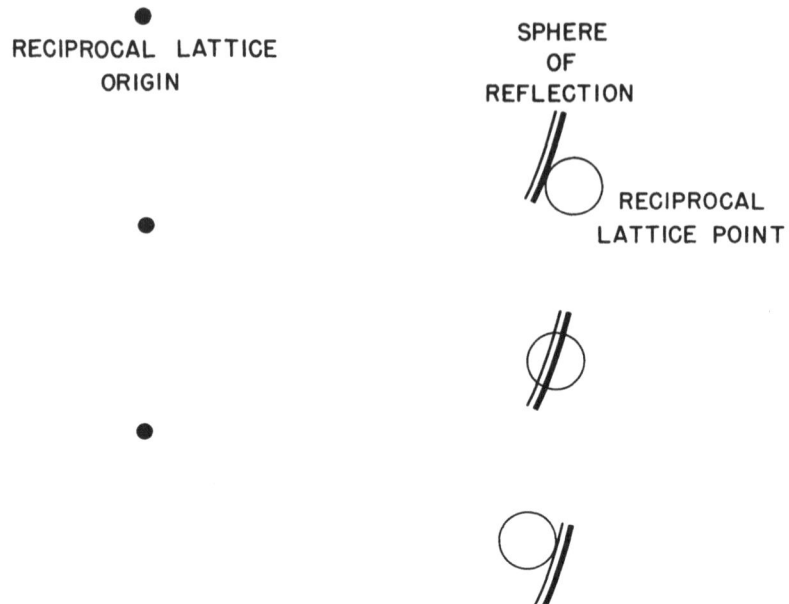

Figure 10. Counter stationary. Reciprocal lattice point moving through sphere of reflection.

CONCLUSION

With care, it is possible to determine the intensity of X-rays diffracted by a single crystal with a precision of about 2%. With data of such reliability available, it is possible to compute structures with considerably less ambiguity than in the past. This is achieved by iteration involving combinations of slightly different atomic coordinates, and of isotropic and anisotropic amplitudes of vibration until best agreement between observed and calculated data is found. Corrections must be made for such effects as dispersion, absorption, and K_α doublet separation. The stationary crystal method is as rapid, convenient, and accurate in operation as any that could be devised for integrated intensity measurement. Ways of further increasing the rapidity and accuracy of this method are possible.

REFERENCES

[1] W. H. Bragg, Phil. Mag., Vol. 27, 1914, p. 881.
[2] K. Lonsdale, Acta Cryst., Vol. 1, 1948, p. 12.
[3] W. Cochran, Acta Cryst., Vol. 3, 1950, p. 268.

[4] L. S. Birks and A. B. Wing, NRL Report 4402, Naval Research Laboratory, Washington, D.C., 1954.

[5] G. A. Jeffrey, G. E. Pringle, and J. R. Townsend, "Equipment for X-ray Crystal Structure Analysis," Final Report from Scaife Radiation Laboratory, University of Pittsburgh, Pittsburgh, Pa., 1955.

[6] H. T. Evans, Jr., Rev. Sci. Instr., Vol. 24, 1953, p. 156.

[7] D. McLachlan, D. F. Clifton, and A. Filler, Rev. Sci. Instr., Vol. 22, 1951, p. 1024.

[8] T. C. Furnas, Jr., and D. Harker, Rev. Sci. Instr., Vol. 26, 1955, p. 449.

[9] W. L. Bond, Abstracts of Communications, Fourth Congress of the International Union of Crystallography, 1957, p. 741.

[10] M. J. Buerger, X-ray Crystallography, Chapter 14, John Wiley and Sons, Inc., New York, 1942.

[11] J. W. M. DuMond and H. A. Kirkpatrick, Phys. Rev., Vol. 37, 1931, p. 136.

[12] W. Gerlach, Physik. Z., Vol. 23, 1922, p. 114.

[13] F. K. Richtmyer, Phys. Rev., Vol. 26, 1925, p. 724.

[14] W. Cochran, J. Sci. Instr., Vol. 25, 1948, p. 253.

SOME IRRADIATION EFFECTS IN NONMETALLIC CRYSTALS

M. C. Wittels and F. A. Sherrill

Oak Ridge National Laboratory,* Oak Ridge, Tennessee

ABSTRACT

Nonmetallic crystals with a variety of structures and chemical compositions have been exposed to fission spectrum neutrons in a core position of the Low Intensity Test Reactor (LITR), Oak Ridge, and examined by X-ray diffraction techniques. Wherever possible, single crystals were employed, and the neutron dosages ranged between 10^{19} and 10^{21} neutrons/cm^2 at a temperature of approximately 85°C. In no case was a special technique used for examining radioactive materials because the problem of heavy induced radioactivity was greatly reduced by working with crystals weighing a milligram or less. Both film and counter measurements were employed.

INTRODUCTION

Nonmetallic crystals, as a group, are exceedingly difficult to classify due to the wide diversity of crystal structures and complexity of chemical compositions. For these same reasons, it is clear why a systematic understanding of the effects of irradiation on this group of crystals is not easily attainable. Although a considerable body of data concerning these effects has been gathered at several laboratories both in this country as well as abroad, no attempt to summarize these findings will be made here. Instead, this report will consist of a brief review only of the results for a small diverse group of crystals irradiated in a core position of the Low Intensity Test Reactor at Oak Ridge.

The materials under consideration are oxides, mixed oxides, and silicates in which the cation components, for the most part, have low thermal-neutron absorption cross sections. This nuclear characteristic together with the refractory nature of the crystals gives the materials a potential reactor utility.

---

*Oak Ridge National Laboratory is operated by Union Carbide Corporation for the United States Atomic Energy Commission.

EXPERIMENTAL METHOD

Conventional X-ray techniques, including the use of Weiss-enberg, Buerger-Precession, and Laue methods on small single crystals were employed. The induced radioactivity was kept at a minimum level so that the standard film methods could be employed without the necessity of extraneous shielding that is often needed to reduce or eliminate background fogging. Similarly, Debye-Scherrer patterns were taken when single crystals were not available. In addition to these normal film methods, a special diffractometer apparatus[1] was employed in studying larger single crystals.

The greater part of these studies was related to the damage produced by fast neutrons, but a few isolated examples of fission fragment and heavy charged particle irradiations were examined as well. All of these irradiations were performed at approximately 85°C in a water-cooled facility.

EFFECTS IN CRYSTALS CONTAINING BERYLLIUM

The remarkable irradiation stability of the moderating material BeO has been extensively reported elsewhere,[2] and will only be briefly mentioned here. Powder samples bombarded with fast fluxes up to $1.8 \cdot 10^{20}$ n/cm² showed no X-ray line broadening, although there is a very small lattice expansion (0.2%) of the c_0 axis, and the a_0 parameter remains unchanged. The mixed oxide $BeAl_2O_4$ (chrysoberyl) is also very stable under fast-neutron bombardment with integrated fluxes up to $8.1 \cdot 10^{20}$ n/cm². Single crystal diffraction patterns remain sharp throughout this range of irradiations. An interesting structural feature of this material is the complete dominance of oxygen ions in the lattice, which are arrayed in a nearly perfect hexagonal close-packed system. This close packing would be expected to result in measurable lattice parameter shifts if atomic displacements were trapped in interstitial positions, and this effect would be enhanced if the large oxygen ion comprised the trapped nuclei. The expansions of the three orthorhombic axes after irradiation with $3.6 \cdot 10^{20}$ n/cm² are 0.43% in a_0, 0.55% in b_0, and 1.05% in c_0, respectively. Laue transmission photographs of $BeAl_2O_4$ after three dosages of irradiation are shown in Figure 1. Note the development of a diffuse scattering pattern at small and moderate angles that is superimposed on the normal Laue pattern. This type of scattering is not of thermal origin, and is commonly observed in

---

[1]Superscripts pertain to references at the end of the paper.

crystals containing irradiation-induced defects. Although the scattering lies outside the zones of selective reflection its high degree of symmetry related to the parent matrix is suggestive of scattering due to inhomogeneous strains developed through the trapping of interstitial atoms at statistical positions of the crystal lattice.

Be_2SiO_4 (phenacite) is also stable after irradiation dosages up to $8.1 \cdot 10^{20}$ n/cm^2, and also exhibits a diffuse scattering pattern in transmission photographs (Figure 2) of heavily irradiated specimens. Unlike $BeAl_2O_4$, however, this crystal has an unusually open structure with open channels roughly parallel to c_0, much like α-quartz. And similar to the anisotropic behavior of quartz[3] under neutron bombardment, Be_2SiO_4 exhibits an anisotropic expansion in the same direction. After irradiation with $4.5 \cdot 10^{20}$ n/cm^2 the a_0 expansion is 1.12% while no change is measurable in c_0. This extreme anisotropic behavior is most likely related to the open channels of the structure as well as the directional binding forces within the lattice, which are more amenable to expansion in the a_0 direction.

As a result of neutron bombardment the least stable crystal structure in this group under discussion is $Be_3Al_2Si_6O_{18}$ (beryl), which is very similar to quartz and phenacite in having open channels parallel to the c axis. This crystal becomes amorphous following bombardments in excess of $6 \cdot 10^{20}$ n/cm^2 when all single crystal reflections disappear and all that remains in the diffraction pattern is a broad diffuse ring characteristic of amorphous matter. Prior to the complete destruction of long-range order, at lower dosages of irradiation, beryl expands anisotropically nearly to the same extent as quartz.[3] After irradiation with $2.8 \cdot 10^{20}$ n/cm^2 the expansion of a_0 is 2.70%, while the distention of c_0 is 0.70%, about a factor of four lower. It is believed that the open c axis channels in this crystal, as in those previously discussed, furnish ideal sites for trapping displacements that result in an anisotropic expansion. This extreme distention results in a highly strained and distorted lattice giving smeared-out X-ray reflections (Figure 3) with successive dosages of irradiation, and a gradual reduction of long-range order.

IRRADIATION-INDUCED AMORPHISM IN SILICATES

The [SiO_4] tetrahedral unit is the basic building block in silicate crystal structures. Since silicates are also common glassy

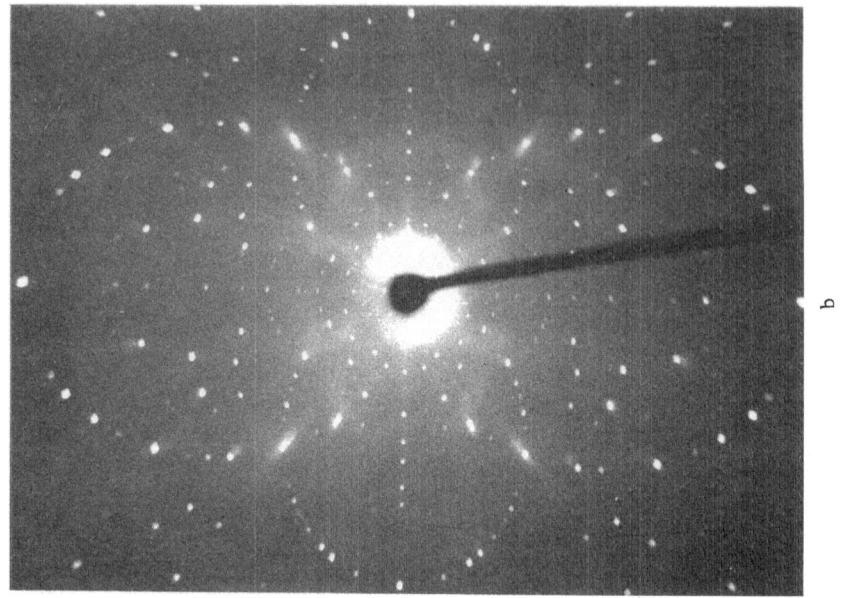

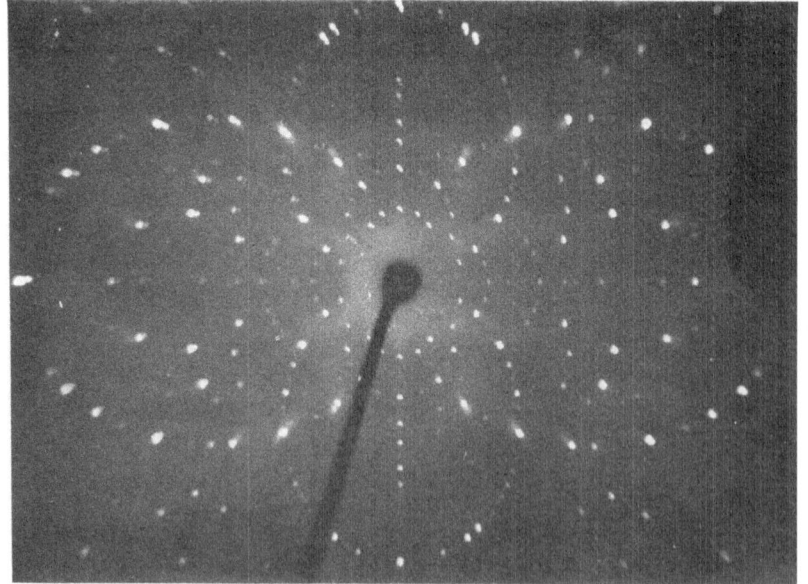

Figure 1. $BeAl_2O_4$ (Chrysoberyl). Laue transmission photographs, X-ray beam parallel to a axis; (a) after irradiation with $2.8 \cdot 10^{20} n/cm^2$, (b) after irradiation with $3.6 \cdot 10^{20} n/cm^2$, and (c) after irradiation with $8.1 \cdot 10^{20} n/cm^2$.

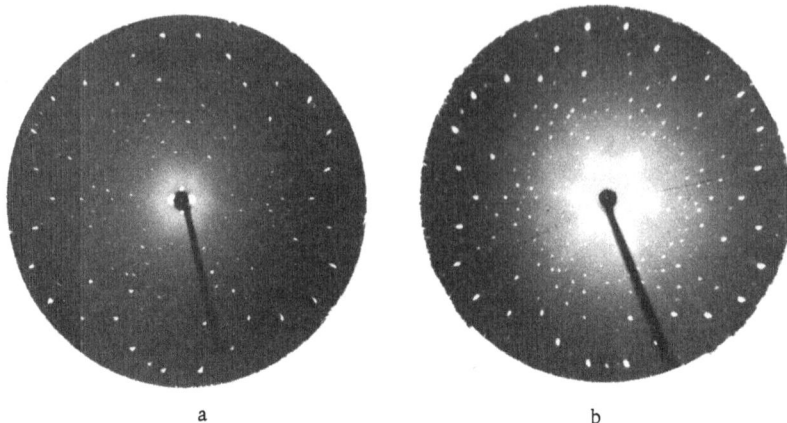

Figure 2. Be_2SiO_4 (Phenacite). Laue transmission photographs, X-ray beam parallel to c axis; (a) after irradiation with $4.53 \cdot 10^{20}$ n/cm^2, and (b) after irradiation with $6.3 \cdot 10^{20}$ n/cm^2.

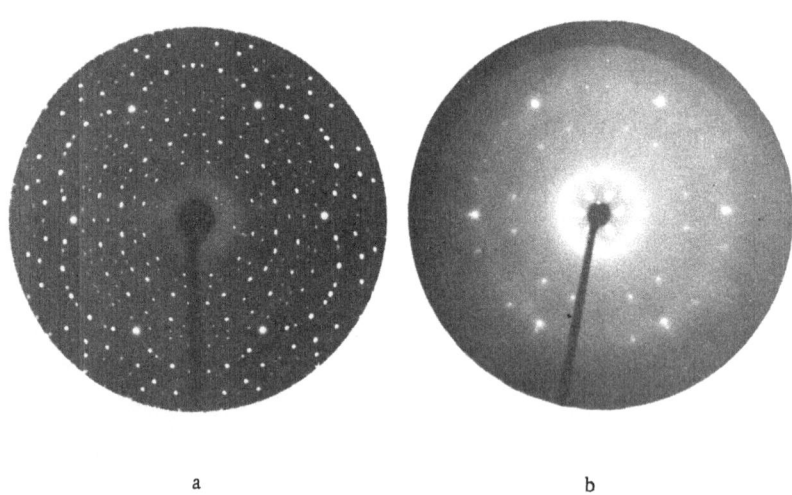

Figure 3. $Be_3Al_2Si_6O_{18}$ (Beryl). Laue transmission photographs, X-ray beam parallel to c axis; (a) unirradiated, and (b) after irradiation with $2.8 \cdot 10^{20}$ n/cm^2.

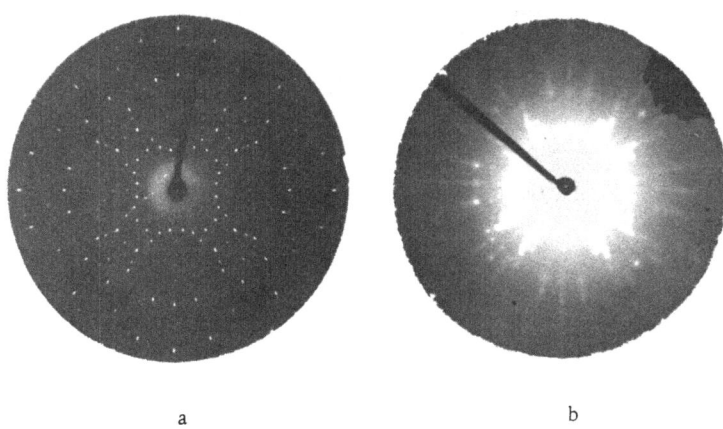

a   b

Figure 4. $Ca_3Al_2Si_4O_{12}$ (Garnet). Laue transmission photographs, X-ray beam parallel to a axis; (a) unirradiated, and (b) after irradiation with $2.53 \cdot 10^{20}$ n/cm^2.

bodies, it is apparent that the rupture of silicon—oxygen bonds and the subsequent distortion of the arrangement of the $[SiO_4]$ tetrahedra resulting from fast-neutron bombardment might tend to produce amorphous silicates. If the process is a progressive one that develops with increasing irradiation dosages one observes the gradual reduction of long-range order, a depression in X-ray peak intensities, and eventually the appearance of the diffuse halo that accompanies the growth of the amorphous material within the skeletal crystal matrix. In the early stages of the process one observes the nonrandom, symmetrical, diffuse scattering that was described in the previous section as arising from inhomogeneous lattice strains and statistically trapped displacements. An example of this effect is shown in Figure 4 of $Ca_3Al_2Si_4O_{12}$ (garnet). Bombardment with $6 \cdot 10^{20}$ n/cm^2 produces amorphism in this crystal. The smearing of X-ray line profiles is exhibited in Figure 5 of $Al_2(OH)_2SiO_4$ (topaz), through a sequence of Weissenberg patterns. Diffractometer measurements on single crystals irradiated with $4.5 \cdot 10^{20}$ n/cm^2 showed axial expansions for $a_0 = 2.87\%$, $b_0 = 0.53\%$, and $c_0 = 2.05\%$. With these large lattice parameter shifts it is not surprising that X-ray reflections are heavily broadened. After irradiation with $6 \cdot 10^{20}$ n/cm^2 this crystal also becomes completely amorphous.

A third sample of radiation-induced amorphism in silicates is $ZrSiO_4$ (zircon), which is completely disordered after irradiation with $6 \cdot 10^{20}$ n/cm^2. Here, also, the progressive growth

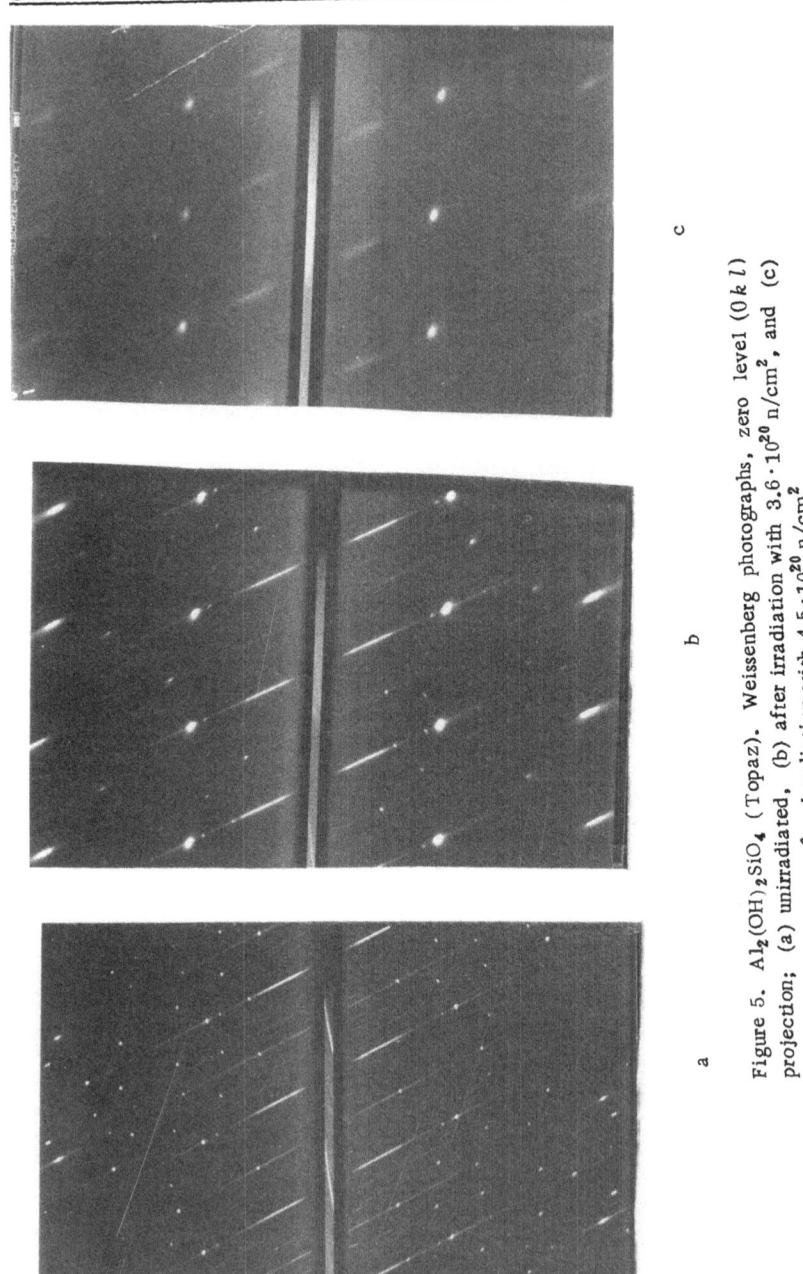

Figure 5. $Al_2(OH)_2SiO_4$ (Topaz). Weissenberg photographs, zero level $(0\,k\,l)$ projection; (a) unirradiated, (b) after irradiation with $3.6 \cdot 10^{20}$ n/cm^2, and (c) after irradiation with $4.5 \cdot 10^{20}$ n/cm^2.

of amorphous regions within a distended crystal matrix is seen after successive lower dosages of irradiation. Annealing a crystal at 1600°C which is only partially amorphous after irradiation with $3.6 \cdot 10^{20}$ n/cm^2 produces recrystallization on the skeletal single crystal matrix to a more perfect structure. This is exhibited in the precession photographs shown in Figure 6.

OXIDES WITH A HIGHER DEGREE
OF RADIATION STABILITY

Single crystals of MgO, Al$_2$O$_3$ (sapphire), and TiO$_2$ (rutile) have been subjected to integrated fast-neutron fluxes in excess of 10^{20} n/cm^2. Preliminary studies indicate that TiO$_2$ is the most stable material of the three, for single crystal diffraction patterns taken after irradiation with $1.1 \cdot 10^{20}$ n/cm^2 show no lattice distortions or imperfections and no lattice parameter shifts within the limits of measurement. Exposure of MgO to $1.8 \cdot 10^{20}$ n/cm^2 produces a small lattice expansion (0.17%), but the over-all lattice perfection is little altered. Al$_2$O$_3$ (sapphire) was irradiated to a higher dosage, $6 \cdot 10^{20}$ n/cm^2, and showed significant lattice parameter shifts, but again single crystal diffraction patterns show that a high degree of perfection is retained. After this dosage, a_0 expands 0.3% and c_0 lengthens 0.45%.

EFFECTS IN CRYSTALS DUE TO
THERMAL-NEUTRON REACTIONS

Crystals which contain fissionable nuclei as constituents of their lattices are interesting materials from a radiation damage viewpoint because of the immense amount of energy which can be dissipated over short distances.

When we employ the natural boron content of the crystal NaMg$_3$B$_3$Si$_6$O$_{27}$(OH)$_4$ (tourmaline) as the fissionable source, amorphism is produced after exposure to $7.8 \cdot 10^{19}$ thermal neutrons/cm^2. The reaction which produces the heavy charged particle is as follows: ^{10}B(n, α)^{7}Li+2.4 Mev, natural cross section 750 barns. The total energy as given is divided between the alpha particle and the recoil nuclei, and the very large cross section makes the reaction attractive for producing damaging effects.

Similarly, Al$_2$O$_3$ (sapphire) single crystals containing natural lithium (0.15% Li$_2$O) as a substitutional replacement for aluminum were exposed to an integrated neutron flux of $1.8 \cdot 10^{20}$ n/cm^2, and were subjected to fast-neutron bombardment as well as charged-particle irradiation due to the following reaction for thermal neutrons: ^{6}Li(n, α)^{3}H+4.8 Mev, natural cross section 70 barns. X-ray transmission patterns of this crystal using

a

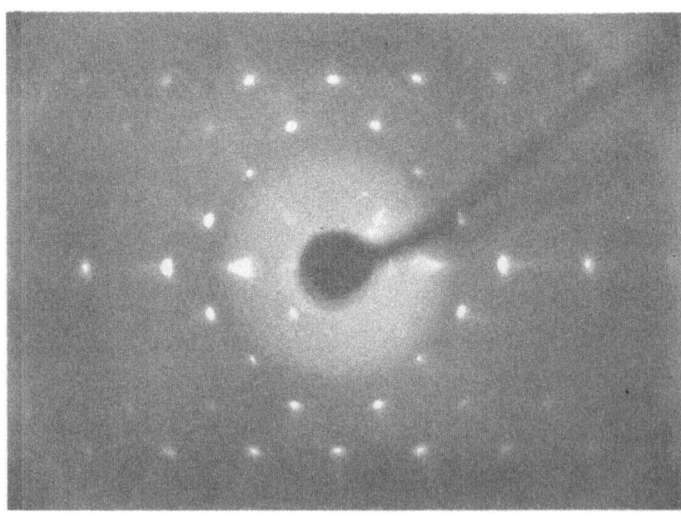

b

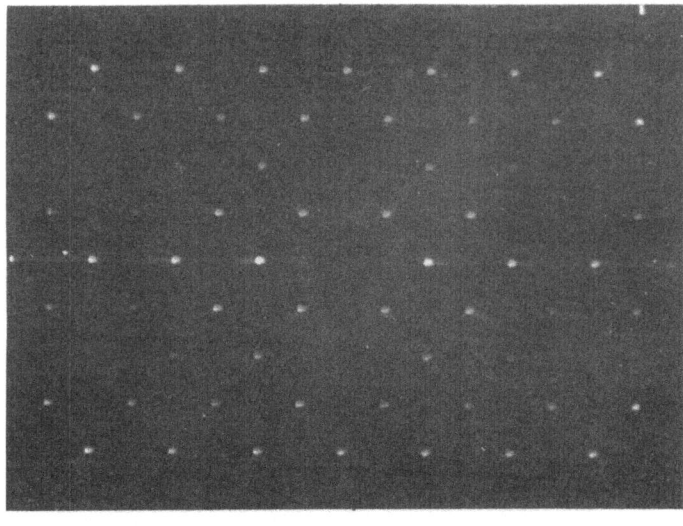

c

Figure 6. ZrSiO$_4$ (Zircon). Precession photographs, zero level $(h\,0\,l)$ projection; (a) unirradiated, (b) after irradiation with $3.6 \cdot 10^{20}$ n/cm^2, (c) same as (b) after annealing at 1600°C.

unfiltered and monochromatic radiation are shown in Figure 7. It is seen that the Laue reflections remain sharp and intense, but that a diffuse scattering pattern containing a symmetry related to the crystal diffraction symmetry is produced as a result of irradiation. Small distentions of the lattice amounting to +0.20% in a_0 and +0.21% in c_0 were measured. Unfortunately, in this case, the effects of fast neutrons and heavy charged particles are superimposed so that the real contributions of the latter were not determined.

An exceptionally clear case for the differentiation of fast-neutron effects and fission fragment effects is seen in the case of natural ZrO$_2$ (baddeleyite). More often than not, natural samples of this oxide contain uranium and thorium impurities in concentrations ranging as high as 1%. Polycrystalline samples of the oxide containing measured concentrations of uranium impurities were measured by the Debye-Scherrer method before and after several irradiations. At the same time, synthetic single crystals and powders of pure oxide were exposed to reactor neutrons for prolonged irradiations. The results[4] of this experiment were that the monoclinic structure of ZrO$_2$ is stable

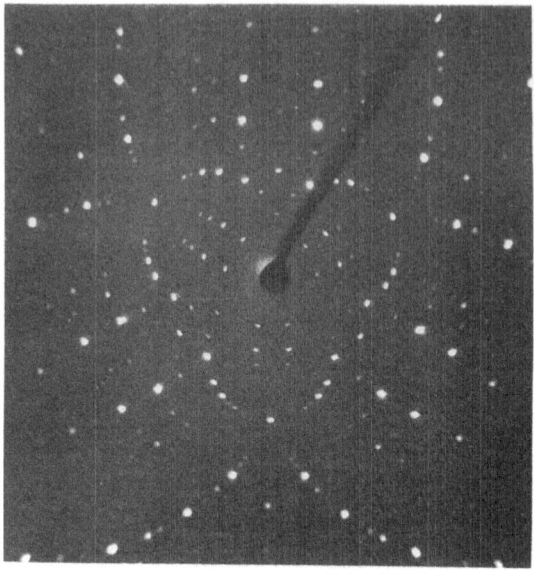

a

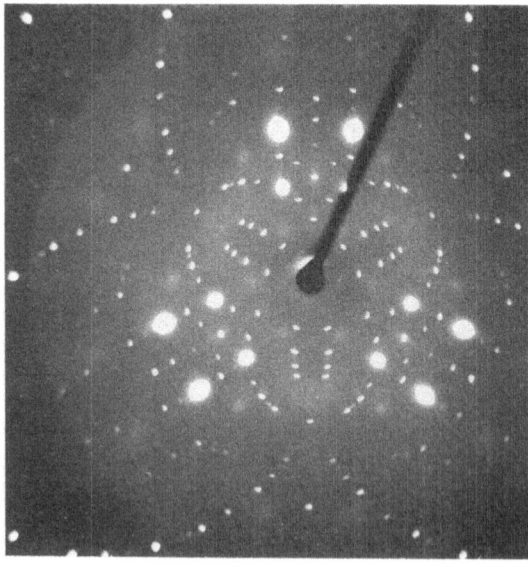

b

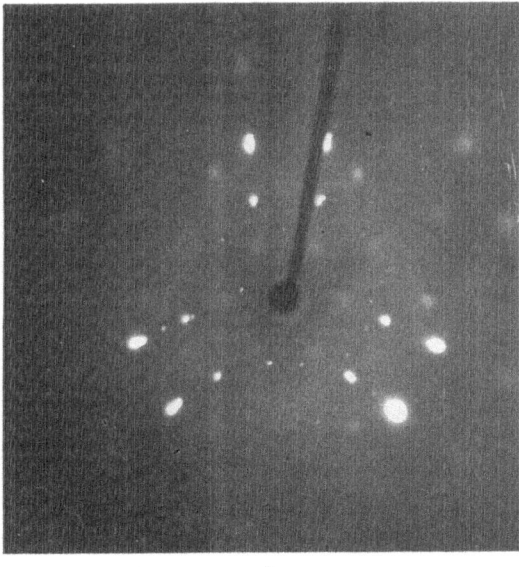

c

Figure 7. Al_2O_3 (Sapphire) containing 0.15% LiO_2. Laue transmission photographs, X-ray beam parallel to c axis; (a) unirradiated, (b) after irradiation with $1.8 \cdot 10^{20}$ n/cm², (c) transmission photograph of (b) using monochromatic Mo K_α radiation.

under fast-neutron bombardments as high as $3.6 \cdot 10^{20}$ n/cm², but that the monoclinic structure transforms to the high-temperature fcc phase upon fission fragment irradiation. The fission reaction producing this crystal transformation is as follows:

$$^{235}U \, (n, \, f) \text{ fission fragments } +160 \text{ Mev},$$
natural cross section 3.92 barns.

Although the cross section for the reaction is small, the amount of energy dissipated by the pair of fission fragments is enormous. By measuring the fraction of transformed material in partially transformed crystals (Figure 8), one is able to determine that the energy dissipated by a pair of fission fragments in ZrO_2 affects $2 \cdot 10^6$ atoms, thereby causing a transformation in a small region of approximately 140 A diameter. The stability of the crystals when exposed to fast neutrons alone indicates that the transformation mechanism is a "fission spike" whereby a small localized region is heated above the transition temperature and rapidly quenched.

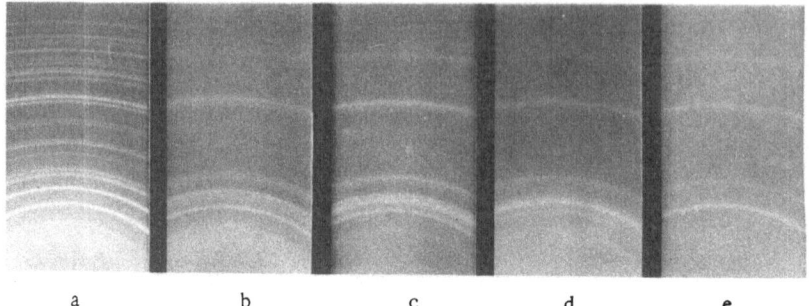

a   b   c   d   e

Figure 8. ZrO_2 (Baddeleyite). Debye-Scherrer photographs of samples irradiated with fission fragments; (a) unirradiated, (b) after irradiation with $7.66 \cdot 10^{14}$ fission fragment pairs/cm^3, 5% transformed, (c) after irradiation with $6.67 \cdot 10^{15}$ fission fragment pairs/cm^3, 50% transformed, (d) after irradiation with $1.07 \cdot 10^{16}$ fission fragment pairs/cm^3, 80% transformed, (e) after irradiation with $3.39 \cdot 10^{16}$ fission fragment pairs/cm^3, completely transformed.

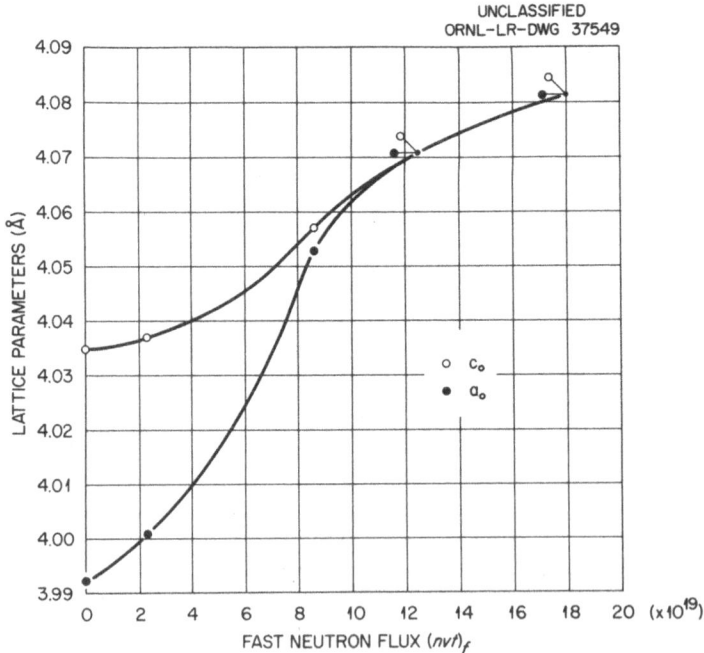

Figure 9. Lattice expansion of $BaTiO_3$ as a function of fast-neutron dosage.

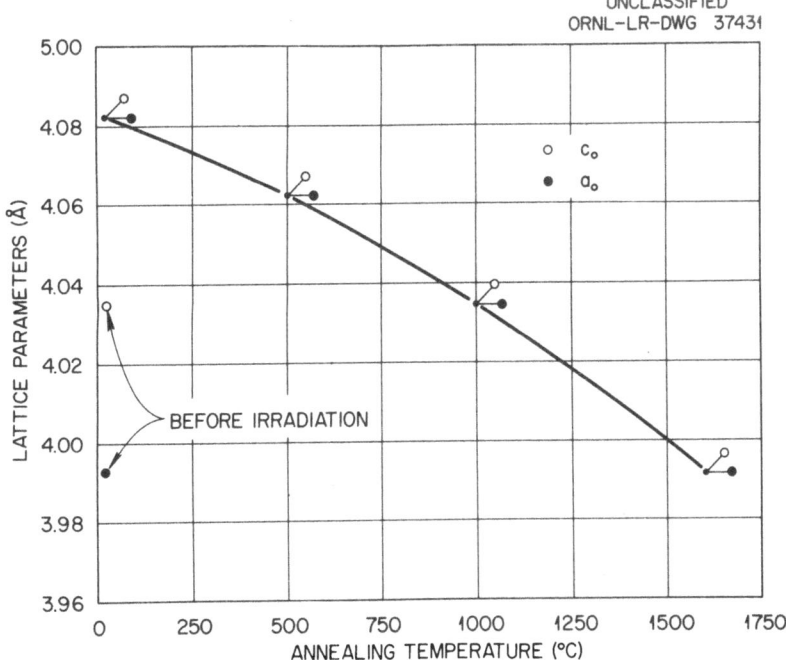

Figure 10. Lattice parameter annealing of irradiated $BaTiO_3$.

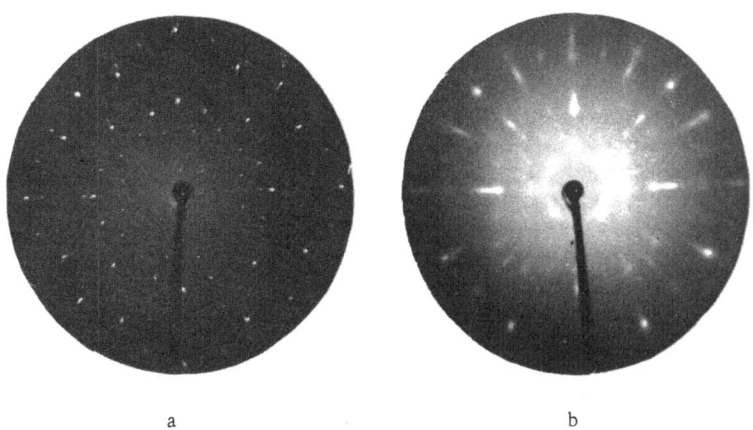

Figure 11. $KNbO_3$, Laue transmission photographs, X-ray beam parallel to c axis.
(a) unirradiated, (b) after irradiation with $4 \cdot 10^{19}$ n/cm².

PHASE TRANSITIONS DUE TO
FAST-NEUTRON IRRADIATION

$BaTiO_3$, $KNbO_3$, and $PbTiO_3$ are perovskite-type crystals that deviate from ideal cubic symmetry at room temperature and transform at higher temperatures to the cubic phase. When exposed to fast-neutron dosages in excess of 10^{19} n/cm^2, these crystals transform[5,6] into the high-temperature cubic modification. The transformation in $BaTiO_3$ proceeds by the anisotropic growth of the tetragonal axes (Figure 9). In unirradiated $BaTiO_3$, at the upper Curie temperature (120°C), thermal atomic motion is sufficient to overcome the weaker restoring force perpendicular to the c axis and into the cubic phase. In the irradiated crystal, at a lower temperature (85°C), the internal strains produced by interstitial atoms are probably similarly relieved by a growth in the weaker directions. The annealing behavior of the transformed crystal (Figure 10) shows the remarkable thermal stability of the irradiated material. After annealing at 1000°C the crystal is still cubic, but with a lattice parameter equivalent to that of the original tetragonal c axis. After annealing at 1550°C, the crystal is still cubic, but the lattice parameter then jumps down to that of the original tetragonal a axis. In effect, the combination of irradiation and annealing results in a transformed crystal with a smaller volume than the original unirradiated crystal. Through the entire irradiation and annealing procedures the single crystal structure is retained.

The results of fast-neutron irradiation in $KNbO_3$ and $PbTiO_3$ are very similar to those for $BaTiO_3$, with the exception that the irradiation-induced phase transition is observed at lower dosages ($4 \cdot 10^{19}$ n/cm^2). $KNbO_3$ is orthorhombic at room temperature but transforms to a tetragonal[7] cell at approximately 228°C and into a cubic structure at approximately 435°C. After irradiation with $4 \cdot 10^{19} n/cm^2$, the Laue transmission photographs (Figure 11) show the distortion induced by the bombardment together with diffuse scattering that was seen from irradiated $BaTiO_3$.[5] Precession, Weissenberg, and powder patterns confirmed the formation of a perovskite-type crystal as a result of neutron irradiation. Annealing these transformed crystals at 1000°C produced only small lattice parameter shifts, but the cubic crystals retained their perovskite-type symmetry.

$PbTiO_3$ is tetragonal[8] at room temperature and transforms into a perovskite-type cubic structure at approximately 490°C. Figure 12 shows the Laue transmission patterns of a crystal irradiated with $4 \cdot 10^{19}$ n/cm^2. Note the exact identity of the two projections indicating the higher symmetry. The annealing be-

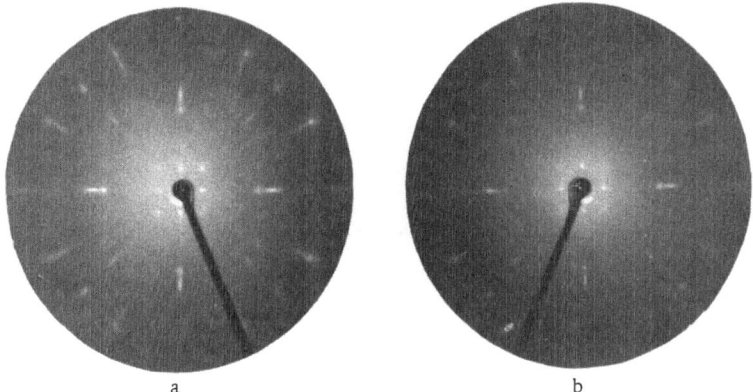

a   b

Figure 12. $PbTiO_3$, Laue transmission photographs, both after irradiation with $4 \cdot 10^{19}$ n/cm^2, (a) X-ray beam parallel to tetragonal c axis, (b) X-ray beam parallel to tetragonal a axis.

havior of irradiated $PbTiO_3$ is similar to that for $KNbO_3$, since the transformed cubic crystal is thermally stable and retains its cubic symmetry.

MACROSCOPIC INTEGRITY OF NEUTRON-IRRADIATED CRYSTALS

None of the crystals discussed in this review suffered any obvious macroscopic damage such as cracks, bubbles, or flaws as a result of fast-neutron irradiation. Although volume changes ranging from a few tenths of a percent to more than 14% were produced, it is notable that natural crystal faces (Figure 13) were left unperturbed. An interesting, if somewhat unrelated, effect is the strain pattern induced in irradiated and annealed Brazilian twinned quartz (Figure 14). This twinned crystal was irradiated with $7.6 \cdot 10^{19}$ n/cm^2, still retaining some long-range order, and then annealed at 950°C for 30 min. Note the development of slotted holes within the twin bands and the associated strains which are visible in polarized light.

DISCUSSION

The structural effects of irradiation upon nonmetallic crystals are highly variable, as determined by X-ray diffraction techniques. Some crystals become amorphous after bombardment with fast-neutron fluxes as low as $1.6 \cdot 10^{20}$ n/cm^2 (quartz), while others display a high degree of perfection even after irradiation dosages as high as $8 \cdot 10^{20}$ n/cm^2. There are no clear-cut

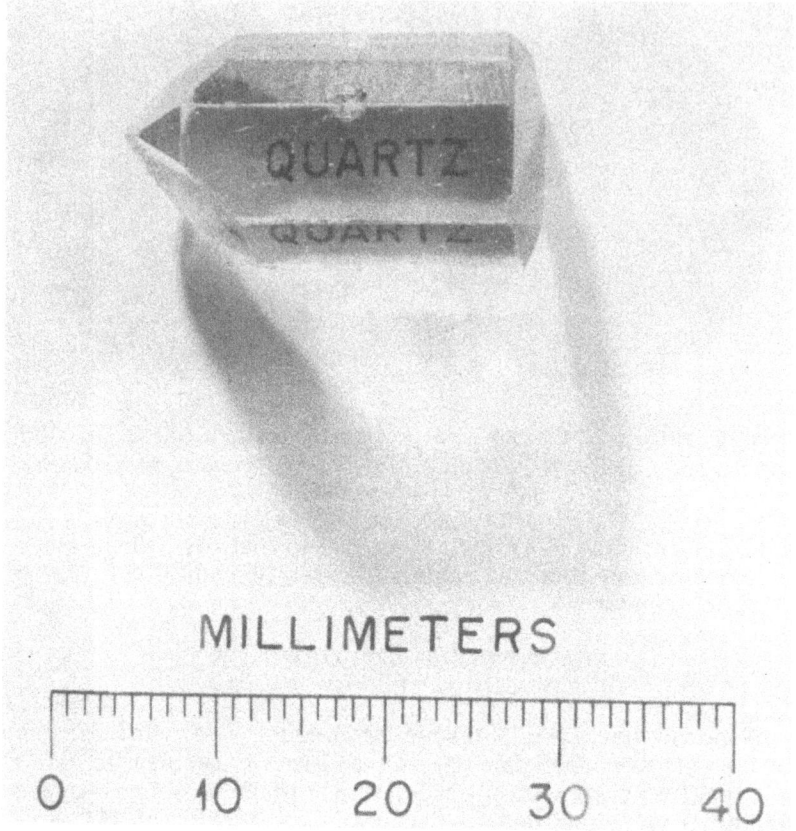

Figure 13. Quartz crystal, amorphous after irradiation with $1.6 \cdot 10^{20}$ n/cm^2, volume increase 14.7%.

answers to explain these wide differences in stability but it would appear that the type of binding, atomic packing in the lattice, and crystal structure all play a role in the ultimate stability of a crystal under fast-neutron bombardment.

Those crystals which contain more ionic binding appear to be more resistant to radiation effects, but this does not seem to be a useful general rule.*

If displacements are produced in an anisotropic crystal, however, the effects which are observed are invariably aniso-tropic as well. As a result, lattice parameter shifts in such

---

*It should be pointed out that this possible explanation was first given by Gold-schmidt and Thomassen[9] in their study of the so-called metamict crystals.

crystals are anisotropic and bombardment-induced defects are often strategically trapped, thus restricting normal thermal atomic motion. From these effects one finds irradiation-induced phase transitions to high-temperature forms, and a very high thermal stability for the transformed crystals as well.

In addition to the effects due to fast neutrons, a few examples of effects due to charged-particle irradiations employing thermal-neutron reactions with the B^{10} nucleus and the Li^6 nucleus have been studied. Another phase transition was found to be due to the energy dissipated by fission fragments in a crystal lattice (ZrO_2), and again the transition was to a more symmetrical high-temperature form. From this analysis it was found that the energy dissipated by a single fission fragment in ZrO_2 affected approximately 10^6 atoms through what is believed to be a "fission-spike" mechanism.

The displacement of atoms in nonmetallic crystals by knock-on collisions produces interstitials, vacancies, and complex aggregates. When the concentrations of these defects is sufficiently high, X-ray reflections are broadened, lattice parameters are distended, and often a diffuse X-ray scattering appears in diffraction patterns of irradiated crystals. It is interesting to note that crystals which suffer volume changes as high as several percent continue to retain macroscopic integrity.

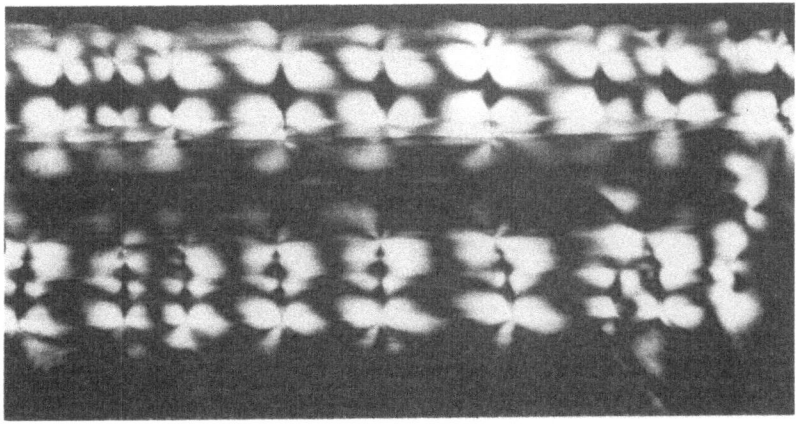

Figure 14. Quartz crystal, Brazilian twins viewed parallel to c axis in polarized light. Irradiated with $7.6 \cdot 10^{19}$ n/cm^2 and annealed for 30 min at 950°C.

REFERENCES

[1] F. A. Sherrill, M. C. Wittels, and T. H. Blewitt, J. Appl. Phys., Vol. 28, 1957, p. 526.

[2] J. R. Gilbreath and O. C. Simpson, Second United Nations International Conference on the Peaceful Uses of Atomic Energy, Paper No. 621, 1958.

[3] M. C. Wittels, Phil. Mag., Vol. 2, 1957, p. 1445.

[4] M. C. Wittels and F. A. Sherrill, Phys. Rev. Letters, August 15, 1959.

[5] M. C. Wittels and F. A. Sherrill, J. Appl. Phys., Vol. 28, 1957, p. 606.

[6] M. C. Wittels, Bull. Am. Phys. Soc., Vol. 4, 1959, p. 285.

[7] E. A. Wood, Acta Cryst., Vol. 4, 1951, p. 353.

[8] Gen Shirane and Sadao Hoshino, J. Phys. Soc. Japan, Vol. 6, 1951, p. 265.

[9] V. M. Goldschmidt and L. Thomassen, Norsk. Ak. Skr., I, 1, No. 5, 1924, p. 58.

X-RAY METHODS FOR DETECTION OF LATTICE IMPERFECTIONS IN CRYSTALS

Volkmar Gerold

Institut fuer Metallphysik am Max Planck Institut
fuer Metallforschung, Stuttgart, Germany

ABSTRACT

Two methods are described for the detection of lattice imperfections of different kinds.

1. Imperfections in Solid Solutions. Normal diffraction methods with monochromatized X-rays are applied to single crystals of supersaturated solid solutions, where preprecipitation states occur. The calculated intensity distribution for a one-dimensional model of clustering shows the influence of atomic distribution and lattice distortion. Results are given on the structure of clusters in aluminum-rich alloys.

2. Detection of Dislocations in Nearly Ideal Crystals. A method first given by Barth and Hosemann is improved and used for the detection of dislocations in crystals of germanium. Use is made of the anomalously low absorption coefficient existing for X-rays which make the Bragg angle with a certain set of lattice planes going through a perfect crystal. The absorption is increased at points where the lattice planes are distorted by a dislocation line, and shadows from these lines can be seen on a photographic plate behind the crystal. The method gives information on the spatial distribution of dislocation lines and the direction of the Burgers vectors.

INTRODUCTION

The latest development of solid state physics is concerned with the influence of lattice defects on the physical properties of crystals. In connection with this the question arises as to the suitability of methods which give direct information on the structure of such defects. For example, the electron microscope gives direct information. A modern method using very thin foils for direct transmission pictures shows the distribution of dislocation lines in such foils. In alloys where precipitation occurs from a supersaturated solid solution, the precipitation in the earliest states including preprecipitation states can be detected.

As the electron microscope technique is limited to very thin foils of about 500 A thickness, the question arises with regard

289

to other methods. In this paper two different X-ray methods are described. The first gives information on the structure of the preprecipitation state in supersaturated solid solutions. The second method describes some experiments on the detection of dislocations by a sort of microradiography.

I. THE DETERMINATION OF THE STRUCTURE OF PREPRECIPITATION STATES

The study of the behavior of quenched metal alloys is of much interest. Quenching from the solid solution region at high temperatures into the two-phase region at low temperatures gives an unstable supersaturated solid solution. During a heat treatment in a distinct temperature region the properties of the material are changed. In many cases an increase of hardness is observed, known as the age-hardening process. This change of properties is connected with the change of the atomic distribution in the matrix: the formation of preprecipitation states and precipitation states. The atomic structure of all these states is of interest.

In many cases the study of X-ray scattering solves the problem. Here only the scattering phenomena from the preprecipitation states may be discussed. In this state the solid solution

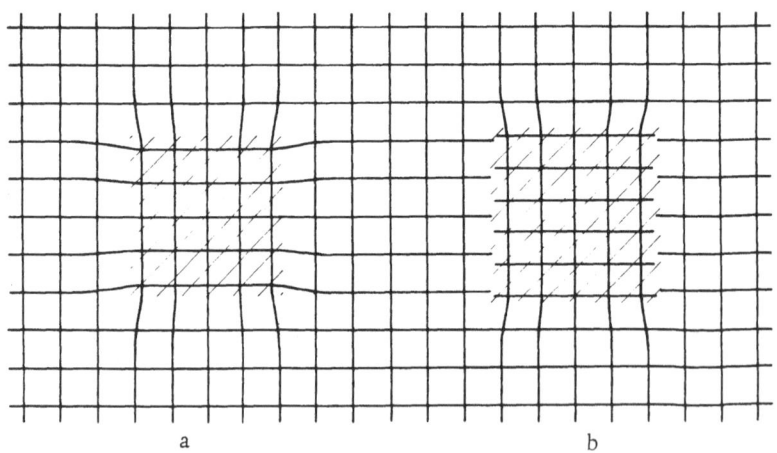

Figure 1. Different states in a supersaturated solid solution. a) Pre-precipitation state. Coherency in all three directions. b) Precipitation state. Coherency has been lost in one direction.

is no longer a homogeneous solution, for at least two different regions can be distinguished. One region is formed by clusters where the concentration of solute atoms is high. The other region surrounds these clusters with a very low concentration of solute atoms. Both regions are part of the matrix, i.e., each atom inside and outside the clusters belongs to a distinct lattice site of the matrix. Due to the different diameters of the atoms this lattice may be distorted elastically as schematically shown in Figure 1a. The lattice is coherent in all three directions. Figure 1b shows another possible state. The coherency between different regions has been lost in one direction due to a different lattice constant in that direction. At the surface between the Regions I and II there is a distortion which is similar to a plastic deformation. In this case Region I appears more like a precipitation and shall not be considered here.

The precipitation state is characterized by:

1. The arrangement of the atoms, i.e., the shape and the size of the clusters, sometimes the ordered structure inside the clusters or the ordered arrangement of different clusters.

2. The lattice distortions inside and outside the clusters.

Both these quantities give a contribution to the background of monochromatic X-ray diffraction which can be distinguished from other contributions (for example, temperature scattering).

One-Dimensional Model of Clustering. In some cases clusters having the shape of plates are formed in supersaturated solid solutions. Two-dimensional clusters of copper atoms are formed in an aluminum-rich alloy Al–Cu during the aging process at room temperature, known as Guinier-Preston zones.[1,2] Other examples of plate-like clusters have been found in Cu–Ni–Fe and Au–Pt, their equilibrium diagram having a miscibility gap. In these alloys the so-called sideband structures occur, first investigated by Bradley.[3] This state is characterized by sidebands on both sides of Debye-Scherrer lines in a small distance. As the different atoms have nearly the same atomic number, the X-ray beam cannot distinguish between the different kinds of atoms. Therefore the sidebands only can be explained by the occurrence of lattice distortions. Daniel and Lipson,[4] Hargreaves,[5] and Tiedema, Bouman, and Burgers[6] gave interpretations for this phenomena. The best explanation is due to the last-mentioned authors.

---

[1] Superscripts pertain to references at the end of the paper.

Now we will start to study the scattering amplitudes from a more general model, where the different atoms have a large difference in their atomic numbers. As we shall see, the intensity formula gives a general insight into the influence of both lattice distortion and atomic distribution.

The model for the clusters is given by Figure 2, similar to the model given by Tiedema et al.[6] It consists of a periodic arrangement of atomic layers; the number of layers forming a period is L. The average distance between the layers is $a_0/2$. The number of layers in a period belonging to the Region I is c_I/L. The rest, or $c_{II}L = (1-c_I)L$ layers form the Region II. The average atomic scattering amplitudes in these regions are respectively f_I and f_{II}. The spacings of the layers are $(1 + \alpha_I)a_0/2$ and $(1 + \alpha_{II})a_0/2$, respectively, where α_I and α_{II} are the distortion constants. The spacing of the layers at the contact surface between I and II is $(1 + \alpha_s)a_0/2$.* Due to the coherence of these regions with the matrix the following relation holds:

$$(c_I L - 1)\,\alpha_I + (c_{II}L - 1)\,\alpha_{II} + 2\alpha_s = 0, \tag{1}$$

because we have $(c_I L - 1)$ layer spacings in the Region I and two

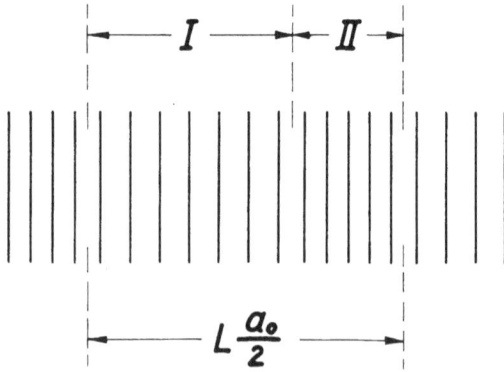

Figure 2. One-dimensional model of periodic clustering in a solid solution.

---

*Tiedema et al. have not introduced a different distance for the contact surfaces. In our calculation the introduction of these spacings leads to simpler equations.

contact surfaces within one period. To get a simple relation, we make the assumption

$$\alpha_s = \frac{1}{2}(\alpha_I + \alpha_{II}).$$ (2)

Equations (1) and (2) give

$$c_I\alpha_I + c_{II}\alpha_{II} = 0.$$ (3)

Now we must calculate the positions $x_n = (n + u_n)a_0/2$ of the layers, where $na_0/2$ is the position in an undistorted lattice and $u_na_0/2$ is the shift from these positions due to lattice distortions. If we number the layers in one period from $-c_IL$ to -1 (Region I) and from 0 to $(c_{II}L-1)$ (Region II) we get the following coordinates:

Region I: $x_n = (n + \alpha_I n + \alpha_I - \alpha_s)a_0/2$ for $-c_IL < n < -1$

Region II: $x_n = (n + \alpha_{II}n)a_0/2$ for $0 < n < (c_{II}L-1)$. (4)

Introducing the reciprocal lattice coordinate $s = h/a_0$, the calculation of the scattering amplitude is given by

$$F(h) = \sum_n f_n \exp[\pi i(n+u_n)h].$$ (5)

We substitute $h = H + g$, where H is an even number and gives the position of the sharp intensity spots from the undistorted matrix. The calculation is given in the appendix. The final result is a periodic arrangement of spots in the background at the positions

$$g_\nu = 2\nu/L, \quad \text{where } \nu = 0, \pm1, \pm2, \pm3, \ldots \ .$$

The amplitude at these positions is

$$F(h_\nu) = \frac{\sin\left[\frac{\pi}{2}c_IL\left(g_\nu + \alpha_Ih_\nu\right)\right]}{\sin\left[\frac{\pi}{2}\left(g_\nu + \alpha_Ih_\nu\right)\right]\sin\left[\frac{\pi}{2}\left(g_\nu + \alpha_{II}h_\nu\right)\right]} \cdot$$

$$\left\{f_I \sin\left[\frac{\pi}{2}\left(g_\nu + \alpha_{II}h_\nu\right)\right] - f_{II}\sin\left[\frac{\pi}{2}\left(g_\nu + \alpha_Ih_\nu\right)\right]\right\}. \quad (6)$$

The integral intensity of these spots is given by the square of equation (6) multiplied by a factor N/L^2, where $N = ML$ is the total number of atomic planes. The weight factor of the number of atoms per layer is omitted. For small values g_ν and αh_ν we get the following approximation for the integral intensity [using equation (3) for the elimination of α_{II}. $c_{II} = 1 - c_I$]:

$$I(h_\nu) = \frac{N}{L^2} \cdot \frac{4 \sin^2\left[\frac{\pi}{2} c_I L \left(g_\nu + \alpha_I h_\nu\right)\right]}{\pi^2 \left(g_\nu + \alpha_I h_\nu\right)^2 \left(g_\nu + \alpha_{II} h_\nu\right)^2} \cdot \left[\left(f_I - f_{II}\right)g_\nu - \overline{f}\frac{\alpha_I}{c_{II}}h_\nu\right]^2 , (7)$$

where $\overline{f} = c_I f_I + c_{II} f_{II}$ is the mean scattering amplitude from the whole lattice.

The approximation is good for periodic structures with large periods L, because then g_ν becomes small. The smaller the number M of periods, the more the background spots are broadened. Finally, when the clusters are not arranged in a periodic manner, neighboring intensity spots overlap and produce an intensity streak. The scattering amplitudes from this streak are given exactly by equation (6) for the positions $g_\nu \neq 0$. The weight factor N/L^2 must be substituted by the number Z of clusters distributed at random in the solid solution. For other positions than g_ν the same equation gives good interpolated values [without the surroundings of $g = -\alpha_{II} h$, where $F(h)$ becomes infinite].

In the case of unperiodic clustering the distortion coefficient α_{II} is no longer a constant. It becomes a decreasing function of the distance from the boundary between Regions I and II. The intensity function is not influenced too much by this deviation from the calculated model. The number L is then an approximated value for the diameter of the distorted area around one cluster, including the cluster (Region I) in the midst and the distorted Regions II on both sides.

The intensity equation (7) may be discussed in detail. It shows in a clear manner the influence of both quantities, the distribution of atoms, and the lattice distortion.

The intensity of the matrix spots is given by $g_\nu = 0$.

$$I(H) = \overline{f}^2 \left(\frac{\sin y}{y}\right)^2 \approx N\overline{f}^2 \exp\left(\frac{-y^2}{3}\right) \qquad (8)$$

where $y = (\pi/2)c_I \alpha_I LH$. This formula is only correct for a strictly periodic structure. The intensity is influenced by the distortion constant α_I and is proportional to $\exp(-H^2)$ as the Debye factor. The value $(c_I \alpha_I L)^2$ is a measure for the mean square atomic displacement as it is the Debye temperature θ_D in the Debye factor.

The intensity outside the matrix spots is of more interest. Several cases may be discussed.

Case I. $f_I \neq f_{II}$, $\alpha_I = \alpha_{II} = 0$.

The intensity formula reduces to

$$I(h\nu) = \frac{N}{L^2} \frac{4 \sin^2\left(\frac{\pi}{2}c_I L g_\nu\right)}{\pi^2 g_\nu^2} \left(f_I - f_{II}\right)^2 . \tag{9}$$

It is proportional to $(f_I - f_{II})^2$ and is a symmetrical function with respect to ν, i.e., to the matrix spot H. The intensity is reduced with increasing $|\nu|$ depending on the values c_I and L. It is independent of the magnitude of H (the dependence on f_I and f_{II} is omitted in this case).

An experimental example cannot be given, since lattice distortion will always be present in plate-shaped clusters. Also in systems where spherical clusters have been found, lattice distortion influences the scattering. This shall be discussed later.

Case II. $f_I = f_{II}$, $\alpha_I \neq 0$.

The first term in the brackets of equation (7) is zero. The intensity distribution around the spot H depends on the value of H. It is nearly proportional to H^2. At small scattering angles there is no intensity contribution. In all cases where $c_I \neq 0.5$, the intensity becomes asymmetric with respect to the matrix spots H. The intensity drops down very rapidly; therefore, in most cases only the first order $\nu = \pm 1$ can be seen.

If $I^{(+)}$ is the intensity on the positive side of a matrix spot ($\nu = +1$ for $H > 0$) and $I^{(-)}$ the intensity on the negative side, we get as a rule:

$$\text{If } c_I < 0.5 \text{ and } \alpha_I < 0 \quad \text{then} \quad I^{(+)} < I^{(-)}.$$
$$\tag{10}$$
$$\text{If } c_I < 0.5 \text{ and } \alpha_I > 0 \quad \text{then} \quad I^{(+)} > I^{(-)}.$$

Examples for this type of periodic clustering are given by the sideband structures in the systems Cu–Fe–Ni[4,5] and Au–Pt[6] as already discussed by Tiedema et al.[6] If the plates are parallel to (200) planes, the formulae discussed here can be applied directly. H is one of the three coordinates of a matrix spot (HKL) of the fcc lattice [the formulae are also valid for odd H in the odd triples (HKL)]. As the concentration of the alloys varies, the fraction c_I is changed too, whereas α_I does not change sign. The experimental value of the ratio $I^{(+)}/I^{(-)}$ is changed in accordance with the theory.

Case III. $f_I \neq f_{II}, \alpha_I \neq 0$.

In this case we have the influence from both the atomic arrangement and the lattice distortion. The intensity formula, equation (7), consists of three terms:

$$I(h_\nu) = I_A + I_D + I_M,$$

where

$$I_A = C^2(h_\nu) \, g_\nu^2 \, (f_I - f_{II})^2,$$

$$I_D = C^2(h_\nu) \left(\frac{\alpha_I}{c_{II}}\right)^2 h_\nu^2 \, \overline{f^2},$$

$$I_M = C^2(h_\nu) \left(\frac{\alpha_I}{c_{II}}\right) g_\nu h_\nu (f_I - f_{II}) \, \overline{f},$$

and

$$C(h_\nu) = \frac{2 \sin\left[\dfrac{\pi}{2} c_I L (g_\nu + \alpha_I h_\nu)\right]}{\pi L \, (g_\nu + \alpha_I h_\nu) \, (g_\nu + \alpha_{II} h_\nu)} \sqrt{N}.$$

The terms I_A and I_D have positive values, whereas the mixed term I_M changes its value with the sign of g_ν. As we have seen, the intensity of the matrix spots is reduced due to the lattice distortion. The lacking intensity is found in the background, mainly in the term I_D. The term I_A gets its intensity from the diffuse monotonic scattering from an ideal solid solution. This diffuse scattering is reduced and would be zero if we would have a total clustering, i.e., in an alloy A–B all A atoms are in the Regions I and all B atoms in the Regions II. The mixed term I_M gives no real contribution to the background. Its main effect is the shift of intensity from one side of a matrix spot *H* to the other side. The direction of the shift depends on the signs of α_I and $f_I - f_{II}$.

Examples for a periodic one-dimensional clustering with $f_I \pm f_{II}$ are not available, but there are different kinds of clusters in alloys where these calculations may be applied. Some alloys which have been investigated in our laboratory shall be discussed here.

The Alloy Aluminum—Copper. In the aluminum-rich alloy, clusters of copper atoms are formed at room temperature, which produce three sets of streaks *(hkl)* parallel to the three axes of the reciprocal lattice. The streaks are asymmetric with respect to the spots *(hkl)*; they are observed only at the positive side of the spots. As we have continuous streaks and no spots in the background, we can assume an irregular distribution of plate-like clusters parallel to the three (200) layers of the matrix.

Figure 3. Radiograms from a single crystal Al+ 2 at. % C.
a) Containing Guinier-Preston zones. b) Without zones.
Intensity streak along (*h* 0 0) through the matrix spots (200)
and (400). Oscillation diagram, oscillation axis 011,
monochromatic copper radiation.

The asymmetry gives information on the sign of I_M and, therefore, on the sign of $\alpha_I (f_I - f_{II})$, which is negative in this case. If the Regions I are clusters of copper atoms, we have $(f_I > f_{II})$ and α_I must be negative.

In former years it has not been possible to determine the structure because the term I_D has not been detected. In pictures made with monochromatic copper radiation, the term I_D has been found experimentally in the immediate surroundings of the matrix spots *(hkl)*, [7] as can be seen in Figure 3. The intensity distribution from one of the three sets of Guinier-Preston zones is shown schematically in Figure 4. The maxima on both sides of the spots are much higher than shown in the figure. In the surroundings of the points $H = 0$ there are no intensity maxima in accordance with the theory where $I_D = 0$ at these points.

The following conclusions can be drawn:
1. The plates of copper atoms forming the Region I are thin. This can be concluded from the long streak going from (000) to (200), which is due to I_A. Its length is reduced by

the term I_M, which diminishes the intensity on the negative
side of (200). The estimation gives a thickness of the order
of one atomic layer for the clusters of copper atoms. The
diameters are about 50 to 100 A.

2. The distorted Region II around the copper plate must be
    very thick. This can be concluded from the intensity dis-
    tribution of I_D around the matrix spots. The relation of the
    intensity is here $I^{(-)} > I^{(+)}$. The intensity is very high and
    mainly due to the terms I_D and I_M. Since $I_M^{(-)} < 0 < I_M^{(+)}$,
    as we have seen, the intensity $I_D^{(-)}$ must be much higher
    than $I_D^{(+)}$. From this we can conclude that c_I must be
    much smaller than 0.5.

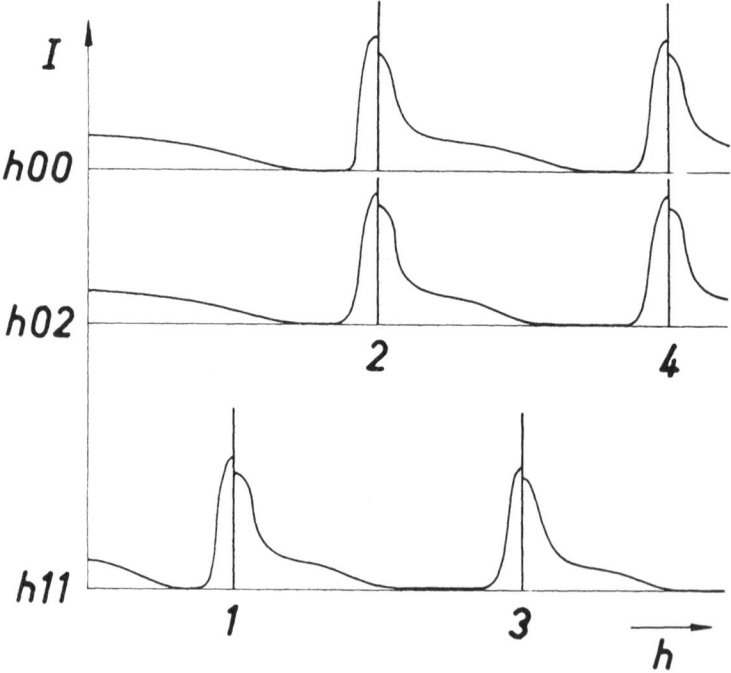

Figure 4. Schematic intensity distribution along different rods ($h\,k\,l$).
The intensity maxima on both sides of the matrix spots are reduced
in the diagram.

Figure 5 gives the model of the Guinier-Preston zones in the alloy aluminum—copper. The lattice distortion reaches much further than shown in the figure. Calculations have given a number of about 15 distorted layers on each side of the copper layer.[7]

Another model for the structure of the zones is given by Toman,[8] consisting of several layers enriched with copper atoms. His results are based on a different intensity distribution at small angles, where the streak also has an intensive maximum. This is in contradiction to the results of other authors.[9] Measurements of the absolute intensity at larger angles cannot be explained by Toman's model, because there are more copper atoms needed than are present in the alloy, but the intensity can be explained by the author's model, where 30 to 50% of the copper atoms are clustered in such zones.[10]

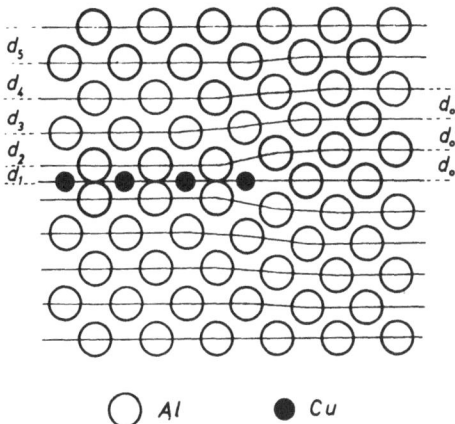

O Al   ●   Cu

Figure 5. Model of the structure of Guinier-Preston zones in Al-Cu section parallel to (002).

Figure 6. Small angle scattering from a foil Al + 5.9 at.% Ag containing clustered zones. The scattering angle for the intensity maximum is about 1°. Monochromatic copper radiation.

The Alloy Aluminum—Silver. In this alloy spherical clusters occur which are enriched by silver atoms. They are not distributed at random, but have a distinct smallest distance from each other. This can be concluded from the small angle scattering, which is a ring as shown in Figure 6. From this intensity distribution, one can calculate the diameter of the cluster (Region I), and the diameter of the surrounding Region II, where no second cluster is to be found.[11] This will not be discussed here.

The same intensity rings as around the spot (000) must be found around each spot *(hkl)*, and if there is no lattice distortion, the intensity must be the same on all sides of the spot. Figure 7 shows the intensity distribution from a coarse-grained sample in the surroundings of the Debye-Scherrer line (400). In Figure 7a the foil contained spherical clusters in the solid solution (after aging for 4 hours at 150°C). Sidebands occur on both sides of the

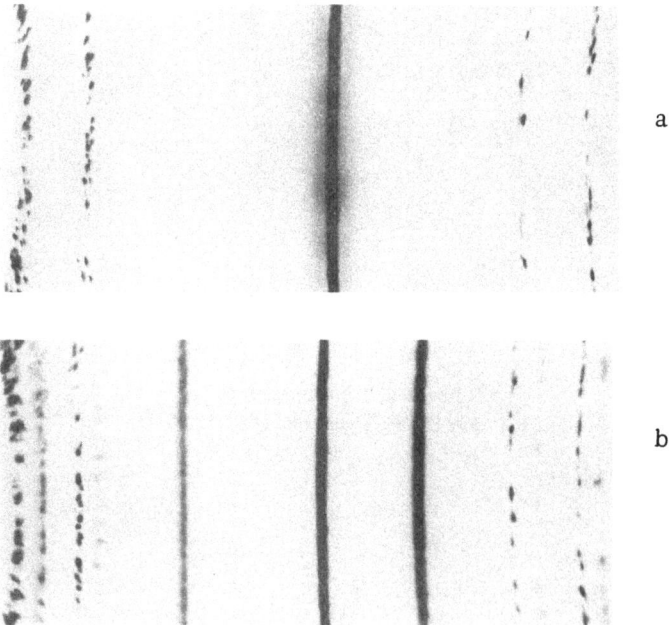

Figure 7. Debye-Scherrer radiogram from a coarse-grained foil Al + 5.9 at.% Ag. Monochromatic CuK_α radiation. a) Solid solution containing spherical zones. Sidebands along the strong line (400). b) Equilibrium state (solid solution + hexagonal precipitation). No sidebands along (400).

intensive line (400), due to the clusters. For comparison, Figure 7b shows a picture from the same foil after a heat treatment at 300°C, where precipitation has occurred. The sidebands have disappeared. In Figure 7a the intensity of the bands is different on the two sides of the line (400). Therefore lattice distortion must have occurred. As the atomic diameters of aluminum and silver atoms are nearly the same, these distortions must be very small.

To get an estimation for the lattice distortion, we use our one-dimensional model of clustering. As α_I is very small in this case, the term I_D in equation (11) can be neglected. The main asymmetry factor is given by the term I_M, which must be positive at the negative side of the spot (400). As $f_I > f_{II}$, the distortion constant α_I must be positive, i.e., the lattice parameter in the spherical silver-rich clusters must have increased a little.

As the diameter of the surrounding Region II is nearly twice the diameter of the cluster,[11] we take $c_I = 0.5$ in our one-dimensional model. Using equation (7) and the value $g_\nu = \pm 2/L$, we get an estimate for the intensities $I^{(+)}$ and $I^{(-)}$ on both sides of the line (400). As α_I is small, the $\sin^2$ term has the value 1, and equation (7) can be reduced to

$$I^{(\pm)} \sim \left[\frac{\pm(f_I - f_{II}) - L\alpha_I h\bar{f}}{\left(\pm 1 + \frac{L}{2}\alpha_I h\right)\left(\pm 1 - \frac{L}{2}\alpha_{II} h\right)} \right]^2 .$$

The ratio $I^{(-)}/I^{(+)}$ is given by

$$\frac{I^-}{I^+} = \left(\frac{1 + Z}{1 - Z} \right)^2 ,$$

where $Z = \dfrac{\alpha_I}{f_I - f_{II}} Lh\bar{f}.$

As the ratio $I^{(-)}/I^{(+)}$ estimated from Figure 7a is about the factor 2, we get $Z \approx 0.17$. Using the values $h = 4$ [matrix spot (400)], $L = 60$ (the diameter of the surrounding Region II is about 120 A, as estimated from the small angle diagram),

$$f_I \approx 0.5 f_{Ag} + 0.5 f_{Al} \approx 2 f_{Al},$$

and

$$f_{II} \approx \bar{f} \approx f_{Al},$$

we find finally $\alpha_I \approx 7 \cdot 10^{-4}$. This value gives only the order of lattice deformation. Calculations with a three-dimensional model show that this value must be reduced by about 50%.

As we can see, the intensity ratio is a function of $\alpha_I/(f_I - f_{II})$. In a first approximation α_I is proportional to $(f_I - f_{II})$, i.e., the distortion constant α_I is proportional to the concentration difference of silver atoms in and outside the clusters. Therefore the ratio $I^{(-)}/I^{(+)}$ should be independent of $(f_I - f_{II})$. Only the absolute intensities are influenced by this factor. This has also been found experimentally. A short heat treatment at 200°C reduces the intensity of the small angle scattering[12] and also of the sidebands for a factor 0.5, because the silver concentration in the clusters is reduced. The ratio $I^{(-)}/I^{(+)}$ of the sidebands remains unchanged by the heat treatment.

The Alloy Aluminum—Magnesium—Zinc. The investigation of the preprecipitation states in this aluminum-rich alloy is very difficult, since at the same time nuclei of metastable precipitation states occur. On radiograms intensity contributions from two different states are present; therefore, the interpretation is sometimes very difficult. Nevertheless, we were able to solve the structure of the preprecipitation states. Two different alloys were investigated:

1. A magnesium-rich alloy containing 4 at.% Mg and 2.7 at.% Zn with a ratio Zn:Mg=10:15.[13]
2. A magnesium-poor alloy containing 1 at.% Mg and 3.3 at.% Zn with a ratio Zn:Mg=10:3.[14]

With the magnesium-rich alloy, superstructure spots have been found[13] with the indices (001), (003), (201), (221), (110), etc., indicating a superstructure of the CuAu I type. The first two indices of the triple are shifted about 15% to smaller values, indicating a tetragonal distortion in the superstructure lattice. The model of a cluster giving such experimental results is shown in Figure 8. It is a MgZn superstructure. The tetragonal distortion is due to the magnesium atoms. An increase of the two lattice constants parallel to the layers of magnesium is produced. Due to the large lattice distortion, the diameter of the clusters is limited to 20 to 25 Å. An investigation of the intensity distribution in the surroundings of the matrix spots was not possible, the intensity was too weak.

In the magnesium-poor alloy the lattice distortion in the clusters is reduced by the substitution of zinc atoms for magnesium atoms. A more complicated structure occurs which is shown in Figure 9. Antiphase domains have been built up in

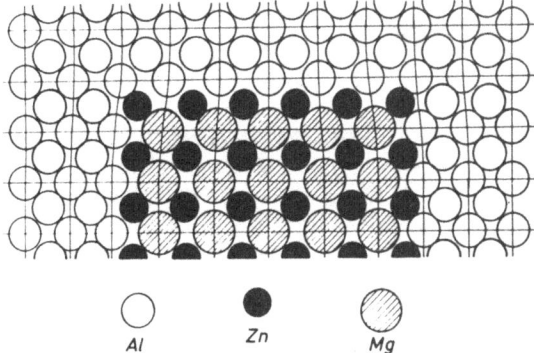

Figure 8. Model of clusters in a magnesium-rich
alloy Al—Mg—Zn. Section parallel to (020).

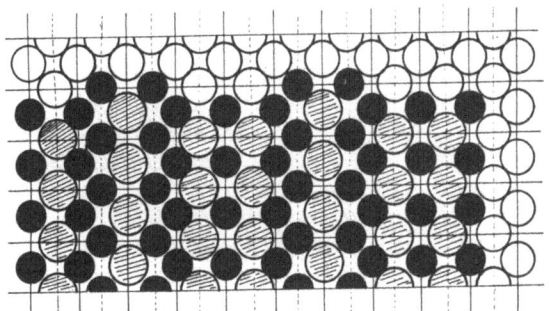

Figure 9. Model of clusters in a magnesium-poor
alloy Al—Mg—Zn. Section parallel to (020).

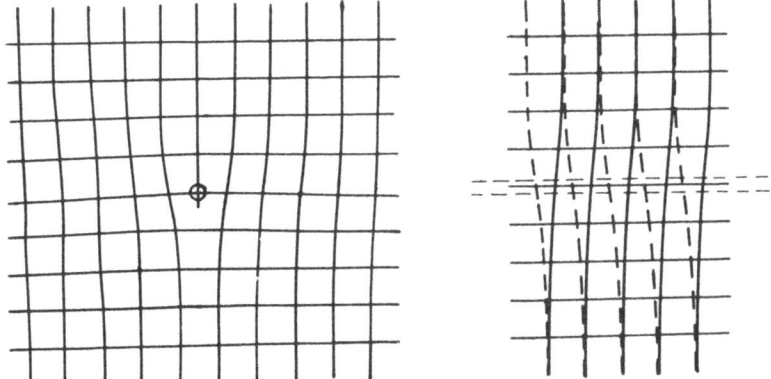

Figure 10. Lattice distortions around an edge dislocation (left) and a
screw dislocation (right).

periodic distances similar to the AuCu II structure. At the domain boundaries all magnesium atoms have been substituted by the smaller zinc atoms. The structure may be described in the following way.

The domain boundaries are parallel to (200), their distances are $2a_0 = 4d_0$, where a_0 is the lattice constant of the matrix and d_0 the distance of the (200) planes. Each domain consists of four (200) planes: one boundary plane and three internal planes. The internal planes have an ordered MgZn structure, where the alternating planes of Mg and Zn atoms are parallel to (002), i.e., perpendicular to the boundary planes. In neighboring domains the position of Mg and Zn atoms has been changed. All boundary planes are pure Zn planes. The structure has the composition Mg_3Zn_5.

The X-ray diffraction photographs from these clusters show two different effects.

1. The superstructure spots $(h'k'l')$ discussed above are split into spots $(h' \pm g', k', l')$, where $g' = 0.25$, indicating the antiphase structure with a period $4a$ in the x direction.

2. The matrix spots (hkl) have side spots with the indices $(h \pm g, k, l)$, where $g = 0.5$, indicating the pure Zn planes at the domain boundaries with a period $2a_0$.

Due to the different sizes of the Zn and Mg atoms, the boundary planes have a smaller thickness than the internal planes. This can be concluded from the different intensities of the side spots on the positive and negative side of a matrix spot. In most cases the spots on the negative side were not observed. Equation (7) was useful in solving the structure. More details are given in the original paper.[14]

II. THE DETECTION OF DISLOCATIONS IN GERMANIUM CRYSTALS BY X-RAYS

Dislocations are linear lattice defects. They are produced during the solidification process and also during plastic deformation. Figure 10 shows two different types of dislocations in a simple cubic lattice. At the left side an edge dislocation is drawn, the dislocation line lying perpendicular to the drawing plane. To the right there is drawn a screw dislocation, the line lying parallel to the drawing plane. In this case the lattice planes perpendicular to the dislocation line form a helicoidal surface around it as schematically shown in the figure.

A dislocation line is characterized by its Burgers vector (generally the shortest translation vector of the lattice) which is parallel to the main shift of atoms (shift from the position in an ideal lattice). In both cases shown in Figure 10 the Burgers vector has a horizontal direction, which is perpendicular to the dislocation line for an edge dislocation and parallel to the dislocation line for a screw dislocation. Generally, there are mixed dislocations present, i.e., their Burgers vectors have an arbitrary angle to the line directions.

If a dislocation line and its Burgers vector lie in a slip plane of the crystal (generally the plane with the largest spacing), this line is able to move through the crystal along the slip plane when a shear stress is acting. The movement of dislocation lines results in a plastic deformation of the crystal.[15] For germanium the slip planes are $\{111\}$; the Burgers vectors lie parallel to $<110>$ directions.

The Experimental Method. An X-ray beam making the Bragg angle with a set of lattice planes is reflected, the reflecting power depending on the deformation of this set of lattice planes. Therefore on microradiographic pictures using the scattered beam an intensity contrast is given between an undistorted region (far away from a dislocation line) and a distorted region in the surroundings of a dislocation. All dislocation lines become visible which deform the set of lattice planes used for the reflection (reflecting plane). As may be seen in Figure 10, most planes are deformed by the line, except the planes having a normal n perpendicular to the Burgers vector b. These planes show no or only a very small deformation. Therefore dislocation lines satisfying the relation

$$b n = 0$$

are not visible in such pictures.

This method has been applied recently by several authors, the reflected beam either leaving the irradiated surface (reflection method) [16,17] or going through the crystal (transmission method).[18-21] Especially in the transmission method, where all dislocation lines from the interior of the crystal are projected on the film, only crystals with a low density of dislocations can be used. Therefore the crystals must be nearly ideal crystals. Semiconductor materials satisfy this condition, because it is possible to grow crystals containing no dislocations. For comparison, a metal crystal has dislocation densities around 10^6 to 10^8 lines/cm^2.

For germanium the normal absorption of X-rays is very high. A transmitted X-ray beam from a copper-target tube running at 30 kv and 20 ma is absorbed nearly totally in a crystal of 1 mm thickness. If the direction of the beam makes the Bragg angle with a set of lattice planes going through the crystal, an anomalously high transmission occurs. The primary and the re-

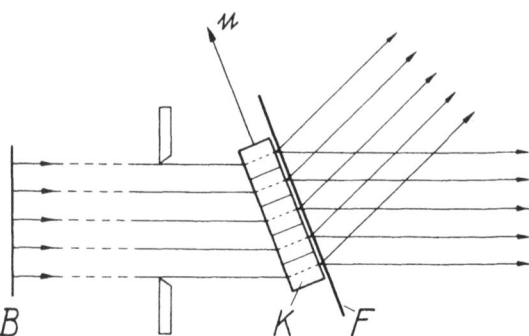

Figure 11. Experimental arrangement (schematic).
B = linear X-ray source, K = germanium crystal,
F = film, n = normal of the reflecting plane.

flected beam form a wave field in the interior of the crystal with a vector of propagation nearly parallel to the reflecting plane.[22] This wave field has a very low absorption coefficient and the intensity can be measured at the reverse side of the crystal lamella. For germanium the best intensities are obtained using (220) planes for the reflection. If a distortion of the reflecting planes has occurred, the absorption coefficient is raised enormously, and the distortion can be detected by intensity contrast.*

Our experimental arrangement is based on an arrangement published by Barth and Hosemann.[20] A linear X-ray source (copper radiation) is used. The crystal lamella is fixed in a position where the characteristic copper radiation has the Bragg angle for a (220) plane as shown in Figure 11. At each point of the irradiated surface a distinct X-ray beam satisfies this condition. All other beams with other directions and other wavelengths are absorbed nearly totally. The transmitting wave field splits up in its two components on the reverse side of the crystal.

Two improvements have been made. As the reflected beam is not exactly parallel, pictures made at a distance from the crystal have poor resolution. Therefore, in our arrangement the film is fixed immediately behind the crystal. This gives a good resolution. Sometimes there was a small distance between film and crystal. Then the contrast lines are doubled due to the split beams.

To get more details from the dislocation distribution in the crystal, the crystal can be turned around the reflecting plane normal n. Now different radiographs can be made using the same reflecting plane, the propagation of the wave field having different directions along this plane.

---

*Crystals with a low absorption coefficient show the reverse contrast.

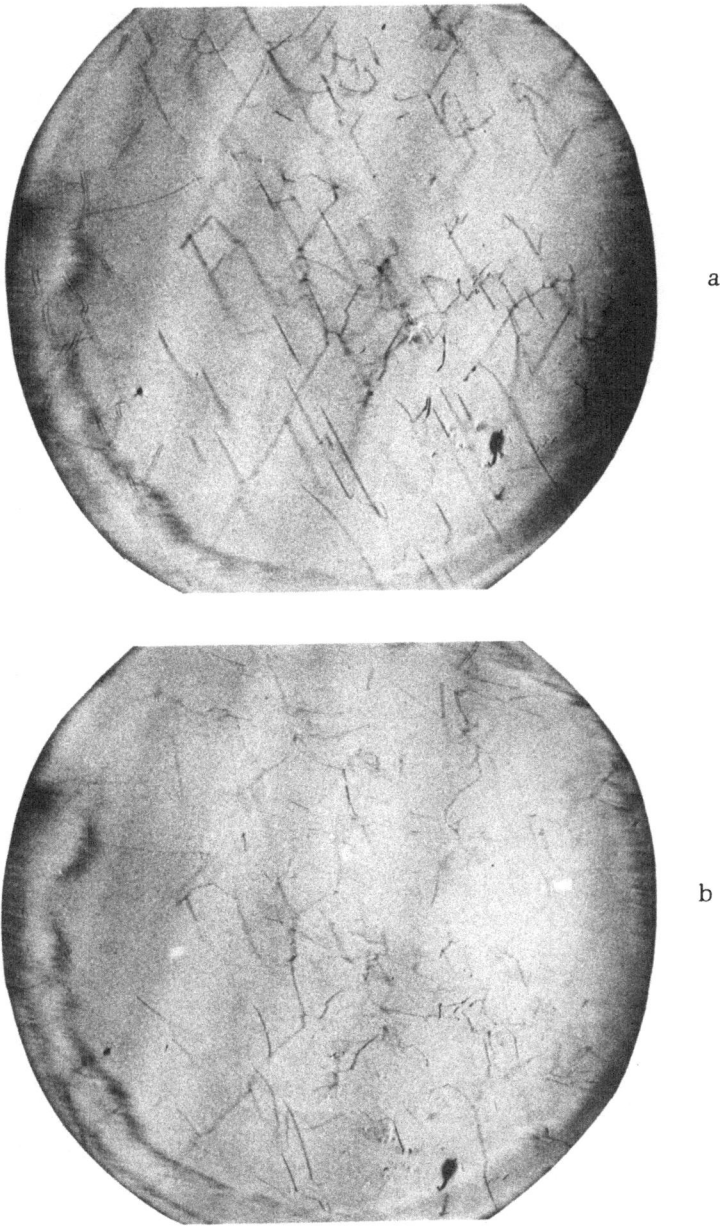

Figure 12. Transmission radiograph from a germanium crystal. Reflecting plane (220). a) Direction of projection [110]. b) Direction of projection [111].

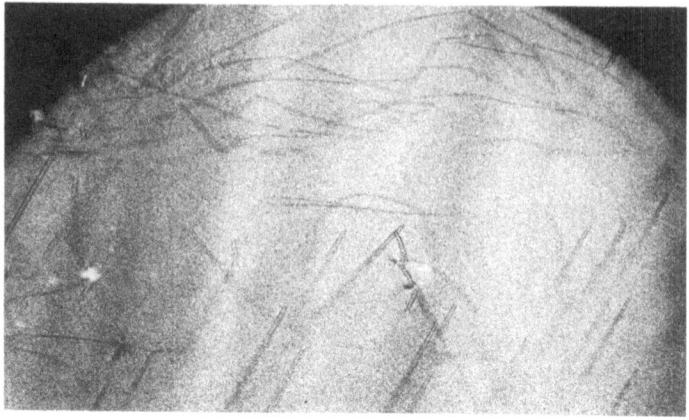

a

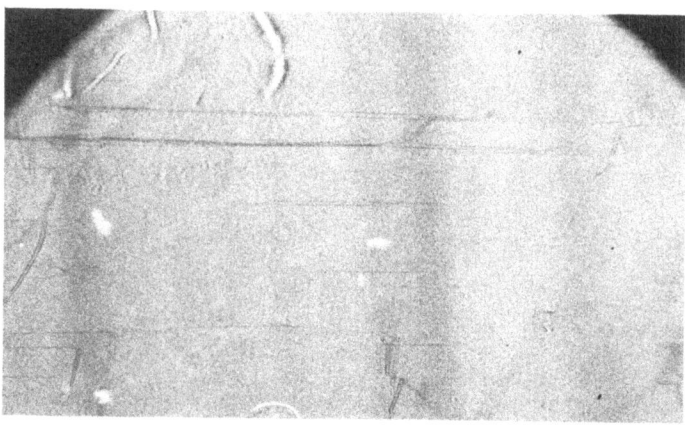

b

Figure 13. Transmission radiograph from a germanium crystal.
Reflecting plane (220). a) Direction of projection [110]. b) Direction of projection ≈[112]. (Circle in Figure 14.)

Experimental Results. Some results have been published
already.[21] The work is still in progress. Two typical radiographs
from a crystal lamella cut parallel to the growing surface of the
crystal are shown in Figure 12. The diameter of the disk is about
5 mm, the density of dislocations is very low.* Both radiographs

*The author is indebted to Dr. H.P. Kleinknecht, Radio Corporation of America,
Somerville, N.J., who has given us the crystal for our experiments.

have been made with the same reflecting plane ($\bar{2}20$), as may be seen in Figure 14. The propagation direction of the wave field is [110] and [111], respectively. The two sets of parallel straight lines in Figure 12a indicate that all these dislocations lie in one of the octahedral planes ($1\bar{1}1$) or ($\bar{1}11$), because the lines are projected on the film in the direction [110] which is parallel to both planes. The dislocations are curved lines in these planes as may be concluded from Figure 12b. Figure 13 shows two radiographs from a lamella cut from the end of the crystal. The reflecting plane was the same as in Figure 12, the direction of projection is indicated in the caption. The dislocation group at the top of the radiograph lies in the ($11\bar{1}$) plane, as indicated by Figure 13b, the dislocations below in ($1\bar{1}1$), as indicated by Figure 13a.

Parallel to each reflecting plane $\{220\}$ there is only one direction <110>. On a radiograph made with any reflecting plane $\{220\}$ all dislocations remain invisible which have a Burgers vector parallel to the one <110> direction parallel to the plane. In changing the reflecting plane, it is possible to determine the Burgers vector from a distinct dislocation line.

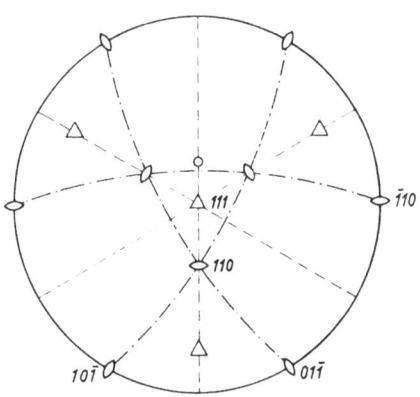

Octahedral planes ·–··–··–··–·

Reflecting planes – – – – – – – –

Figure 14. Stereographic projection from the crystal lamellae. Surface normal [111].

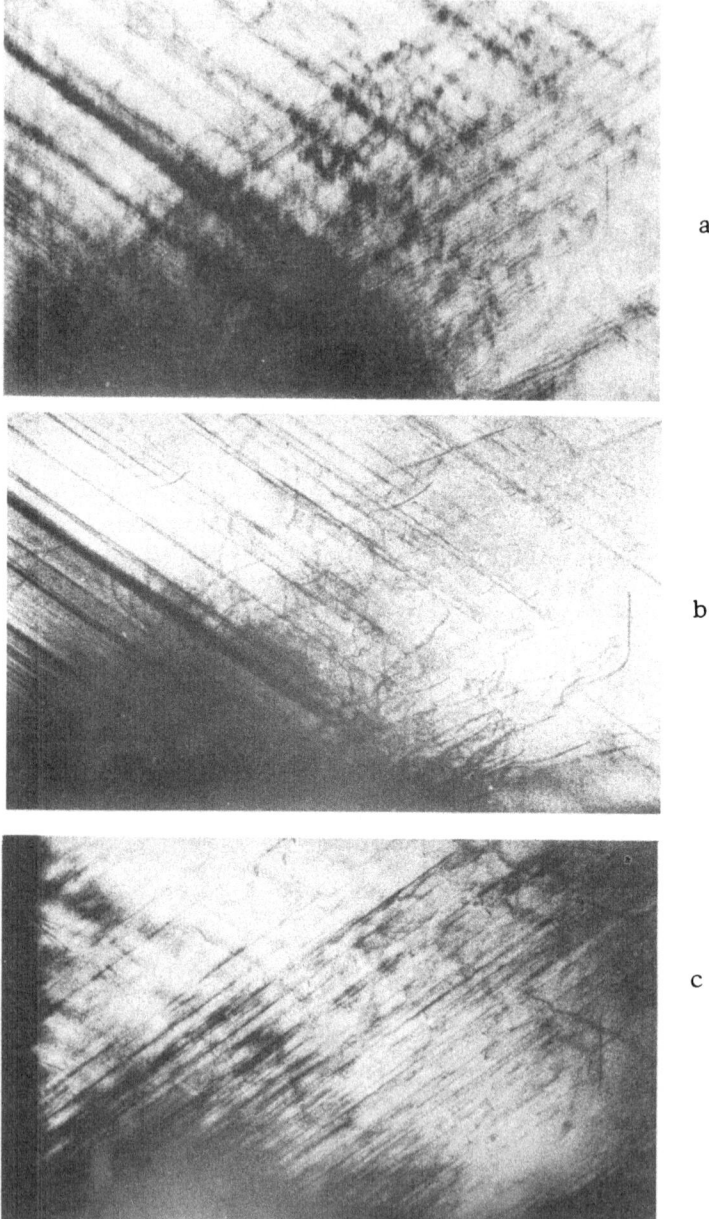

a

b

c

Figure 15. Transmission radiograph from a plastically deformed germanium crystal. a) Dislocations in both slip systems (A and B) are visible. b) Only dislocations from the slip system A are visible. c) Only dislocations from the slip system B are visible.

This may be proved on two groups of dislocations, each group having a distinct Burgers vector. Such groups can be produced by plastic deformation. The crystal was pressed at high temperatures, where germanium becomes plastic. When the pressure was very low, no macroscopic deformation occurred. Only some groups of dislocations were introduced into the crystal from the loading surface. The crystal was cut in such a manner that two different groups of dislocations are introduced, lying in different slip planes and having different Burgers vectors (double slip). Figure 15 shows three radiographs from the same crystal area made with different reflecting planes. In Figure 15a, both groups of dislocations can be observed, whereas in Figures 15b and c, only the one is visible. The Burgers vectors determined in this way are in agreement with the theory of plastic deformation.

In the same way, the Burgers vectors from single dislocations can be determined in undeformed crystals. It comes out that about 80 to 90% of all dislocations lie in octahedral planes with the Burgers vector in the same plane as the dislocation line. All these lines can move and have been introduced into the crystal during the growth by slip from the growing surface due to thermal stresses. Moreover, in the octahedral plane parallel to the growing surface no dislocations are present. The rest of the dislocation lines are real grown-in dislocations and cannot move because they do not lie in octahedral planes.

ACKNOWLEDGMENT

The author is grateful to Dr. H. Schmalzried and Mr. H. Haberkorn for their assistance with the experiments described in Part I, and to Mr. F. Meier for his assistance with the experimental mounting and taking the radiographs shown in Part II. The author would like to express his appreciation to the Deutsche Forschungsgemeinschaft for the support of this program.

APPENDIX

Substituting $h = H + g$, where H is an even number, equation (5) is changed to

$$F(h) = \sum_n f_n \exp \pi i \ (ng + u_n h). \tag{a}$$

Substituting n and u_n from equation (4) and assuming a periodic arrangement with M periods of L layers ($N = ML$ is the total number of layers), we get

$$F(h) = \left\{ f_I \exp\left[\pi i(\alpha_I - \alpha_s)h\right] \sum_{n=c_I L}^{-i} \exp \pi i n(g + \alpha_I h) + \right.$$

$$\left. + f_{II} \sum_{n=0}^{c_{II}L-1} \exp \pi i n(g + \alpha_{II}h) \right\} \sum_{m=0}^{M-1} \exp \pi i m L g. \qquad (b)$$

The sum over m has the result

$$\frac{\sin \dfrac{\pi}{2} LMg}{\sin \dfrac{\pi}{2} Lg} ,$$

where the phase factor is neglected. For large M this term is noticeable only at the positions

$$g_\nu = 2\nu/L, \qquad \nu = 0, \pm 1, \pm 2, \ldots . \qquad (c)$$

The sum over n is given by the general formula

$$\sum_{n=0}^{N-1} \exp inx = \frac{\sin(Nx/2)}{\sin(x/2)} \exp (N-1)ix/2. \qquad (d)$$

The first factor in equation (b) is given by

$$f_I \exp\left[\pi i(\alpha_I-\alpha_s)h\frac{\pi i}{2}(c_I L + 1)(g + \alpha_I h)\right] \frac{\sin\left[\dfrac{\pi}{2} c_I L(g + \alpha_I h)\right]}{\sin\dfrac{\pi}{2}(g + \alpha_I h)} +$$

$$+ f_{II} \exp\left[\frac{\pi i}{2}(c_{II}L - 1)(g + \alpha_{II}h)\right] \frac{\sin\left[\dfrac{\pi}{2} c_{II}L(g + \alpha_{II}h)\right]}{\sin\dfrac{\pi}{2}(g + \alpha_{II}h)} . \qquad (e)$$

Both terms are multiplied by

$$\exp\left[-\pi i(\alpha_I - \alpha_s)h + \frac{\pi i}{2}(c_I L + 1)(g + \alpha_I h)\right] .$$

Then in the first term the exponential factor becomes 1, and in the second term it becomes

$$\exp \frac{\pi i}{2}\left\{\left[2\alpha_s - \alpha_I - \alpha_{II} + (c_I\alpha_I + c_{II}\alpha_{II})L\right]h + (c_I + c_{II})Lg\right\},$$

which reduces to $\exp\left[(\pi i/2)Lg\right]$ due to relations (2) and (3). Substituting $g = g_\nu = 2\nu/L$ from equation (c), the factor

$$\sin\left[\frac{\pi}{2}c_{II}L(g + \alpha_{II}h)\right]\exp\left[\frac{\pi i}{2}Lg\right]$$

can be changed in $-\sin\left[\dfrac{\pi}{2}c_I L(g + \alpha_I h)\right]$.

The final result is given by equation (6).

REFERENCES

[1] A. Guinier, Nature Vol. 142, 1938, p. 569.

[2] G. D. Preston, Proc. Roy. Soc. Vol. A 167, 1938, p. 526.

[3] A. J. Bradley, Proc. Phys. Soc. Vol. 52, 1940, p. 80.

[4] V. Daniel and H. Lipson, Proc. Roy. Soc. Vol. A 181, 1943, p. 368 and Vol. A 182, 1944, p. 378.

[5] M. E. Hargreaves, Acta Cryst. Vol. 4, 1951, p. 301.

[6] T. J. Tiedema, J. Bouman, and W. G. Burgers, Acta Met. Vol. 5, 1957, p. 534.

[7] V. Gerold, Z. Metallkde. Vol. 45, 1954, p. 599.

[8] K. Toman, Acta Cryst. Vol. 10, 1957, p. 187.

[9] V. Gerold, Acta Cryst. Vol. 11, 1958, p. 230.

[10] K. Toman, private communication, 1959.

[11] B. Belbeoch and A. Guinier, Acta Met. Vol. 3, 1955, p. 370.

[12] V. Gerold, Z. Metallkde. Vol. 46, 1955, p. 623.

[13] H. Schmalzried and V. Gerold, Z. Metallkde. Vol. 49, 1958, p. 291.

[14] V. Gerold and H. Haberkorn, Z. Metallkde. Vol. 49, 1959, in press.

[15] For details see, for example, Cottrell, Dislocations and Plastic Flow in Crystals, Oxford, Clarendon Press (1953).

[16] J. B. Newkirk, Phys. Rev. Vol. 110, 1958, p. 1465.

[17] U. Bonse, Z. Physik Vol. 153, 1958, p. 278.

[18] A. R. Lang, Acta Cryst. Vol. 12, 1959, p. 249.

[19] G. Borrmann, W. Hartwig, and H. Irmler, Z. Naturforschg. Vol. 13a, 1958, p. 423.

[20] H. Barth and R. Hosemann, Z. Naturforschg. Vol. 13a, 1958, p. 792.

[21] V. Gerold and F. Meier, Z. Physik Vol. 155, 1959, p. 387.

[22] G. Borrmann, G. Hildebrand, and H. Wagner, Z. Physik Vol. 142, 1955, p. 406.

TEXTURES IN EXTRUDED URANIUM

R. B. Russell

Nuclear Metals, Inc., Concord, Massachusetts

ABSTRACT

By means of crystallographic (inverse) pole figures for extruded uranium rods and tubes, the effect of extrusion variables on the texture can be examined. The principal variables are: prior texture, billet and liner temperatures, reduction in area, and ram speed. The final texture can be deeply affected by the prior texture, since two extrusions made under the same conditions but with different prior textures can be markedly different.

In general, an increase in extrusion temperature has an effect similar to an increase in reduction ratio. Ram speed has no important effect between 13 and 100 in./min except that a slow ram speed allows more time for recrystallization to occur.

The effect of temperature on texture is the development of a strong 110 axial texture for relatively high temperature (above about 525°C), a strong 010 axial texture for relatively low temperatures (below 400°C), and a mixture of 110 and 010 axial textures at intermediate temperatures, although these textures may be somewhat complicated by 130 axial recrystallization textures.

The variation of axial texture along the extrusion length and parallel to the radius of extruded tubes may be large.

A 900°C gamma-phase extrusion is shown to have a mild 100, 010, 021, 001 texture.

INTRODUCTION

The purpose of this investigation was to find what textures are produced by the extrusion of uranium rods and tubes under various conditions of prior billet condition and extrusion variables, as a prelude to studying the effect of texture on mechanical, physical, and chemical properties. For example, some work has already been done on the prediction of thermal expansion, based on preferred orientation data supplied by the modified Harris analysis.[1]

The effect of cold work on extruded uranium can be seen from Table I. We observe that as the amount of strain at room temperature is increased, the coefficient of thermal expansion

---

* This work was performed under United States Atomic Energy Commission Contract No. AT(30-1)-1565.

[1] Superscripts pertain to references at the end of the paper.

is decreased. Because of the anisotropic nature of a single crystal of uranium in thermal expansion, this can only mean that the amount of b component in the texture is increased as the strain is increased, because the coefficients of thermal expansion of the a and c crystallographic directions are large positive numbers, whereas the coefficient for the b direction is a small negative number. The difference in texture between the as-extruded tube and a portion of the tube wall strained 17.5% axially may be seen in the inverse or crystallographic pole figures for the axial direction shown in Figures 1a and 1b. It is seen that the amount of 010 pole density has increased substantially in the strained section compared with the extruded section.

Table I.   Mean Linear Coefficients of Thermal Expansion
Parallel to the Axial (Extrusion) Direction for
Tube As-Extruded and Cold-Worked by Axial Strain

% Extension in 2 in.	Coefficient of thermal expansion $\times 10^6 \times °C$ (25-300°C)		
	Cycle 1	Cycle 2	Cycle 3
0	15.64	17.61	– See Fig. 1a
1.6	11.70	15.09	15.16
1.7	10.78	16.44	13.84
3.9	5.78	13.38	–
7.1	3.55	10.98	12.06
9.1	4.17	11.83	13.21
13.1	7.55	8.75	8.20
17.5	–	5.08	4.12 See Fig. 1b
18.0	3.25	5.83	7.57

The effect of strain in the axial, radial, and tangential directions is shown in Figures 1, 2, and 3, respectively. The similarity between these figures and those predicted by Calnan and Clews[2] is reasonably good. The axial pole figure for as-extruded uranium (Figure 1a) corresponds to these authors' pole figures for a tension texture resulting from a large amount of slip and some twinning, leading to strong 110 and weak 010 textures, respectively, whereas the cold-worked axial texture (Figure 1b) corresponds to their predicted tension texture for less slip and more twinning, leading to weak 110 and strong 010 textures.

The r a d i a l pole figure for the as-extruded metal (Figure 2a) agrees with their prediction of a compression texture for twinning and a large amount of slip leading to 150-152 texture

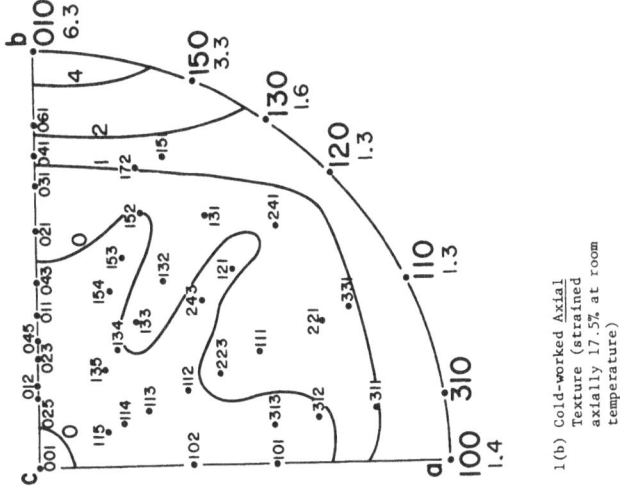

1(b) Cold-worked Axial
Texture (strained
axially 17.5% at room
temperature)

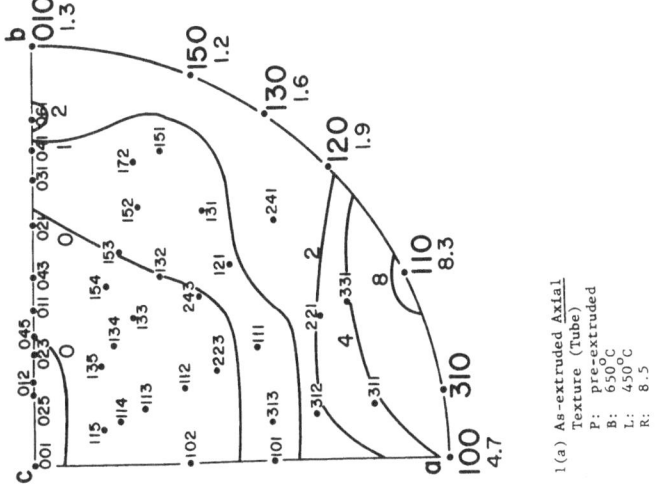

1(a) As-extruded Axial
Texture (Tube)
P: pre-extruded
B: 650°C
L: 450°C
R: 8.5

Figure 1.

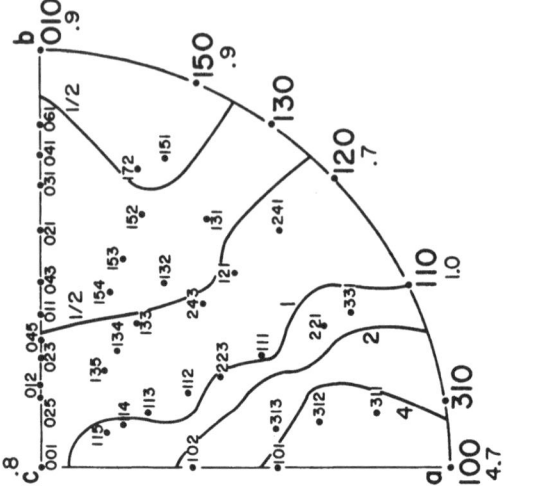

2(b) Cold-worked Radial
      Texture

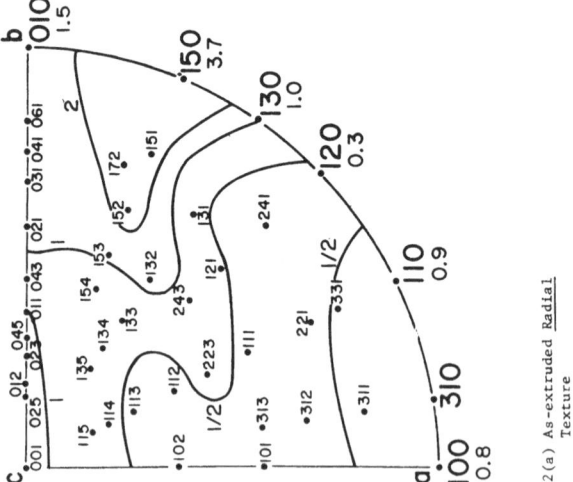

2(a) As-extruded Radial
      Texture

Figure 2.

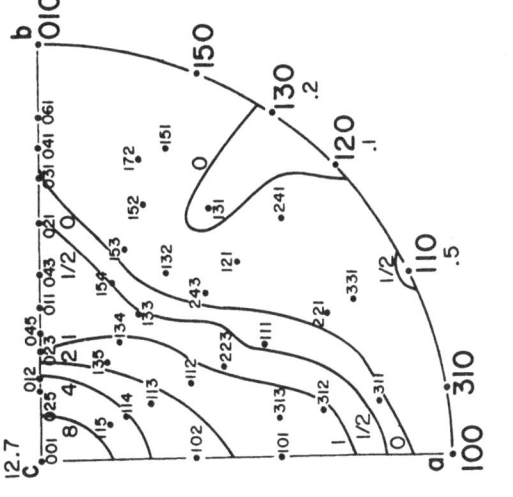

3(b) Cold-worked <u>Tangential</u>
Texture

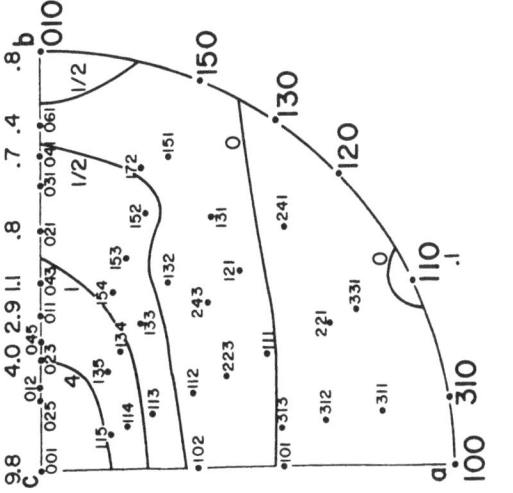

3(a) As-extruded <u>Tangential</u>
Texture

Figure 3.

concentrations. The radial pole figure for the cold-worked metal (Figure 2b) agrees moderately well with their compression textures for twinning and some slip leading to a 101 texture.

The t a n g e n t i a l textures for the as-extruded and strained rods are shown in Figures 3a and 3b, respectively. In each case, the principal texture is 001, except that the as-extruded tangential texture (Figure 3a) extends toward 010, whereas the cold-worked tangential texture (Figure 3b) extends toward 100.

A later investigation on a series of axial strains from 2 to 16% at 400°, 300°, and room temperature introduced into previously extruded uranium showed that slip is more prominent at 400° than at 300°C, where twinning becomes important. At room temperature, twinning is still more important, as shown by microexamination and the increase of pole densities toward the 010 pole.

EFFECT OF EXTRUSION VARIABLES ON ROD TEXTURES

The effect of extrusion variables on alpha-extruded uranium is difficult to demonstrate clearly, for the simple reason that the factors that produce a given texture are themselves variable and sometimes difficult to determine. The texture depends principally on the following:

    a.   prior texture,
    b.   billet temperature,
    c.   liner temperature,
    d.   reduction in area, and
    e.   ram speed.

These variables, which in themselves are often indeterminate — such as actual extrusion temperature — have regularly been observed to react with each other, so that two different textures may be produced by extrusion at the same temperature as a result of having different prior textures. Furthermore, the texture of the starting billet (prior texture) may vary along its length as well as in the radial and circumferential directions. These variations are caused by differences in temperature and degree of working of the original billet during its prior history. To avoid some of these difficulties, some of the experiments were conducted with triple beta-quenched uranium billets so as to have an unbiased or randomly oriented set of grains which would not, therefore, mask the individual effect of given extrusion variables.

In general, however, we find that an extrusion variable which tends to increase the temperature induces an increasing 110 pole component, and that a variable which tends to decrease the temperature induces an increasing 010 component.

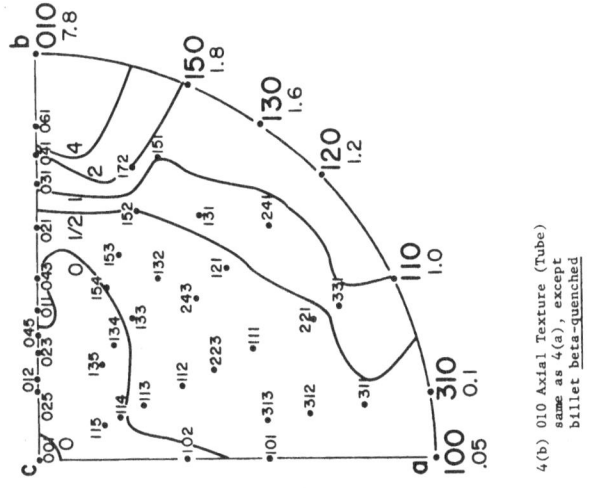

4(b) 010 Axial Texture (Tube)
same as 4(a), except
billet beta-quenched

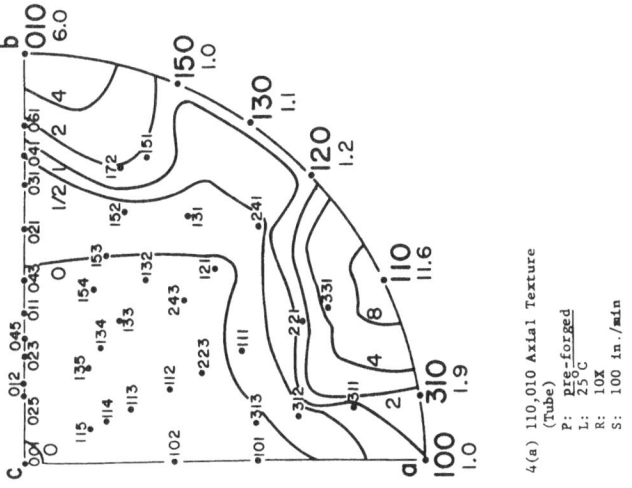

4(a) 110,010 Axial Texture
(Tube)
P: pre-forged
L: 25°C
R: 10X
S: 100 in./min

Figure 4.

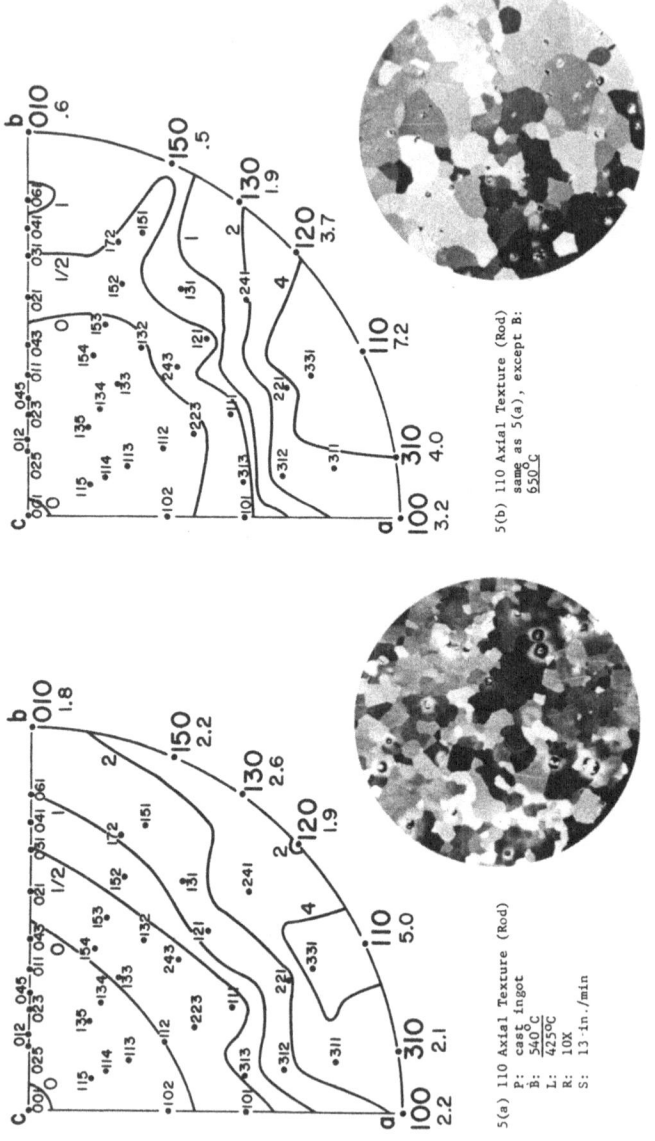

5(a) 110 Axial Texture (Rod)
P: cast ingot
B: 540°C
L: 425°C
R: 10X
S: 13-in./min

5(b) 110 Axial Texture (Rod)
same as 5(a), except B:
650°C

Figure 5.

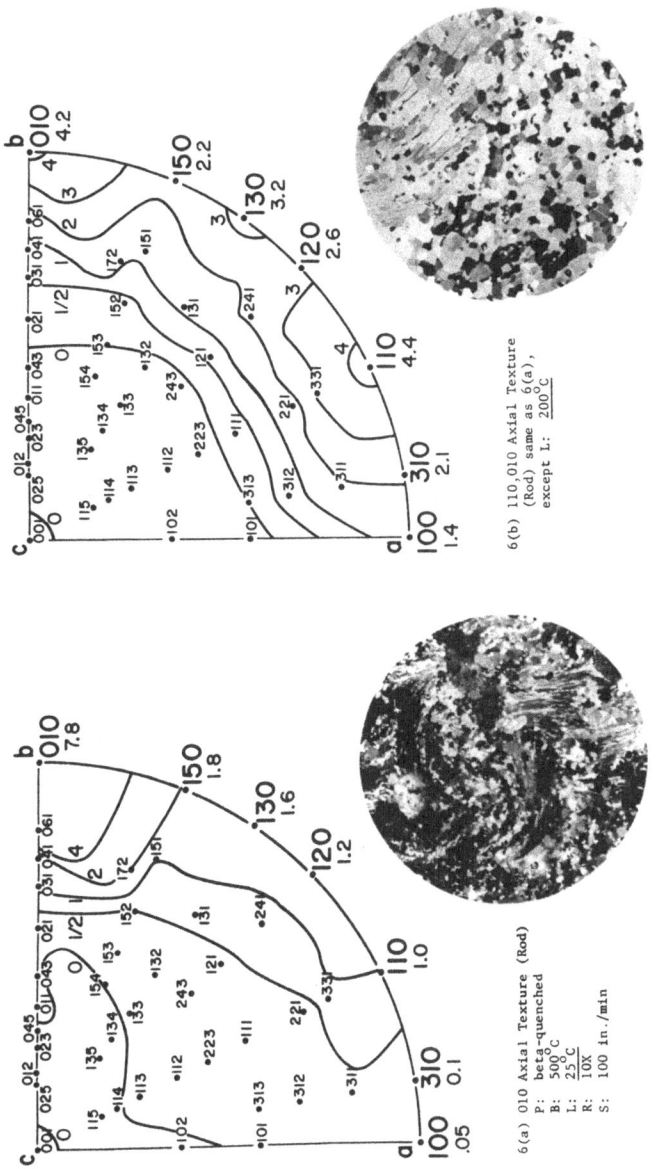

6(b) 110,010 Axial Texture
(Rod) same as 6(a),
except L: 200°C

6(a) 010 Axial Texture (Rod)
P: beta-quenched
B: 500°C
L: 25°C
R: 10X
S: 100 in./min

Figure 6.

Prior Texture. A case illustrating the effect of prior texture may be seen in Figure 4, showing the textures produced at about 500°C from an as-forged billet, Figure 4a, and from a beta-quenched billet, Figure 4b. We can see that an appreciable amount of 110 texture was inherited from the forged billet, whereas with an unbiased (random) texture at this temperature, a predominantly 010 texture is produced.

Billet Temperature. The effect of billet temperature on the texture of rod is demonstrated in Figures 5a and 5b for 540° and 650°C, respectively. We observe that the 100-120 region is increased, the 010 region decreased for the higher temperature, and that the grain size is about 60 μ, compared with about 20 μ for the lower temperature.

Liner Temperature. The effect of liner temperature on texture is as important as billet temperature and affects the texture similarly. The textures of two rods extruded at about 500°C (billet) but with a liner at room temperature and another at 200°C are seen in Figures 6a and 6b. The 010 region is seen to predominate in the rod from the cold liner and the microstructure shows a large amount of distortion. In the rod from the 200°C liner, the 110 component is considerably enhanced and the microstructure shows more recrystallized and equiaxed grains. There is a noticeable increase in the recrystallization 130 component in the latter rod. This has been found to arise from the recrystallization axial mode

$$010 \longrightarrow 130$$

which has often been found to change the texture of uranium.

Reduction. The effect of increasing the reduction ratio is similar to that of increasing the temperature as shown in Figures 7a and 7b. The grain size increases only slightly, from 20 to 30 μ, at the higher reduction but there is a substantial increase in the region around 110.

Ram Speed. Ram speed changes between 13 and 100 in./min have not been found to alter the texture very substantially except in the amount of recrystallization permitted. Figures 8a and 8b illustrate this difference for an isothermal 480°C (billet and liner at the same temperature) rod extruded at 13 in./min and 100 in./min. The decrease of 130 axial texture as a result of increasing the ram speed is apparent. Evidently recrystallization has not had enough time to develop to the extent that it does in the slower extrusion.

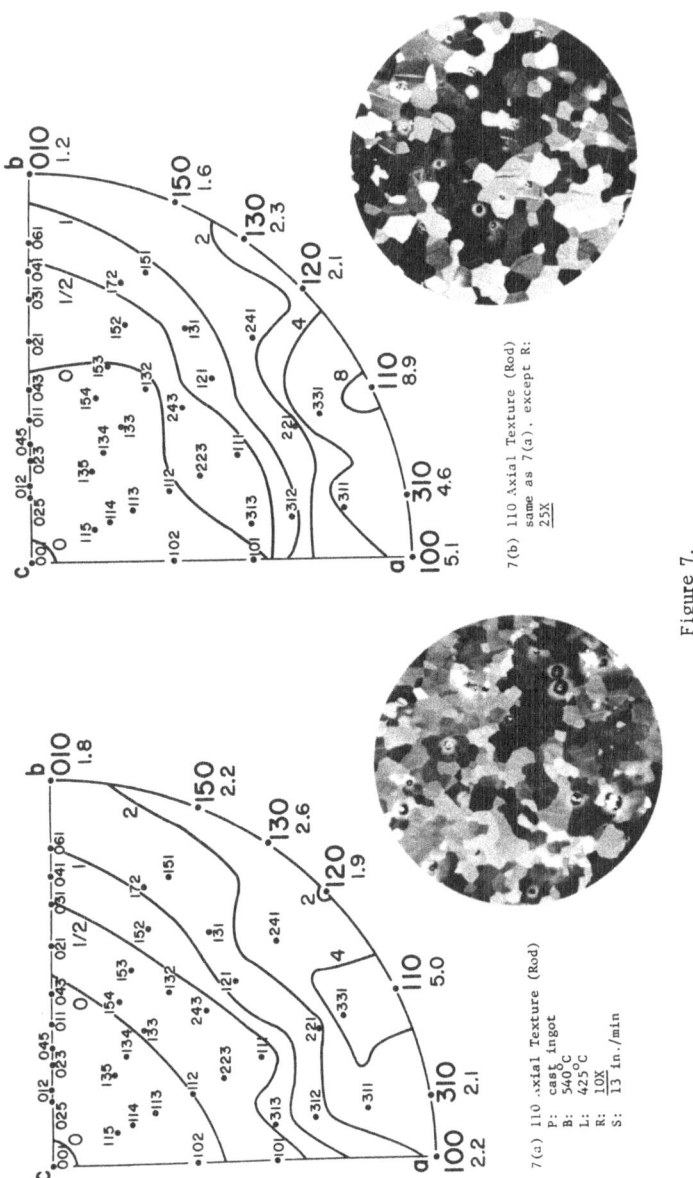

7(b) 110 Axial Texture (Rod)
same as 7(a), except R:
25X

7(a) 110 Axial Texture (Rod)
P: cast ingot
B: 540°C
L: 425°C
R: 10X
S: 13 in./min

Figure 7.

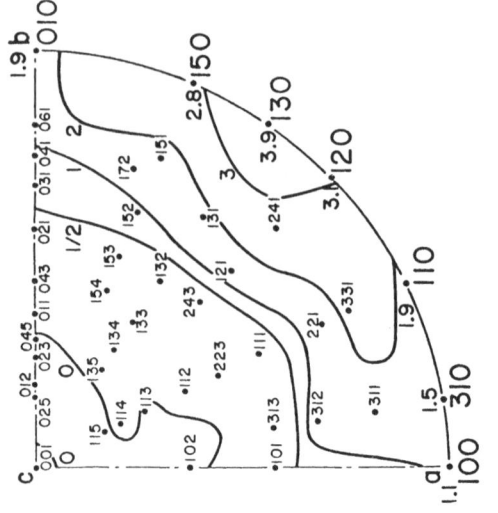

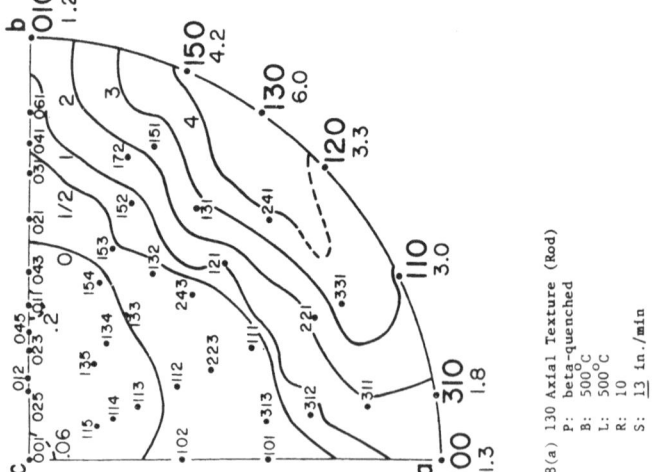

8(a) 130 Axial Texture (Rod)
P: beta-quenched
B: 500°C
L: 500°C
R: 10
S: <u>13 in./min</u>

8(b) 130 Axial Texture (Rod)
same as 8(a), except S:
<u>100 in./min</u>

Figure 8.

VARIATION OF TEXTURE WITHIN SINGLE EXTRUSIONS

Change of Texture Along the Rod Length. If a warm billet is placed in a cool liner and immediately extruded, the first part of the rod to be extruded can be expected to represent that part of the billet which has had the least time to cool to the liner temperature; the last of the rod to extrude, that part of the billet having had the most time to cool to the liner temperature. Accordingly, one can usually expect that the front of the rod has a texture developed by a relatively higher temperature than that at the rear of the rod. In Figures 9a, 9b, 9c, the pole figures for the front, middle, and rear of an extruded rod, this effect is clear, since the 110 pole density decreases markedly from front to rear whereas the 010 pole density, developed by a low temperature, increases from front to rear.

Change of Texture with Radius. The change of axial and tangential texture in the radial direction has been investigated only in tubes. The change in axial texture within a $1\frac{3}{16}$-in.-diameter tube with a $\frac{5}{16}$-in.-thick wall is shown in Figures 10, 10a, 10b, and 10c, which show the inside, middle, and outside thirds of the total wall thickness. Since this tube was extruded from a preextruded billet in which the texture variation in the radial direction was not determined, it is not possible to explain the "warmer" texture on the inside wall and the "cooler" texture on the outside, but if this pregradient had been known, the texture gradient could probably have been explained by some scheme involving the relative heat capacities of the billet with respect to those of the mandrel and liner. We can see, however, that the texture gradient along the radius can be large.

The texture in the tangential direction for the same tube is primarily 001 whose pole density was found to have decreased about 30% from inside to outside.

EXTRUSION OF URANIUM IN THE
BETA AND GAMMA PHASES

Extrusion of uranium in the beta phase has not been successful. A rod usually extrudes with a violently fluctuating pressure and the final surface is badly cracked and barklike. In some cases where the effective extrusion temperature reaches the gamma temperature (over about 770°C), this part of the extrusion will be recognized by the detachment of a long, thin, smooth rod, undoubtedly caused by the fact that gamma uranium is very much more plastic than beta uranium.

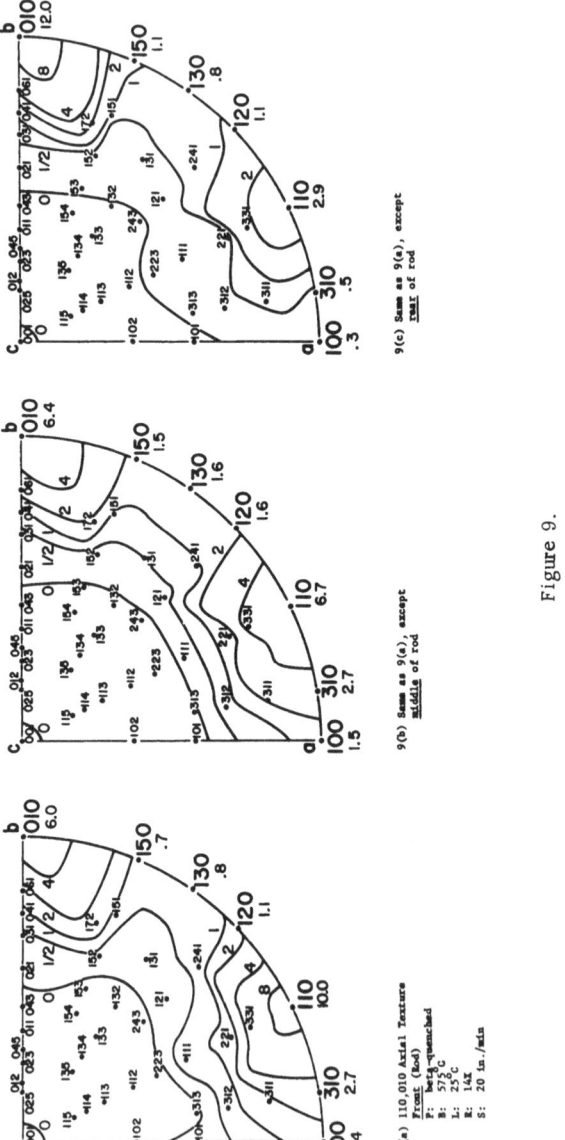

Figure 9.

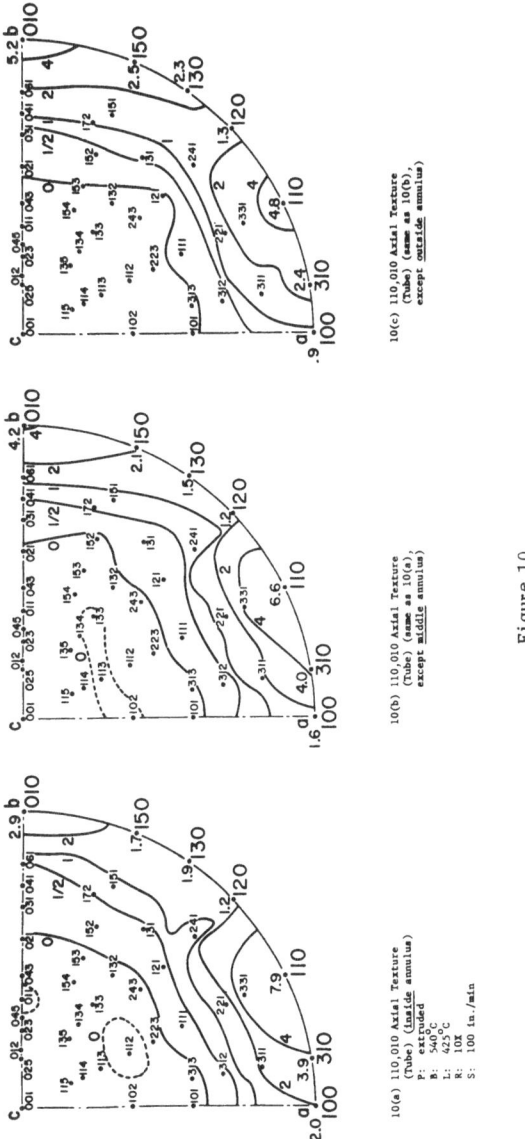

Figure 10.

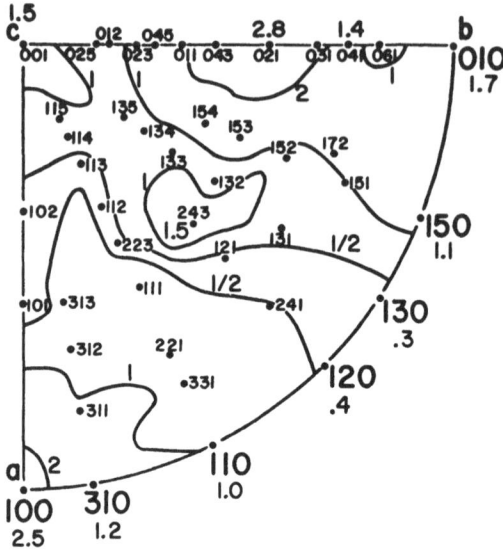

11 100,010,021,001 Axial
Texture (Rod)
P: as-cast
B: 900°C
L: 480°C
R: 25X
S: 140 in./min

Figure 11.

Figure 11 shows the relatively mild texture developed at the front of a gamma-extruded rod made from a billet at about 900°C, at a reduction of 25 times. The texture can generally be described as 100, 010, 021, 001.

REFERENCES

[1]R. B. Russell, Trans. Am. Soc. Metals, Vol. 52, 1960 p. 781.
[2]E. A. Calnan and C. J. B. Clews, Phil. Mag., Vol. 43, 1952, p. 93.

PRECISION X-RAY STRESS ANALYSIS OF URANIUM AND ZIRCONIUM

Lt. B. J. Wooden, Lt. E. C. House, and R. E. Ogilvie

Massachusetts Institute of Technology
Cambridge, Massachusetts

ABSTRACT

The feasibility of using X-ray diffraction methods to measure residual stresses in uranium and zirconium (Zircaloy-2) was investigated. A precision method was developed for the determination of diffraction peak positions and the precision associated therewith. The statistical tables of Fisher and Yates were used to determine what order polynomial provided the best least squares fit within the known precision of the observed data. It was found that a second-order polynomial provided an adequate regression. With the aid of a desk calculator less than 5 min calculation time is required to determine the peak position to a precision of $\pm 0.01°$.

The stress constant for uranium was determined to be 1308 ± 110 psi/0.01° shift in $\Delta 2\theta$ for copper radiation on the (116) planes at $2\theta = 158.3°$. The stress constant for Zircaloy-2 was determined to be 430 ± 1 psi/0.01° shift in $\Delta 2\theta$ for chromium radiation on the (10.4) planes at $2\theta = 156.4°$.

* * *

A great deal of work has been done in the past few years investigating mechanical failure of materials, particularly in attempting to correlate the material strength and fatigue endurance to the physical and metallurgical characteristics of the material. Most investigators believe the residual surface stresses to be an important parameter in this regard. In addition to fatigue, several applications of stress measurement techniques are applicable in the field of reactor design which include the following:

1. Evaluation of stresses produced by local hot spots in fuel rods.
2. Determination of the stress produced by welding and other fabrication techniques in fuel elements.
3. Determination of the proper heat treatments to remove residual stresses in the clad materials.

Several methods have been devised to measure residual stresses. Of these, only the X-ray diffraction method is non-destructive. However, the X-ray diffraction technique has never become generally accepted because of the lack of precision of the reported results. It is the firm belief of the authors that the method itself is quite reliable, but that this reliability can only be obtained by treating the experimental procedure and data with great care.

In order to determine the stress in a specimen the general procedure is first to determine the stress constant for the material if it is not already available. This constant arises from the combination of Bragg's law and classical elasticity theory.

$$\sigma = \frac{E}{1 + \nu} \cdot \frac{\Delta\theta}{\sin^2 \psi}$$
$$= K \Delta\theta \ .$$

where σ is the surface stress in the direction of ψ, E is Young's modulus, $\Delta\theta$ is the difference in the diffraction line position for the normal and oblique exposures, ν is Poisson's ratio, and K is the stress constant.

The material is stressed a known amount and the $\Delta\theta$ value measured. The slope of a straight line through several such points is the desired stress constant. With the stress constant known, the $\Delta\theta$ value can be measured in any similar material and the surface stress evaluated using the known stress constant. The $\Delta\theta$ values of 0.1° to 0.3° cover the elastic range of most materials, and it is necessary to determine $\Delta\theta$ to ±0.01°.

A precision method of diffraction peak position was developed. By use of the orthogonal polynomials in conjunction with the tables of Fisher and Yates the best least square polynomial was selected to pass through seven points spaced 0.05° apart across the peak, with the center point as near the maximum as could be determined by a slow scan across the peak. The criterion for fitting was the near equality of the mean square deviation from the fitted curve and the statistically determined variance of the observed data. A parabola was found to satisfy this criterion.

Uranium and Zircaloy-2 specimens were examined with copper, cobalt, and chromium radiations in order to determine the appropriate diffraction lines to be used. Since at a constant stress a higher-angle line will result in a greater observed value of $\Delta\theta$. the best high-angle lines were selected. These lines were:

Uranium — copper radiation; 2θ 158.3°; (116) planes.
Zircaloy-2 — chromium radiation; 2θ 156.4°; (10.4) planes.

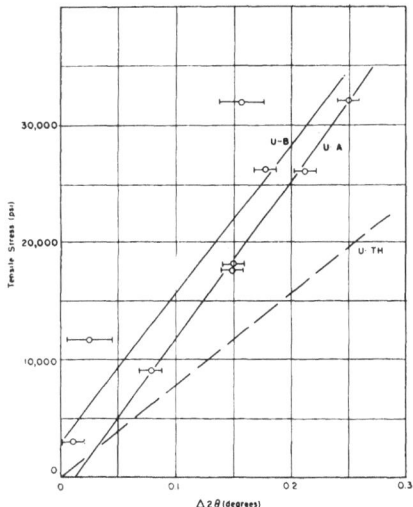

Figure 1. Calibration curve for uranium.

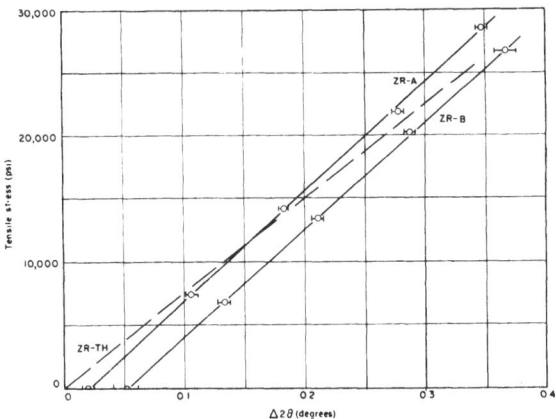

Figure 2. Calibration curve for Zircaloy-2.

A General Electric XRD-3 diffraction unit was used through-out. The specimens were vacuum annealed at 500°C for 8 hr to reduce the fabrication residual stresses to as low a value as possible. Type A-7 SR-4 strain gauges were mounted on both sides of the irradiated area, and the strains measured on a Type

Table I. X-ray Diffraction Data

$i \rightarrow$	1	2	3	4	5	6	7	
$2\theta \rightarrow$	156.50°	156.55°	156.60°	156.65°	156.70°	156.75°	156.80°	j
Time for	36.22	33.68	34.57	34.75	34.94	35.58	36.40	1
16.384	36.51	35.50	34.25	34.74	35.23	35.20	36.69	2
counts	36.56	35.42	34.87	34.27	34.83	36.22	36.49	3
(in sec)	36.91	35.64	35.34	34.69	34.36	35.71	36.26	4
Y_i	36.55	35.56	35.01	34.61	34.84	35.68	36.46	

In this array, $m = 4$ and $n = 7$.
Z' values ($Z' = hZ$) are taken from the Fisher and Yates tables for $n = 7$.

Calculation of Variance

i	y_i	y_i^2	$z_1' y_i$	$z_2' y_i$	$z_3' y_i$	$z_4' y_i$	$z_5' y_i$
1	36.55	1335.9025	−109.65	+182.75	−36.55	+109.65	−36.55
2	35.56	1264.5136	−71.12	0	+35.56	−248.92	+142.24
3	35.01	1225.7001	−35.01	−105.03	+35.01	+35.01	−175.05
4	34.61	1197.8521	0	−138.44	0	+207.66	0
5	34.84	1213.8256	+34.84	−104.52	−34.84	+34.84	+174.20
6	35.68	1273.0624	+71.36	0	−35.68	−249.76	−142.72
7	36.46	1329.3316	+109.38	+182.30	+36.46	+109.38	+36.46
	248.71	8840.1879	−0.20	+0.20	+0.04	−2.14	−1.42

$$\bar{y} = \frac{\Sigma y_i}{n} = 35.53 \qquad\qquad \bar{y}^2 = 1262.3809$$

L Baldwin strain indicator, Ser. J53470. The strains were read before and after each run and the average of the readings used. The specimens were stressed in a four-point bending device to insure a uniform stress over the irradiated area. Alignment of the specimens in the goniometer was held to 0.001 in.

The results for uranium are plotted in Figure 1. The stress constant was determined to be 1308 ± 110 psi/0.01°. This value was determined by a weighted least square correlation with the predicted linear variation.

The results for Zircaloy-2 are plotted in Figure 2. The stress constant was determined to be 430 psi/0.01°.

The results correlated quite well with the predicted linearity of stress as a function of $\Delta 2\theta$. In each case the plotted curve does not pass directly through the origin. The intercept of the curve with the ordinate axis indicates the residual stress in the specimen. The deviation of the theoretical and experimental values of the stress constant is probably due to the anisotropy of the material constants.

Table II. Analysis of Variance

Degree of fit	Sum of squares (ss)	Residual $\sum(y_i)^2 - \sum(ss)_i$	Average residual
0	$n\overline{y}^2 = 8836.6663$	3.5216	
1	$\dfrac{\left[\sum(z'_{1\,i}\,y_i)\right]^2}{\sum(z'_{1\,i})^2} = 0.0014$	3.5202	0.503
2	$\dfrac{\left[\sum(z'_{2\,i}\,y_i)\right]^2}{\sum(z'_{2\,i})^2} = 3.4648$	0.0554	0.008
3	$\dfrac{\left[\sum(z'_{3\,i}\,y_i)\right]^2}{\sum(z'_{3\,i})^2} = 0.0003$	0.0551	0.008
4	$\dfrac{\left[\sum(z'_{4\,i}\,y_i)\right]^2}{\sum(z'_{4\,i})^2} = 0.0297$	0.0254	0.0036
5	$\dfrac{\left[\sum(z'_{5\,i}\,y_i)\right]^2}{\sum(z'_{5\,i})^2} = 0.0240$	0.0014	0.0002

The correlation between the observed and calculated precision is excellent. The two curves for Zircaloy-2 were run independently of each other by the authors, and are a good indication of the reproducibility of the method.

APPENDIX

In order to obtain the maximum amount of information from the data, a preliminary study of the statistics and errors involved is essential. A regression analysis of a typical peak was accomplished using the orthogonal polynomials. For the data of

Table I the residuals of Table II were calculated. The conclusions verified the parabolic method of peak determination. With some not too involved mathematics one may determine the peak position directly from the data by evaluating

$$x_{max} = -\frac{\sum y_i z'_{1i}}{\sum y_i z'_{2i}} \cdot \frac{\sum (z'_{2i})^2}{\sum (z'_{1i})^2} \cdot \frac{h_1}{2h_2},$$

where the z's and h's are constants which depend upon the number of points used across the peak. These constants may be found in the tables of Fisher and Yates. In order to obtain the actual peak position the x_{max} value is multiplied by the point spacing and added or subtracted from the value of the center point as appropriate.

Equally important to the statistical location of the peak is the error involved in the location. This may be found by assuming Poisson statistics for each individual count and applying the law of propagation of errors.

The error, ϵ, in x_{max} was determined to be

$$\epsilon = \frac{\delta}{\left(\sum y_1 z_{2i}\right)^2} \left[\pi^2 \sum \left(z'_{1i}\right)^2 + (x_{max})^2 \sum \left(z'_{2i}\right)^2\right]^{\frac{1}{2}},$$

where δ is the error in each point and the z's and h's may be determined as noted previously.

As an example, for the data of Table I,

$$x_{max} = -\frac{(-0.2)}{17.06} \cdot \frac{84}{28} \cdot \frac{1}{2} = 0.0176,$$

$$\delta = 0.137,$$

and

$$\epsilon = \frac{0.137}{(17.06)^2} \left[\frac{9}{4} \cdot 28 + 0.003 \cdot 84\right]^{\frac{1}{2}} = 0.00406.$$

Utilizing a spacing of 0.05° with the center point at 156.65°,

$$2\theta = 156.65° + \left(x_{max} \pm \epsilon\right) 0.05 = 156.651° \pm 0.002.$$

The error is reported as 0.002° since this value is the quoted error for the goniometer.

INFLUENCE OF GONIOMETRIC ARRANGEMENT AND ABSORPTION IN QUALITATIVE AND QUANTITATIVE ANALYSIS OF POWDERS BY X-RAY DIFFRACTOMETRY

J. Leroux and M. Mahmud

Department of National Health and Welfare
Ottawa, Canada

ABSTRACT

A theoretical approach of some factors influencing the intensity diffracted by a polycrystalline material at a definite Bragg angle has been confirmed by experimental data obtained with a high-angle Norelco diffractometer. The factors mainly considered are the focusing arrangement of the goniometer which delineates the geometric shape of the volume fraction of the sample being irradiated and the absorption of the latter.

Tests have been performed (a) with quartz mixed in samples covering a large range of absorption coefficients, (b) with two different K radiations using copper and molybdenum targets, (c) with three different angles of beam divergence of 1, $\frac{1}{2}$, and $\frac{1}{4}°$, respectively, and (d) with the sample packed and leveled in a copper grid with openings of about $350\,\mu$. As this test with copper grids was to demonstrate that each opening was acting like a volume of sample irradiated by a beam of extremely small divergence (0.1°), it also shows that more accuracy in the measurement of the exact Bragg angle can be reached in these conditions as compared to that obtained with conventional sample holders.

A general equation, suitable for qualitative and quantitative analysis when these factors have to be taken into consideration, is proposed.

*   *
*

It is well known that the total intensity diffracted and reflected by a powder at a definite Bragg angle 2θ is largely influenced by the linear absorption coefficient of the powder.

In terms of quantitative analysis, this influence is made evident in Figures 1 and 2 where the intensity of quartz is plotted against its concentration in mixtures of beryllium oxide and titanium dioxide. The two external curves called "ideal" have been calculated from the "ideal" expression

$$x = \frac{I_1}{\left(I_1\right)_0} \left(\frac{\mu_s^*}{\mu_1^*}\right) , \tag{1}$$

which has been derived from an expression previously proposed by Klug and Alexander,[1] giving the relationship between the concentration, the intensity diffracted, and the mass absorption coefficient of the sample.

Equation (1) was the base of the theoretical considerations which yielded to the direct quantitative analysis by an absorption-diffraction technique published by our laboratory in two previous papers.[2,3]

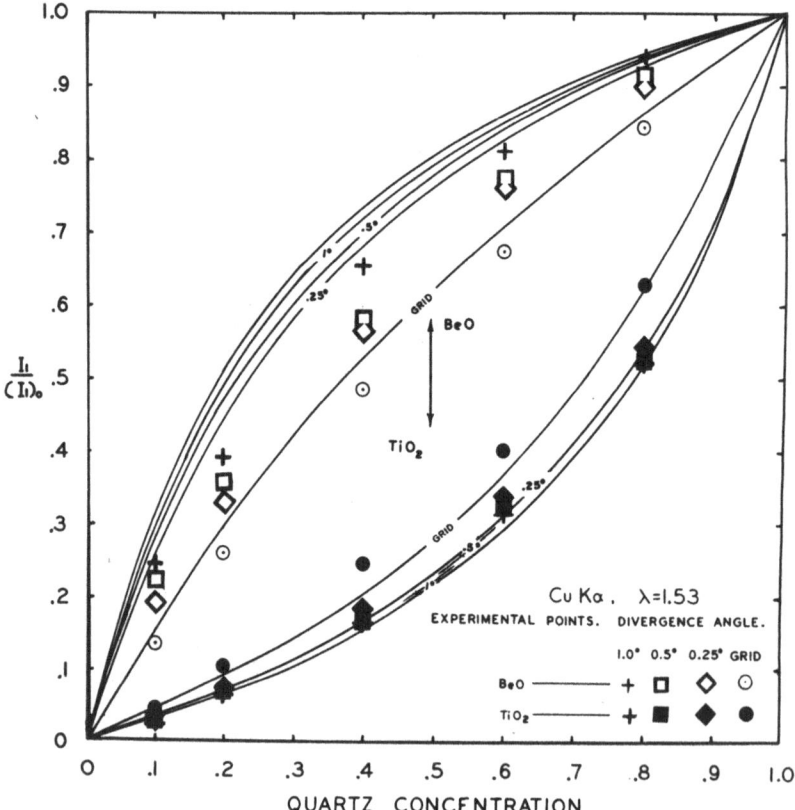

Figure 1. Calculated and experimental intensity ratios as functions of quartz concentration for copper K_α radiation.

---

[1] Superscripts pertain to references at the end of the paper.

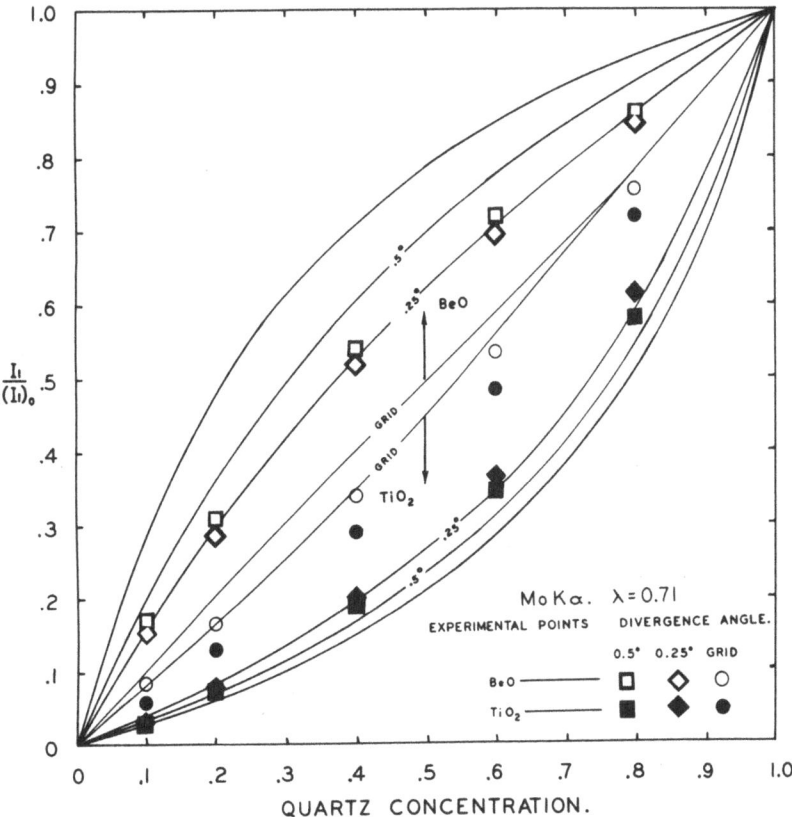

Figure 2. Calculated and experimental intensity ratios as functions of quartz concentration for molybdenum K_α radiation.

In the original paper the technique was described with the use of Cu K_α radiation, while in the second paper reference was made to Mo K_α radiation. In both cases, the final expression became

$$x = \frac{I_1}{\left(I_1\right)_0} \left(\frac{\mu_s^*}{\mu_1^*}\right)^C, \tag{2}$$

the new constant C being slightly different for the molybdenum radiation than that found originally for the copper radiation.

Equation (2) is quite accurate when appropriate collimator slits are used, although it is only an empirical expression which was derived from Figure 3 where the average slope of the concentration lines determines the value of C.

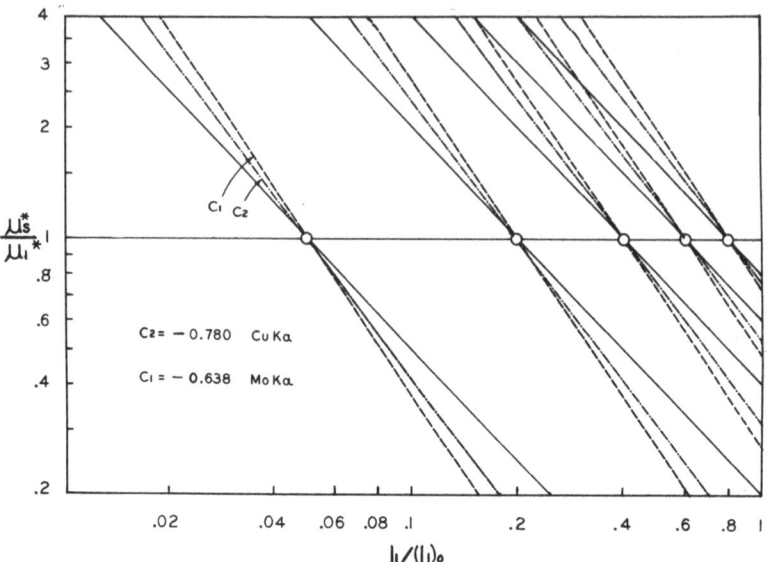

Figure 3. Diffraction and absorption ratios for ideally divided mixtures (full lines) and actual mixtures of beryllium oxide and titanium dioxide with. quartz for copper and molybdenum K_α radiations, respectively (dotted lines).

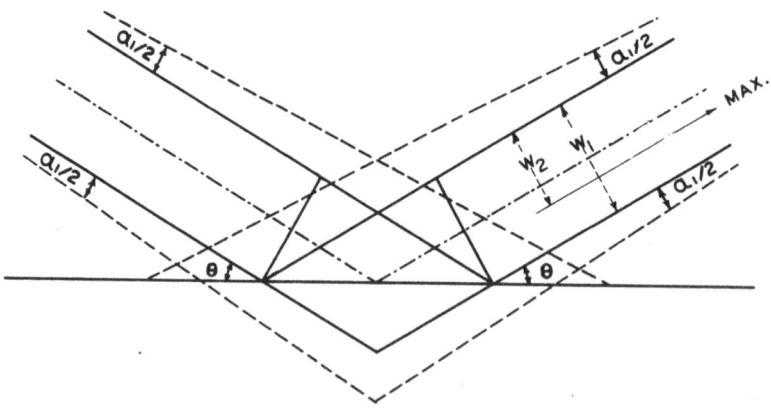

Figure 4. The path of the incident and diffracted beam of divergence through the flat sample is considered as parallel within bold lines.

A few preliminary tests made with different slit openings led to the conclusion that the departure from the "ideal" ratios was not fortuitous. This departure coincided with the size variations of the volume fraction of the sample irradiated by a beam of different widths. Because of the parafocusing geometrical arrangement of usual diffractometers, this volume fraction is a triangular prism from which all crystallites properly oriented are diffracting in a beam considered as parallel within an "effective width" generally smaller than the projected width of the receiving slit of divergence α_1, as illustrated in Figure 4.

Assuming the existence of such a prism, the integration from zero to infinity, as evaluated in many textbooks for the intensity reflected by a flat powder sample, yields equation (1), which is inaccurate for the geometry of an X-ray diffractometer. It is the purpose of the present paper to demonstrate that the departure of the experimental ratios from the ideal values is largely due to the geometric shape of the volume fraction of the sample irradiated, and consequently to the absorption coefficient of the latter, neglecting all other factors. Being given a flat sample, of mass absorption coefficient μ_s^* and apparent density ρ_s, irradiated by a beam of intensity I_0, it can be shown that the intensity I of the parallel beam diffracted and transmitted through the effective width W_1 from the triangular prism by the analyte of concentration x_1 and apparent density ρ_1 is

$$I = \frac{I_0 x_1 W}{2\mu_s^* \rho_1}\left[1 - \frac{1}{2\mu_s^* \rho_s \csc(2\theta)W}\left(1 - e^{-2\mu_s^* \rho_s \csc(2\theta)W}\right)\right]. \quad (3)$$

Replacing the linear absorption coefficient term $2\mu_s^* \rho_s \csc(2\theta)W$ by ψ_s, this equation may be written

$$\frac{I}{I_0} = \frac{x_1 W}{2\mu_s^* \rho_1}\left[1 - \frac{1}{\psi_s}\left(1 - e^{-\psi_s}\right)\right]. \quad (4)$$

This general equation, which applies to the geometrical conditions of a diffractometer, shows that the intensity is dependent on all the factors involved in the linear absorption coefficient ψ_s, including the Bragg angle, because of the integration within a solid triangular prism of limited size.

Comparative values or different families of curves can be readily computed or built for I_1/I_0 ratios of equation (4) by means of Figure 5, where the function in brackets of the right-hand member has been plotted against its argument.

APPLICATION TO QUANTITATIVE ANALYSIS

In the method of analysis by the absorption-diffraction technique, the intensity diffracted by a sample is always compared

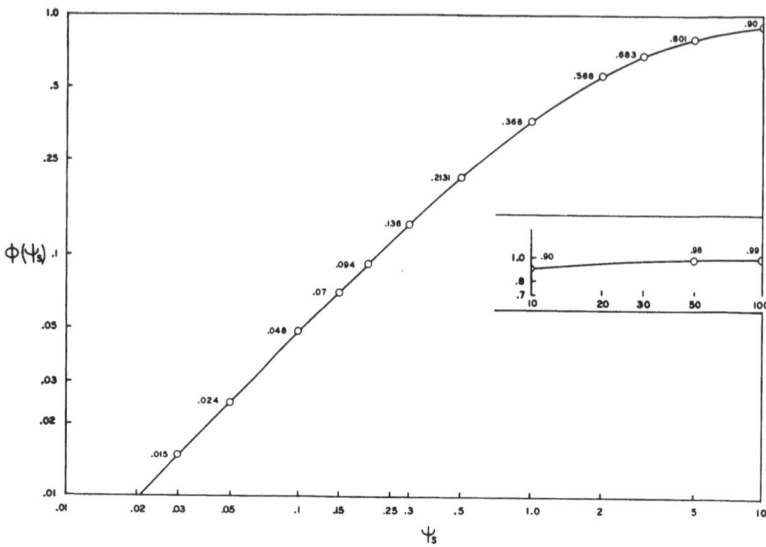

Figure 5. Relationship between ψ_s and $\phi(\psi_s)$ as calculated from equation (4).

with the intensity diffracted by the sample made up of the pure compound. From equation (4), the ratio of the intensities becomes

$$\frac{I_1}{(I_1)_0} = x_1 \frac{\mu_1^*}{\mu_s^*} \frac{\phi(\psi_s)}{\phi(\psi_1)} , \qquad (5)$$

where we recognize again the expression for the ideal ratio corrected this time by the factor $\phi(\psi_s)/\phi(\psi_1)$ instead of the less rational exponent C originally derived by experimental graphics.

Tests have been performed (a) with quartz mixed in samples covering a large range of absorption coefficients; (b) with two different K_α radiations using copper and molybdenum targets; (c) with three different angles of beam divergence of 1, $\frac{1}{2}$, and $\frac{1}{4}°$, respectively; and (d) with the sample packed and leveled in a copper grid (60 mesh copper wire). This grid was used in order to overcome the difficulty arising from the critical alignment of slits significantly smaller than $\frac{1}{4}°$ and the weakness of the intensity which can be provided by such slits. By packing and leveling the powder in such a grid, a multifold intensity diffracted from each grid opening over a large area was expected to satisfy approximately equation (4) calculated for an equivalent "effective width" smaller than 0.1°.

Figures 1 and 2 show the quartz concentration against calculated values of diffraction ratios for "ideal" conditions, for "effective widths" of 1, $\frac{1}{2}$, $\frac{1}{4}$° and for grids (0.05° with Cu K_α and 0.025° with Mo K_α). Experimental points obtained by using corresponding divergent slits have been plotted with symbols described in the caption of each graph.

By transferring to Figures 6 and 7 the experimental points (not indicated here for sake of simplicity) on corresponding calculated concentration curves plotted as functions of calculated diffraction ratios and effective widths, respectively, it is possible to evaluate the average "effective width" obtained with each set of divergence slits. This average "effective width" value is met by the straight line which intersects the concentration curves in the vicinity of each corresponding experimental point. The values have been transferred to Table I. By inspection of these graphs, we observe:

1. All the results calculated from equation (4) indicate that the smaller the effective width, the closer it brings the diffraction lines toward linearity; that is, to a point where penetration is so small that the intensity becomes merely proportional to the concentration of the analyte. As could be predicted from equation (4), the rate of change of the ratios with the value of the effective width is much faster with the molybdenum radiation because of its larger power of penetration.

2. Similarly, the experimental diffraction ratios show a rate of change comparable to that of the calculated ones and in the same direction.

3. By using slits larger than 1 and 0.5° for copper and molybdenum radiations, respectively, the experimental diffraction ratios remain the same. This indicates that, even in the case of extreme beam divergences, the bulk of the radiation contributed by the sample is governed by a law at least comparable to that

Table I. Comparative Values, in Degrees, of the
Slit Divergence α_1 and Corresponding Average Effective
Width α_2 Determined by Means of Figures 6 and 7 for
Copper K_α and Molybdenum K_α Radiations, Respectively

1 – Copper	α_1	$\geqslant 1$	0.5	0.25	0.05 (Grid)
	α_2	0.194	0.13	0.094	0.036
	α_2/α_1	0.20	0.25	0.38	—
2 – Molybdenum	α_1	—	0.5	0.25	0.025 (Grid)
	α_2	—	0.277	0.2159	0.038
	α_2/α_1	—	0.555	0.86	—

expressed by equation (4), which was strictly derived for a solid triangular prism.

4. One notices in Table I that the ratios α_2/α_1 for both radiations tend toward unity with decrease in the width of the divergence slits used. This phenomenon helps to confirm the fundamental assumption just discussed, that is, the bulk of the intensity takes place from the prism defined strictly by the "effective width": the smaller the slit, the less is the divergence and consequently more agreement is found between the calculated and experimental diffraction ratios when equation (4) is set with a W equal to the degree of the slit. On the other hand, the reason why the agreement is better with the molybdenum radiation is easily explained by a slight difference in the alignment of the goniometer for each radiation. In fact, the order of magnitude of the degree values involved in this investigation is within the limits of accuracy obtainable in the zero setting of the instrument for which no device has yet been developed for actual commercial diffractometers.

5. For the same reasons of critical alignment just mentioned, the effective width computed from Figures 6 and 7 is not, for copper K_α radiation, twice the value found for the molybdenum K_α radiation as it should be in the case of the grid. Nevertheless, these results demonstrate clearly that the use of grids provides, at least qualitatively, the equivalent effect of a very fine beam, because practically the diffraction takes place only in the top layers of the flat powder sample.

APPLICATION TO QUALITATIVE ANALYSIS

Many authors[4,5] have discussed the displacement of line peaks toward lower 2θ angular values in diffractometry. The treatment of this phenomenon is rather complex and we wish to point out how equation (4) permits one to evaluate, to a certain extent, the influence of absorption only.

If an X-ray spectrum were made of the total intensity diffracted from the triangular prism of Figure 4, then, obviously the recorded intensity along W_1 (receiving slit) would start from 0 at the top and would increase exponentially along W_1 to reach its maximum value at the level of the sample surface.

As a first approximation, one can say that, when scanning the goniometer toward zero degree, the maximum peak should show at an angle $2\theta - \Delta(2\theta)$ for which the intensity reflected through the upper portion of W_1 is half the value of the total intensity diffracted through the whole effective width.

If we call W_2 the upper portion of W_1 collecting half the integrated intensity, we obtain from equation (4)

$$\frac{I_{W_2}}{I_{W_1}} = \frac{1}{2} = \frac{W_2}{W_1}\frac{\phi(\psi_2)}{\phi(\psi_1)} \ . \tag{6}$$

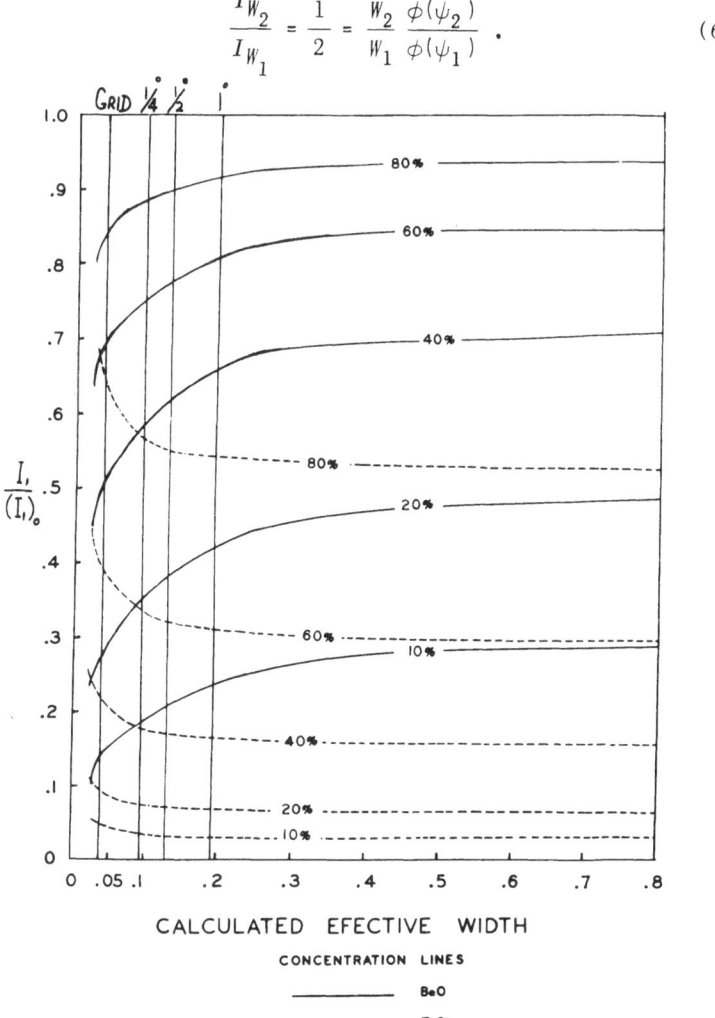

Figure 6. CuK_α radiation. The values of the average effective widths corresponding to each set of divergence slits are represented by the straight lines which intersect the concentration curves in the vicinity of the corresponding experimental points (not indicated in the figure).

If β designates the number of degrees it takes for the goniometer to scan a complete diffraction line over the background, then the ratio $\Delta(2\theta)/\beta$ should be approximately proportional to $(W_2/W_1 - 0.5)$.

The influence of the absorption factor, independently of all other factors, has been experimentally verified by comparing the positions of the peak intensities recorded on a strip chart for a compound packed alternatively in an ordinary sample holder and in a copper grid. The compound was L-glutamic acid ($C_5O_4H_9N$) and was selected for its very low mass absorption coefficients and its well resolved peak lines for lattice spacings lying between 3 and 5 A.

The peak positions obtained with the grid sample holders, being considered as closer to the real 2θ values, because of the low influence of the absorption factor in this case, were taken as points of reference for the $\Delta(2\theta)$ displacements taking place with the ordinary slides. Table II gives comparative results obtained with the two types of sample holders and with both radiations.

Table II. Peak Lines Obtained from L-Glutamic Acid
with an Aluminum Slide and with a Grid

Cu K_α			Mo K_α		
Al slide	Grid	$\Delta(2\theta)$	Al slide	Grid	$\Delta(2\theta)$
28.47	28.53	−0.06	−	−	−
26.585	26.64	−0.055	12.125	12.15	−0.03
23.55	23.65	−0.10	10.675	10.775	−0.10
20.11	20.16	−0.05	9.156	9.21	−0.054
18.22	18.26	−0.04	8.233	8.325	−0.092
Average	$\Delta(2\theta)$−0.06		Average	$\Delta(2\theta)$−0.069	

The average $\Delta(2\theta)$ as calculated from equation (6) was found to be −0.13 and −0.18° for Cu K_α and Mo K_α radiations, respectively. The equation was calculated with the W values 0.2159 and 0.194° previously derived for both radiations using 1 and 0.5° slits and with the same goniometer settings. The apparent density of the sample was 1.252. The mass absorption coefficient values were 7.432 and 1.07 for copper and molybdenum radiations, respectively. Average 2θ values of 23 and 10° were chosen for the two radiations.

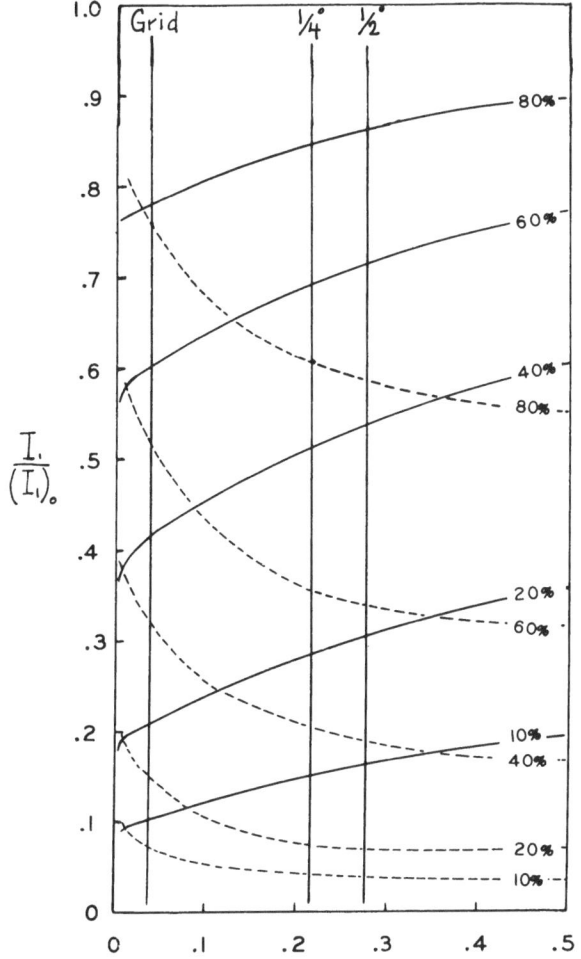

CALCULATED EFECTIVE WIDTH

CONCENTRATION LINES

——————— BeO

---------- TiO_2

Figure 7. Molybdenum K_α radiation. The values of the average effective widths corresponding to each set of divergence slits are represented by the straight lines which intersect the concentration curves in the vicinity of the corresponding experimental points (not indicated in the figure).

These last results indicate that the effect of absorption is quite significant, independently of all other factors, in the displacement of the peak values toward lower angular 2θ values.

CONCLUSIONS

From a practical point of view, this investigation leads to the following conclusions.

1. For quantitative analysis, if the resolution of the peak lines is not a problem, it is preferable to use the maximum slit opening. Under these conditions slight misalignment of the goniometer from the zero axis can be compensated for by an effective width of the slit large enough to permit reliable analysis, as was demonstrated in Table I.

2. Results obtained in the second part of this investigation strongly suggest that a qualitative analysis approaching the accuracy of the film techniques could be obtained by diffractometry if a sample holder based on the principle of the grid described before was developed. Basically, such a sample holder could be made of a nonfluorescing and high-absorbing material. A series of grooves having a section of 0.25 mm^2 could be traced on the surface of the holder into which the powder could be uniformly packed and leveled.

    It is believed that, with these conditions, a much higher degree of surface uniformity could be reached as compared to the grid, thus permitting lattice spacing determinations with a minimum of correction factors.

REFERENCES

[1] L. Alexander and H. P. Klug, Anal. Chem., Vol. 20, 1948, p. 886.
[2] J. Leroux, D. H. Lennox, and K. Kay, Anal. Chem., Vol. 25, 1953, p. 740.
[3] J. Leroux, Norelco Reporter, Vol. IX, No. 5, Sept.-Oct., 1957.
[4] H. S. Peiser, H. P. Rooksby, and A. J. C. Wilson, Technical Editors, X-Ray Diffraction by Polycrystalline Materials, Institute of Physics, London, 1955.
[5] H. P. Klug and L. Alexander, X-Ray Diffraction Procedures, John Wiley and Sons, 1954.

THE APPLICATION OF X-RAY DIFFRACTION
TO MEDICAL PROBLEMS

Jonathan Parsons

Department of Physics
Edsel B. Ford Institute for Medical Research
Detroit, Michigan

ABSTRACT

The use of the X-ray diffraction Debye-Scherrer powder technique for the identification of crystalline material often found in pathologic tissue is discussed. Cases cited include: heart calcification, lung silicosis, thorium dioxide in liver and spleen, and granulomas of the skin and pericardium. The routine analysis of kidney stones using powder cameras as well as an X-ray diffractometer will be described.

* *
*

Crystalline substances are frequently observed by pathologists during routine examination of autopsy and biopsy tissue specimens. The methods available in the past for definitely determining their nature have been limited and often dependent upon the experienced guess of the examining doctor. Good as this may be, a definite and sensitive method of analysis is very desirable. X-ray diffraction powder camera analysis had been used for years, by the author, for analyzing the gross concrements of the body, namely, bone, teeth, kidney, and gall stones. The ability of diffraction analysis to identify small amounts of material encouraged its application to the microscopic crystals in fresh tissue specimens.

A fine needlelike sliver of fresh or formalin-fixed tissue is cut from the area where the crystals are suspected or observed. This sliver of tissue is then introduced into a funnel end of special thin-walled glass capillaries.* Figure 1A shows one of these capillary tubes as received; Figure 1B, as mounted through cylindrical pins; and Figure 1C, with funnel end removed, after the insertion of the tissue sliver. After sealing with a drop of cement, the mounted tissue (Figure 1C) is placed in a powder camera for a 2-5 sec exposure on copper radiation (nickel filter). The following separate examples illustrate our application of X-ray diffraction powder analysis to medical problems.

---

* Available from Caine Sales Co., 3020 N. Cicero Avenue, Chicago 41, Illinois.

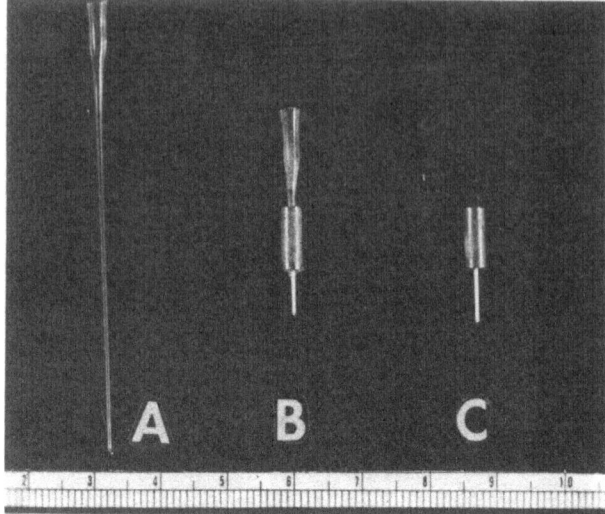

Figure 1. A) Glass capillary tube. B) Glass capillary
tube in pin mount, with specimen. C) Mounted specimen
tube with funnel broken off.

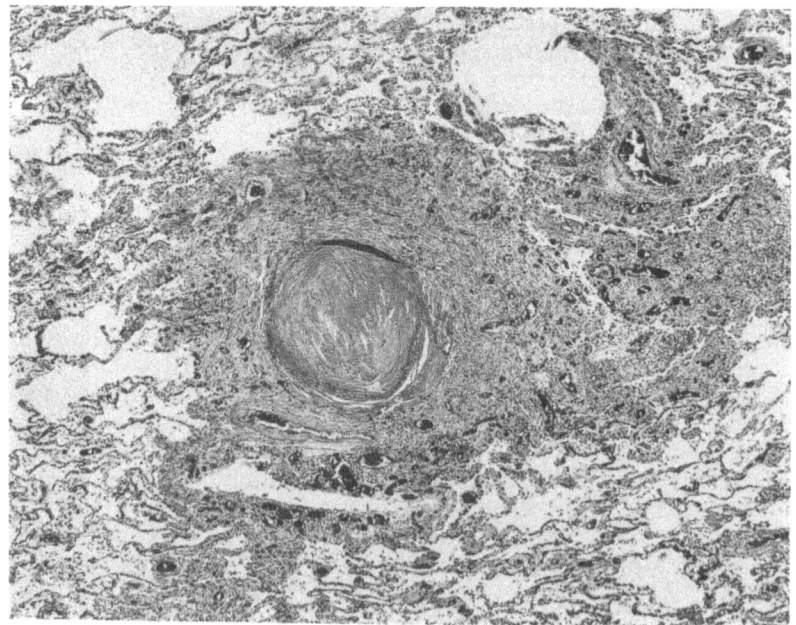

Figure 2. Silicotic nodule, lung, H and E stain, 35×.

Figure 2 is a photomicrograph (35×) of lung tissue from a pottery factory worker who died of pneumonia. Figure 3 shows the area just outside of the large nodule at 730× magnification. The profusely scattered birefringent crystals in the tissue are very clearly seen in the polarized light. The X-ray diffraction pattern obtained from this area is shown in Figure 4B. Upon examination this material was found to be a combination of commercial kaolin and silica in the form of quartz (Figure 4A and C).

A number of years ago, radiologists used thorium dioxide (thorotrast) as an opaque agent in making diagnostic X-ray examinations. This radioactive material cannot be readily eliminated from the liver and spleen and is no longer used for this purpose. Figure 5 shows the small dark crystals of thorotrast appearing as circles around a central vein in the photomicrograph (225×) from a section of liver. Figure 6 shows the patterns from spleen and liver tissue sections as compared with standard thorotrast.

Figure 7 is a photograph of a patient's arm showing a tattoo design which became ulcerated in the red area. The photomicrograph of tissue from this area is shown in Figure 8. Clumps of dark brown crystals were identified (Figure 9) as mercuric sulfide or cinnabar, a pigment used by the tattoo artist in producing the red coloring.

Autopsy following the death of a man from renal failure revealed his pancreas to be greatly enlarged and containing cysts which were filled with calculi or stones (Figure 10). Upon crushing one of these stones and making a powder diffraction pattern, as shown in Figure 11, the nature of these calculi was established as calcium carbonate in the form of calcite.

Figure 12 is a photomicrograph of a lung section from a worker in a barium sulfate processing plant. Numerous crystals can be seen both in the body of small cells and outside them in larger form. Microincineration was carried out on a microscopic slide. The powder residue was subjected to X-ray diffraction analysis; the resulting pattern is shown in Figure 13. The crystals were definitely identified as barium sulfate.

One of our earliest successes at analyzing crystals in tissue was the case of crystals observed during autopsy in the thickened portion of the heart covering, called the pericardium. The individual had died of heart disease. Figure 14 is a gross photograph of a portion of this pericardium. The tissue specimen shown in the photomicrographs and for analysis was taken from the central white fibrotic area immediately outside of the horizontal line of demarcation between light and dark tissue. Figure 15 shows the birefringent crystals at a magnification of 375× in

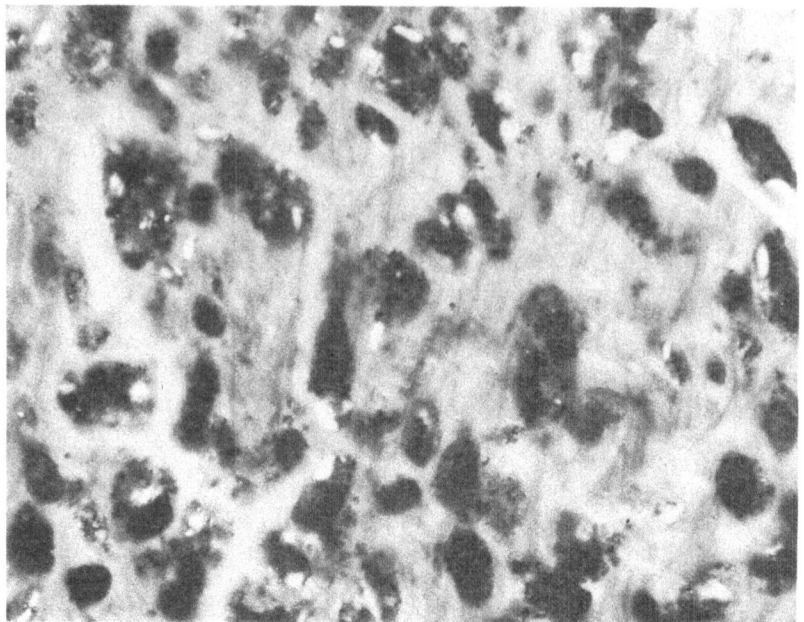

Figure 3. Crystals in lung nodule, partially polarized, H and E stain, 730×.

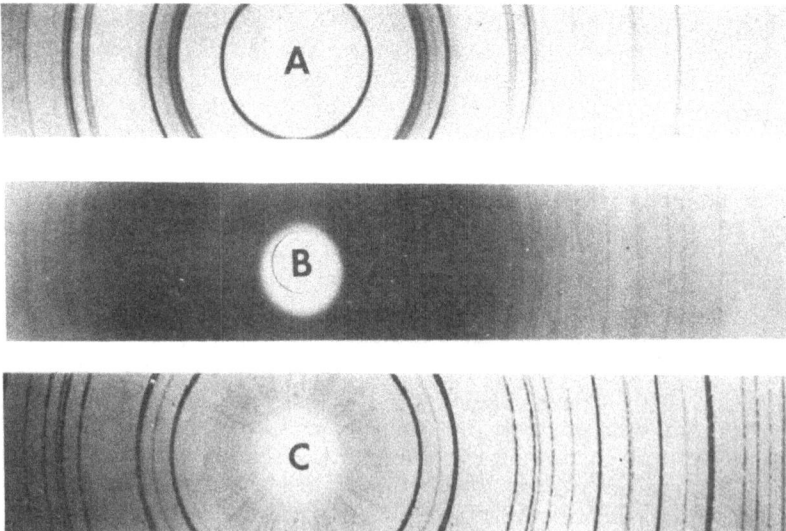

Figure 4. X-ray diffraction patterns. A) Kaolin. B) Lung tissue. C) SiO_2 (quartz).

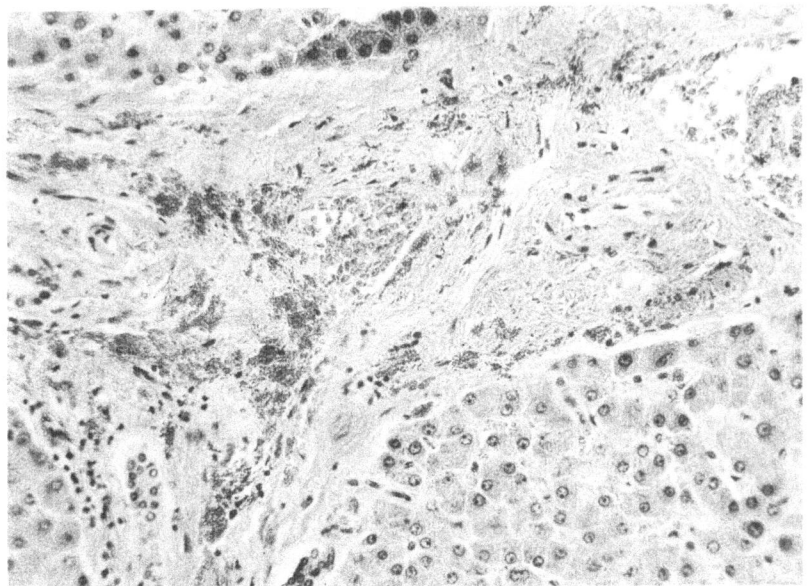

Figure 5. Thorotrast deposits in liver, H and E stain, 225×.

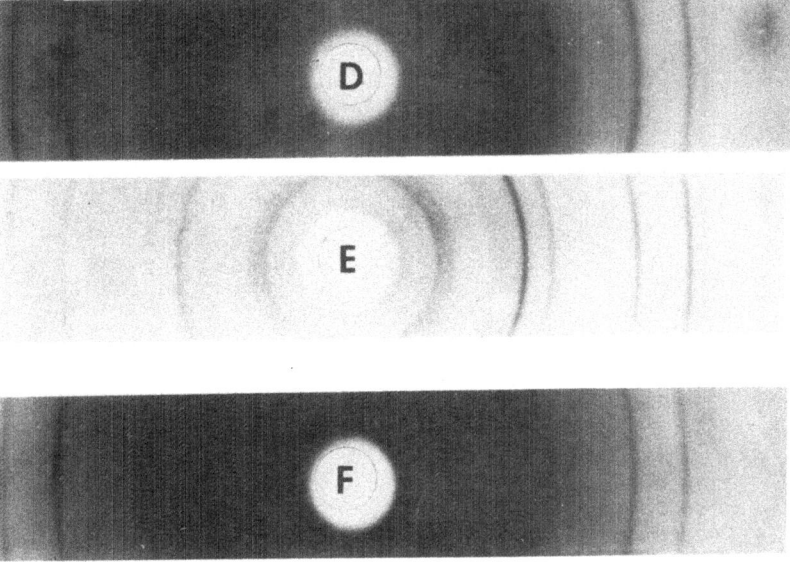

Figure 6. X-ray diffraction patterns. D) Sample of tissue from spleen.
E) Standard thorotrast. F) Sample of tissue from liver.

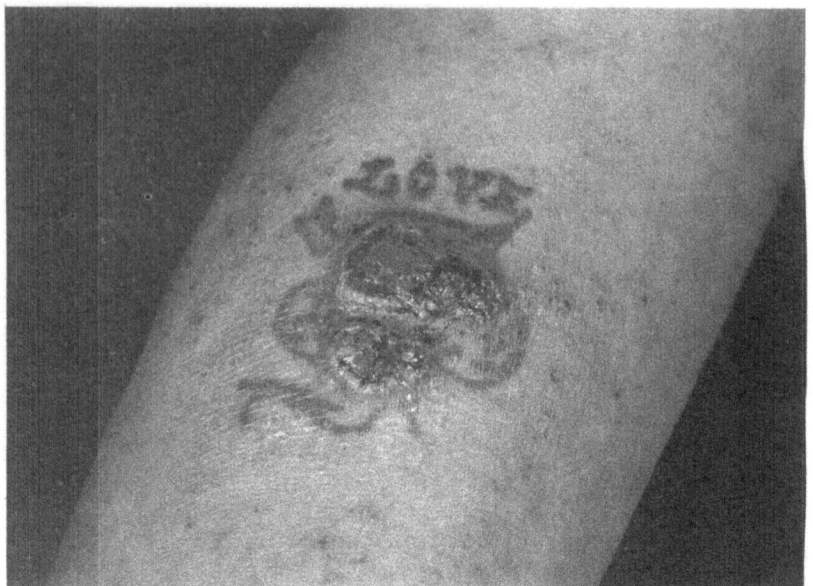

Figure 7. Tattoo mark, ulcerated in central red portion.

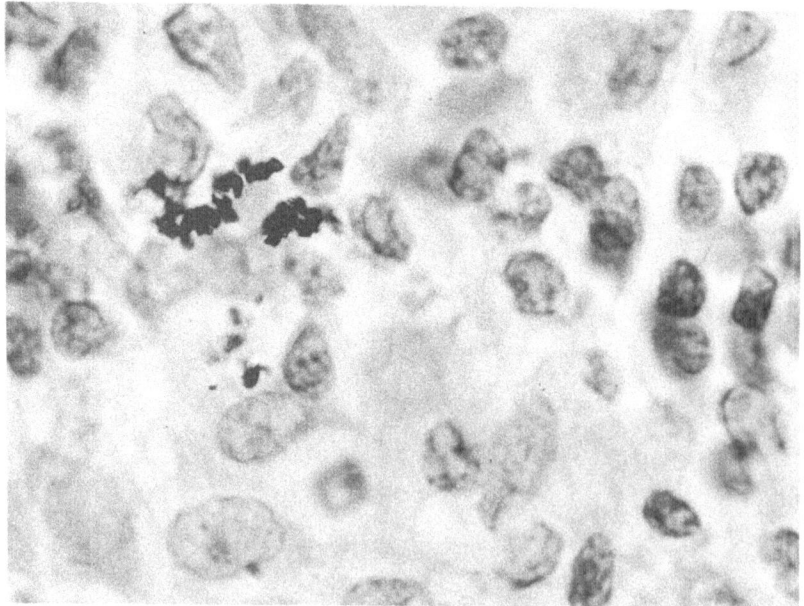

Figure 8. Mercuric sulfide crystals in ulcerated tattoo, H and E stain, 1100×.

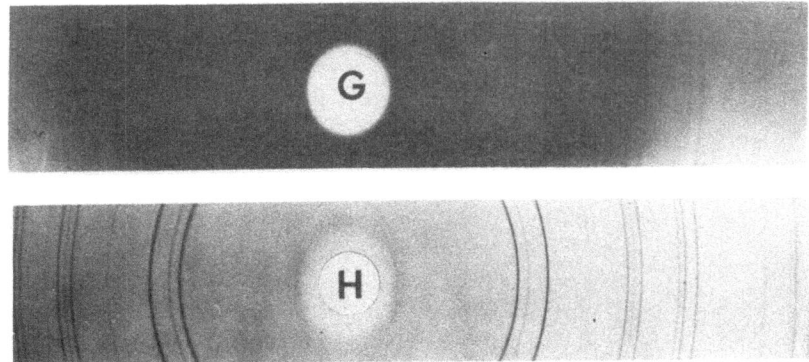

Figure 9. X-ray diffraction patterns. G) Sample of ulcerated tattoo tissue.
H) Mercuric sulfide.

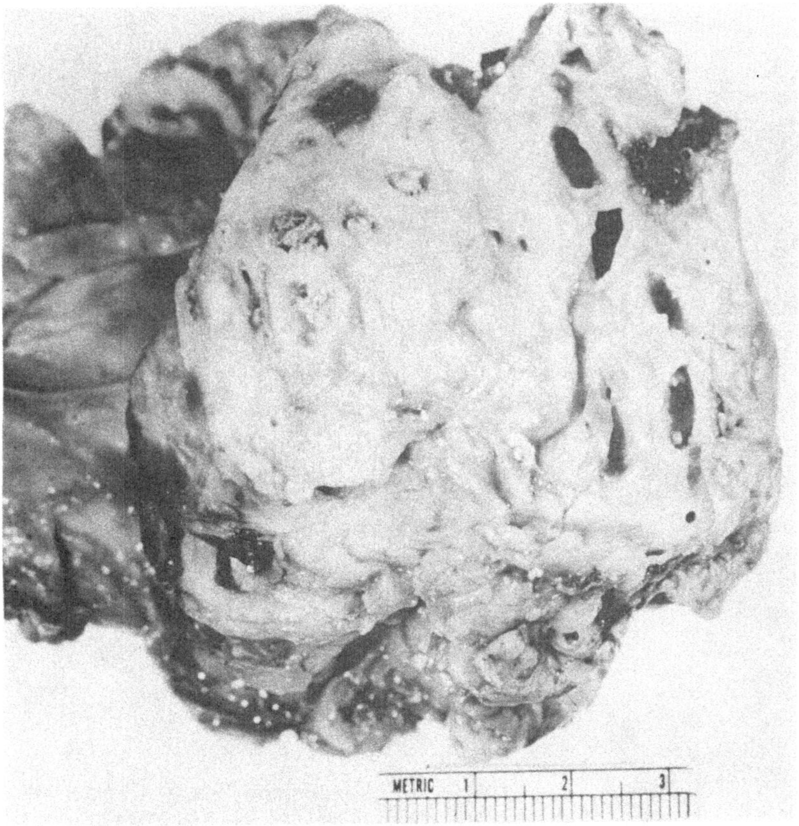

Figure 10. Gross photograph of fibrotic pancreas with stones.

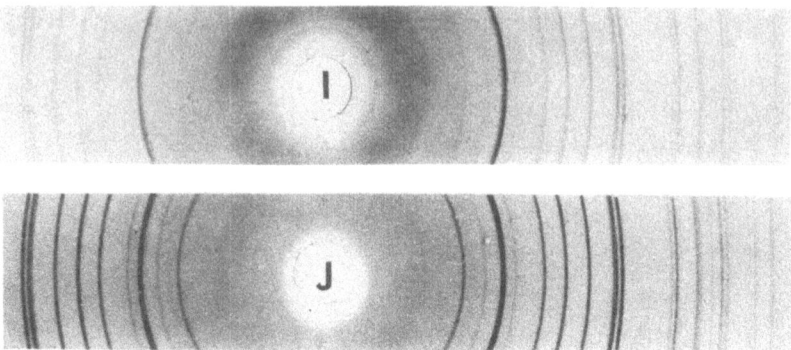

Figure 11. X-ray diffraction patterns. I) Powdered pancreatic stone. J) $CaCO_3$ (calcite).

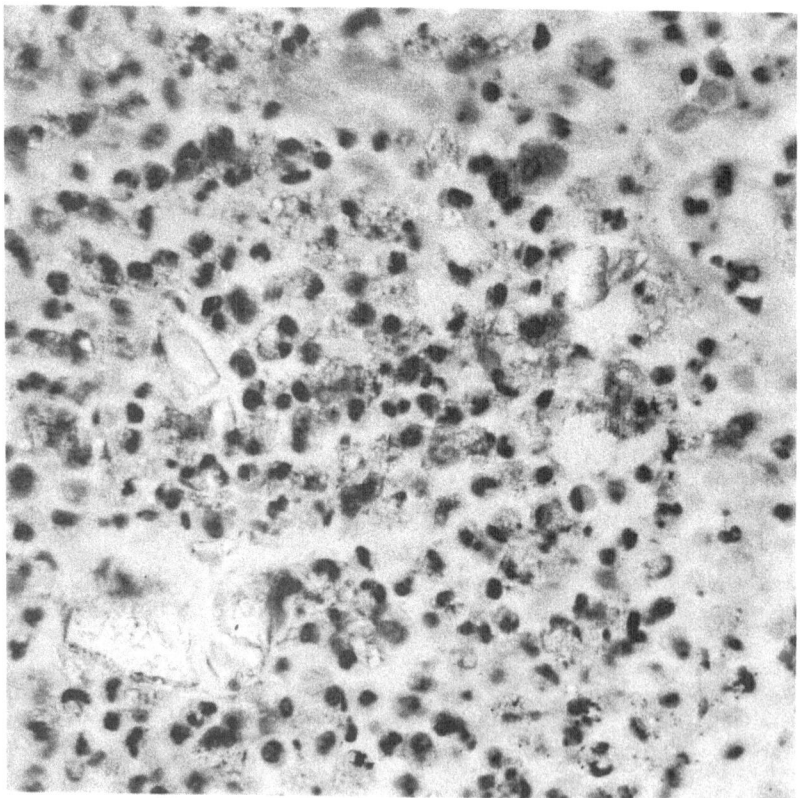

Figure 12. Barium sulfate crystals in lung, H and E stain, 900×.

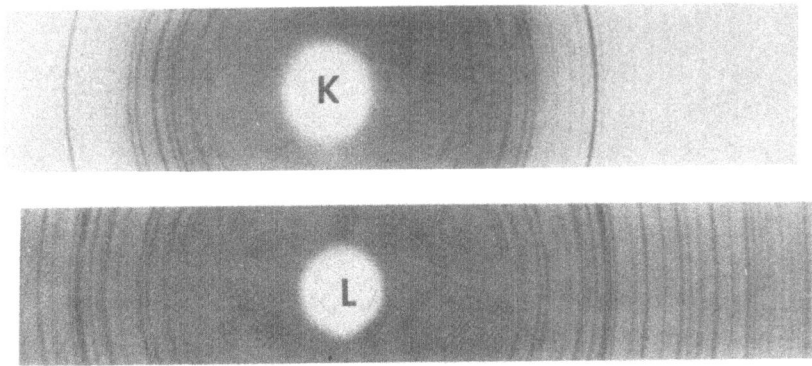

Figure 13. X-ray diffraction patterns. K) Ashed lung. L) Barium sulfate.

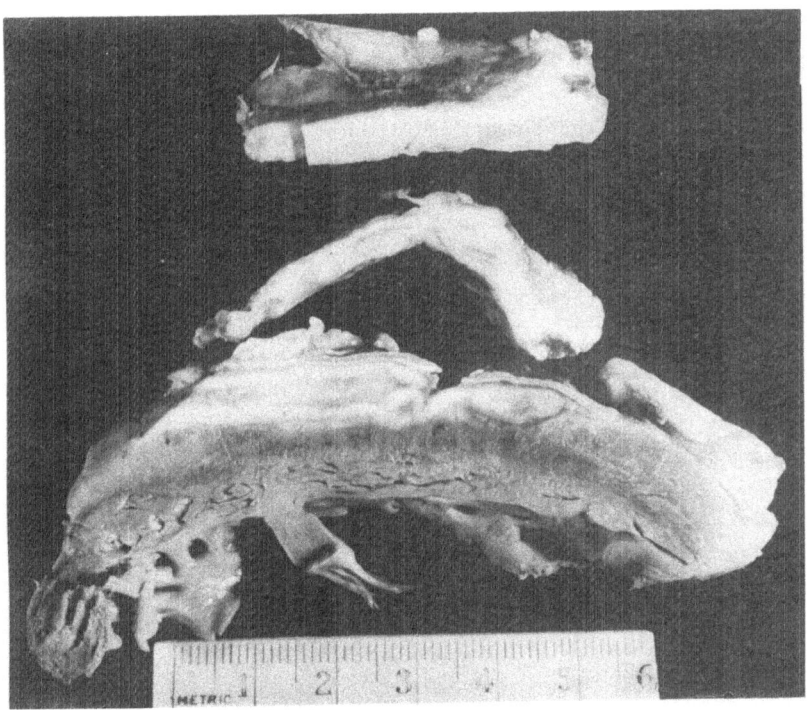

Figure 14. Gross photograph of pericardium (heart sac).

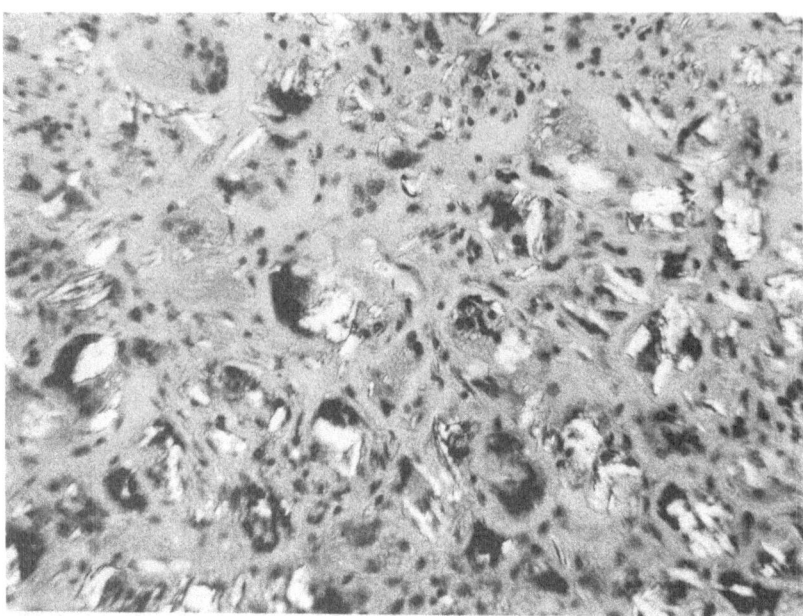

Figure 15. Talc crystals in pericardium (heart sac), partially polarized light, 375×.

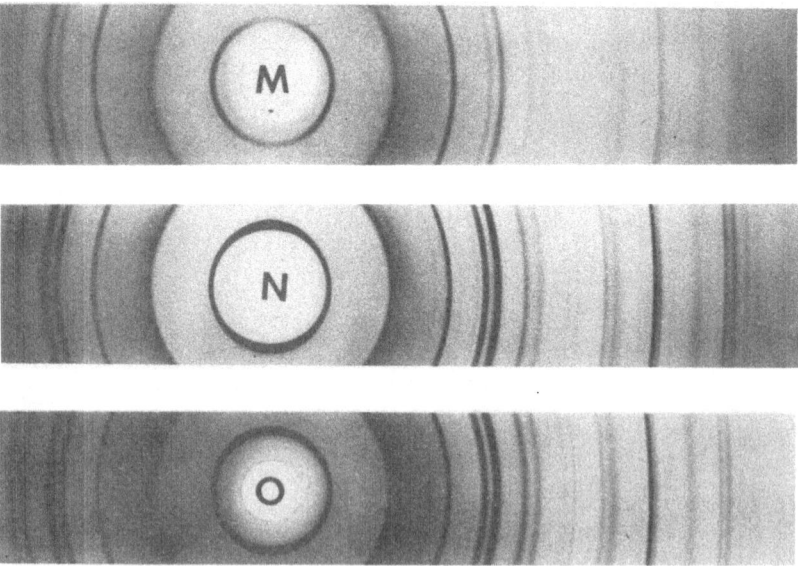

Figure 16. X-ray diffraction patterns. M) Sample of tissue from pericardium. N) Ashed tissue sample from pericardium. O) Talc.

partially polarized light. Figure 16M and N, shows the diffraction pattern before and after the tissue was ashed. In this case ashing did not change the crystalline structure. A study of the diffraction lines using the X-ray powder data file indicated that magnesium silicate (talc) matched the data very well. A standard pattern from talc is shown in Figure 16O. The way in which talc got into the pericardium was a puzzle to the pathologists and doctors interested in this case. It was later learned from the man's widow that talcum powder had been introduced in an experimental surgical procedure several years before death in an effort to relieve his condition by activating the small blood vessels. The small visible scar had been completely overlooked at autopsy.

Some disease conditions involve a calcification of the heart muscle. Figure 17 compares a diffraction pattern from such a heart condition with a standard apatite pattern. Apatite is the complex calcium phosphate mineral which is also the inorganic matrix for bone and teeth.

Of the types of kidney stones which the writer has analyzed by X-ray diffraction, the most common type is composed of calcium oxalate. Figure 18 is a photograph of a stone composed of a base of dark hard calcium oxalate monohydrate with platelike crystals of calcium oxalate dihydrate protruding edgewise at various angles. The patterns for these two hydrates are shown in Figure 19. Figure 20 shows a 20-mm kidney stone composed of cystine, a rather uncommon type of stone. Figure 21 shows the patterns for cystine and uric acid, another more common crystalline component of kidney stones.

Gout is a disease characterized by excess amounts of uric acid in the blood, and by the formation of chalky deposits of sodium acid urate in the cartilages of the joints. Figure 22 demonstrates the finding by X-ray diffraction of sodium acid urate associated with the small bones of the ear. Other symptoms of gout were later found in this case.

Figure 23 shows the X-ray analysis of an internal biliary fistula stone as being composed of cholesterol. This stone worked its way from the ducts associated with the gall bladder through the intestinal wall.

Figure 24 shows other types of kidney stone X-ray diffraction film patterns. X-ray diffractometer kidney stone charts for the 2θ range of 20 to 40° are illustrated in Figure 25. The use of the diffractometer is desirable when enough specimen is available to fill the cup, since a scan over a limited 2θ range allows much quicker identification. A rotating specimen cup was used

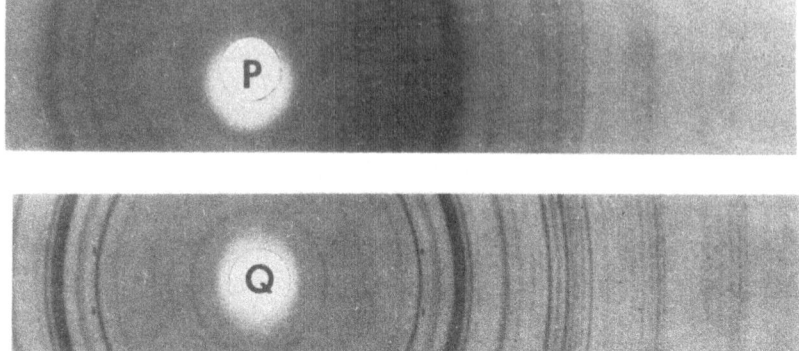

Figure 17. X-ray diffraction patterns. P) Sample of tissue from calcified heart. Q) Apatite.

Figure 18. Calcium oxalate kidney stone. Bulk of stone composed of the monohydrate with calcium oxalate dihydrate platelike crystals protruding from edges.

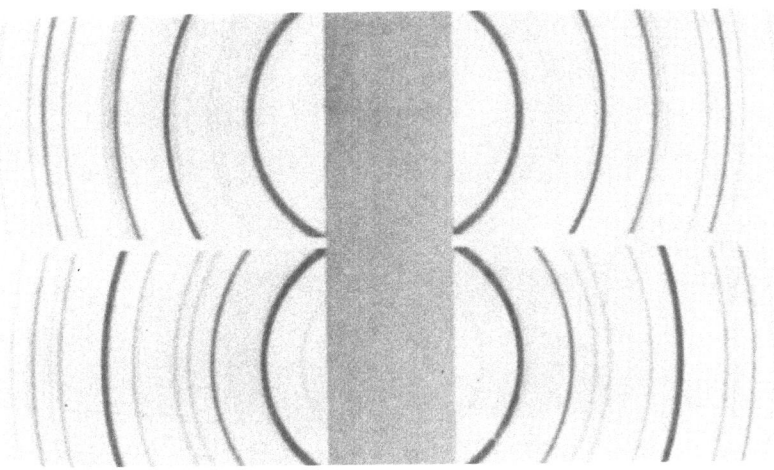

Figure 19. X-ray diffraction patterns of calcium oxalate kidney stones.
Top pattern—monohydrate. Bottom pattern—dihydrate.

Figure 20. Cross section of pure cystine kidney stone (scale in mm).

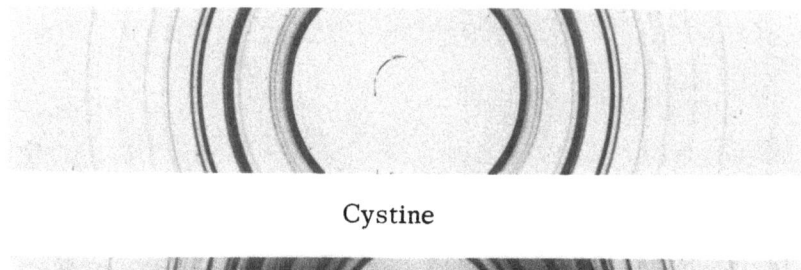

Cystine

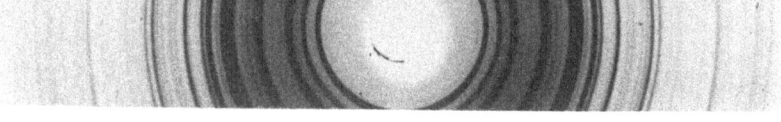

Uric Acid

Figure 21. X-ray diffraction patterns of cystine and uric acid kidney stones.

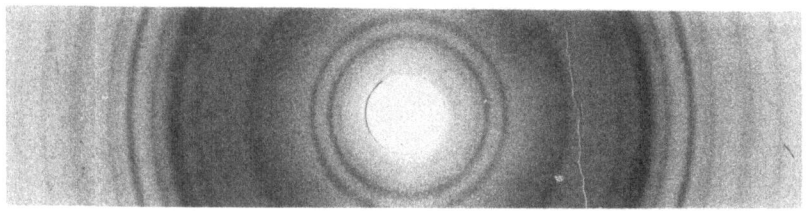

Sodium Acid Urate (chemically prepared)

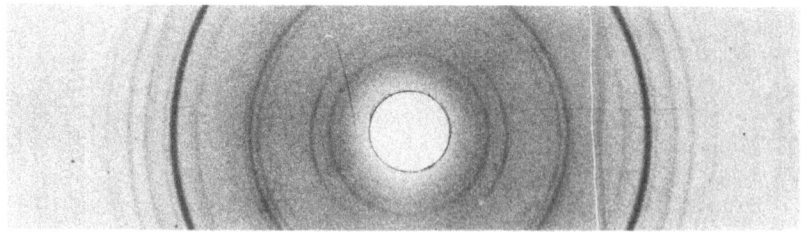

Deposit from Auricular Tophus

Figure 22. X-ray diffraction patterns. Top—standard sodium acid urate. Bottom—powdered deposit from the ear.

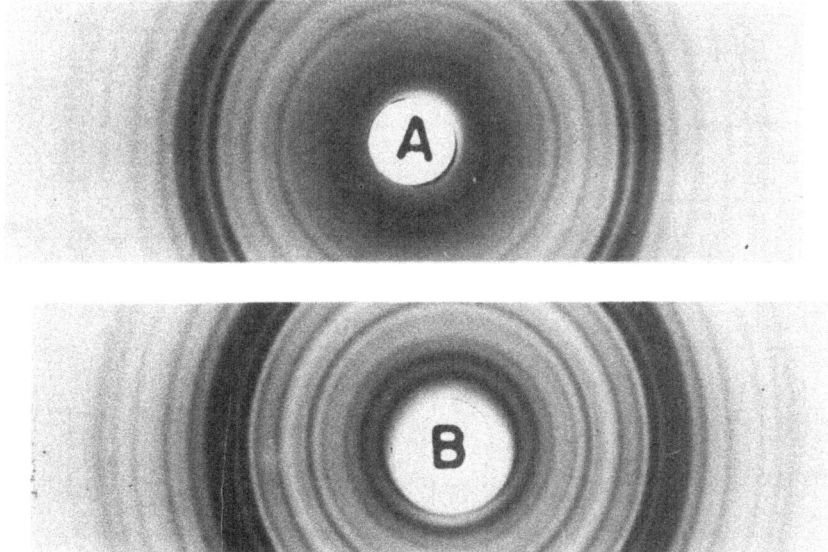

Figure 23. X-ray diffraction patterns. A) Internal biliary fistula stone.
B) Standard cholesterol.

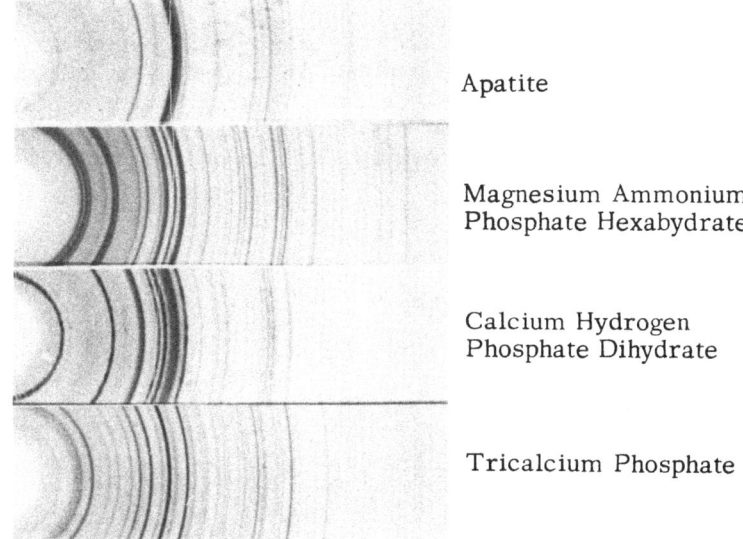

Apatite

Magnesium Ammonium
Phosphate Hexabydrate

Calcium Hydrogen
Phosphate Dihydrate

Tricalcium Phosphate

Figure 24. X-ray diffraction kidney stone patterns.

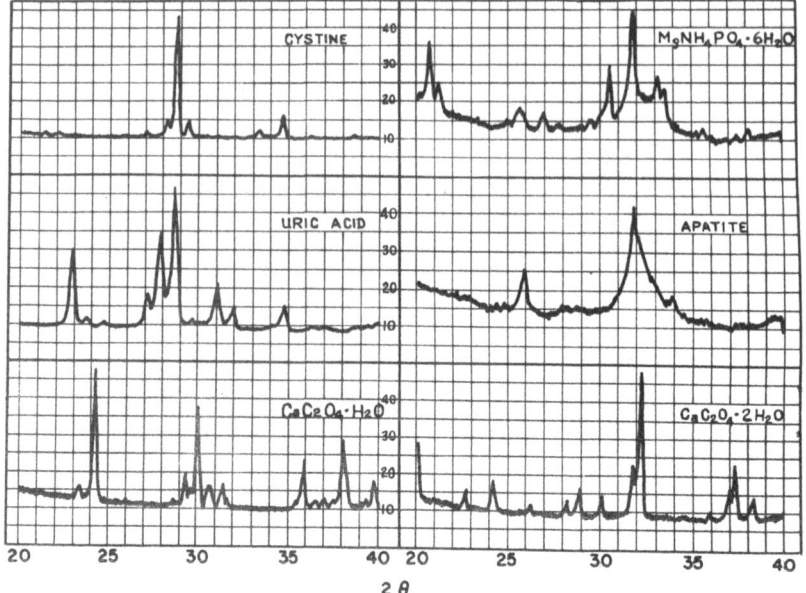

Figure 25. X-ray diffractometer kidney stone charts.

and is practically a necessity in order to minimize the effects of preferred orientation.

The intention of the author has been to illustrate the use made of X-ray diffraction in the Edsel B. Ford Institute laboratory. It is hoped that the presentation of these brief unrelated examples has been of interest to a nonmedical audience, in showing how a valuable tool of the industrial laboratory can help solve medical problems.

A HIGH-TEMPERATURE STUDY OF PHASE
PRECIPITATION IN SUPERALLOYS

John F. Radavich

Physics Department, Purdue University
Lafayette, Indiana

ABSTRACT

Many of the iron- and nickel-base superalloys exhibit brittle properties on heat treatment, welding, or other fabrication processes at temperatures of about 2000°F or higher. Studies have been carried out by means of electron microscopy, electron diffraction, and X-ray diffraction and fluorescence analysis of the precipitation in the metal and in an isolated form.

Results of the electron microscope study of the surface of the metal show a grain boundary constituent to be present which increases in amount as the temperature is increased. Studies on the isolated residue of such samples show a very thin "featherlike" film to be located at the grain boundaries and enclosing the grains. Electron diffraction, X-ray diffraction, and X-ray fluorescence analysis studies of the thin films indicate that they are a TiC phase with very little alloying elements in solution.

At temperatures above 2000°F the thin film becomes quite thick and tends to force the grains apart. It is believed that this form of the TiC phase promotes the severe embrittling nature of these alloys at high temperatures. Suitable heat treatment at lower temperatures causes the TiC film to agglomerate and the grain boundaries become "tight," and a more ductile condition results.

INTRODUCTION

The type and amount of phases produced in iron-base superalloys by aging reactions determine the physical properties of the alloy. A large number of phase studies have been carried out on the two more common iron-base alloys, A-286 and V-57, using electron microscopy, electron diffraction, and X-ray and fluorescence analysis.[1,2,3,4] In such studies emphasis was placed on the nature of the phase precipitation at temperatures below 2000°F after solution heat treatments.

This study is part of a comprehensive investigation of the phases precipitating in V-57 and A-286 alloys at temperatures

---

[1] Superscripts pertain to references at the end of the paper.

of 2000°F or higher. It is in this temperature range that commercial heat treatments, welding, and forging procedures occur.
It is believed that reactions occurring at 2000°F, or higher, determine in part the embrittlement found after welding or forging
as well as aging response at the lower temperatures.

This study involved the use of the electron microscopy,
electron diffraction, and X-ray diffraction techniques as well
as use of X-ray fluorescence analysis of the phases formed at
the high temperatures.

The results to be reported in this paper are divided into
three parts:

    1. Structures found in embrittled forgings.
    2. Structures found in cracked welds.
    3. Structures found after furnace heat treatments.

PROCEDURE

All the samples were given 3/0 paper polishes and then
electropolished in a perchloric alcohol solution. The surfaces
were then examined in the as-polished condition both by optical
and electron microscopy. The samples were then etched for 3
to 5 sec in modified Marbles' reagent or electrolytically in 10%
HCl—methanol solution and reexamined by electron microscopy.
It was found that the resolution of the optical microscope was
too low to show the structures produced by the high-temperature treatments.

After the surfaces were examined by the usual surface rep-
lication technique, the samples were chemically digested in a
10% HCl—methanol solution using 15 v and 1.2 amp for 1 to 3 hr.
The residue was allowed to settle, and then collected for X-ray
and X-ray fluorescence analysis. In many cases the residues
were examined optically to estimate the volume of the various
phases present as seen by X-ray diffraction.

RESULTS

1. Embrittled Forging. Samples of forgings which had
cracked upon the initial blows were examined by electron microscopy after an optical examination showing nothing. Modified
Marbles' reagent etch or electrolytic etching produced a knife-
edge attack at some of the grain boundaries. Figure 1 shows the
precipitation at the grain boundaries with some evidence of fragments of a film being pulled off by the replica. Figure 2 shows
the grain boundary after electrolytic etching wherein the matrix
was dissolved away, and the grain boundary precipitation was

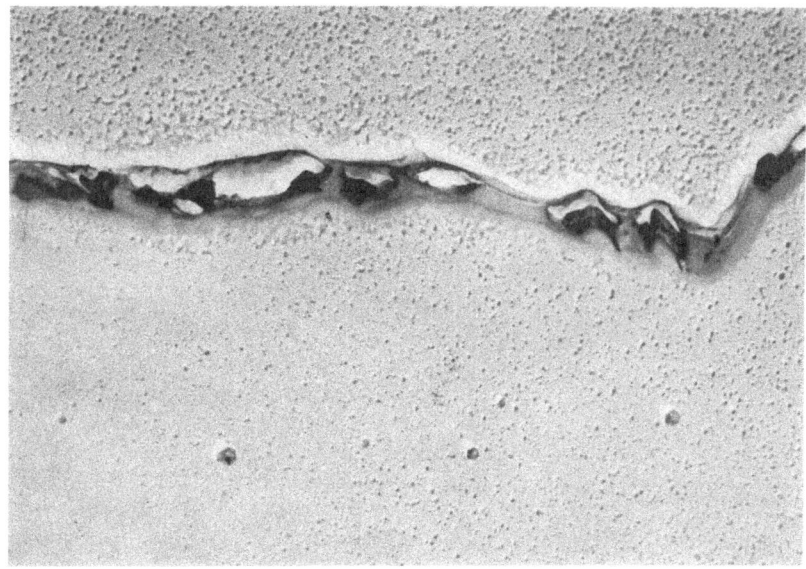

Figure 1. Grain boundary precipitation in V-57 forging etched in modified Marbles' reagent—8500×.

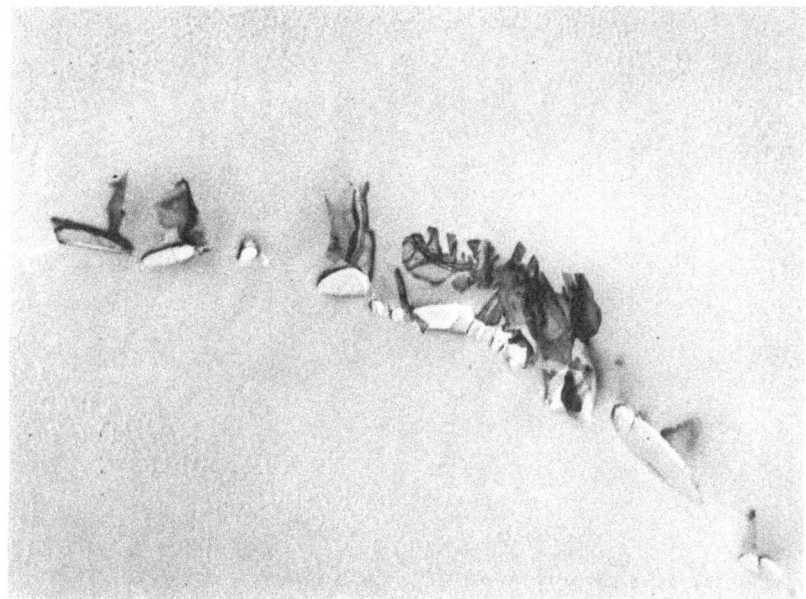

Figure 2. Grain boundary film in V-57 forging electrolytic etchant—8500×.

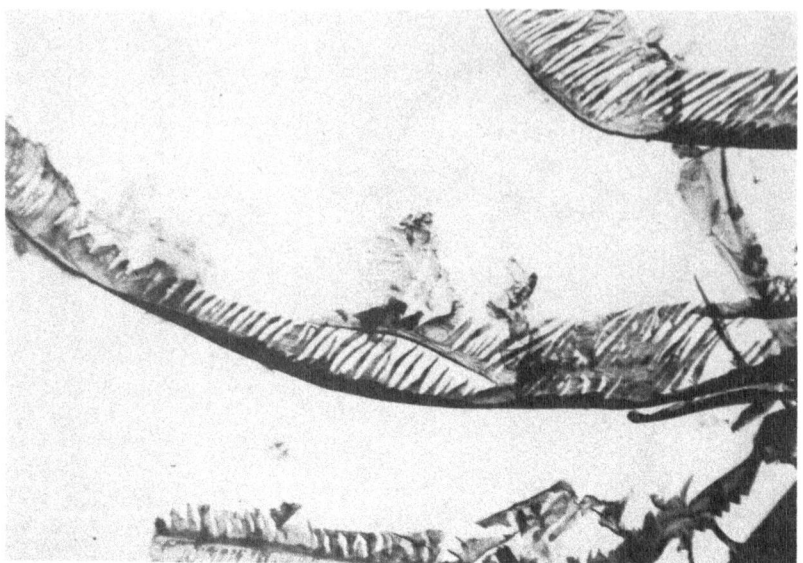

Figure 3. Isolated grain boundary films of V-57 forgings—5000×.

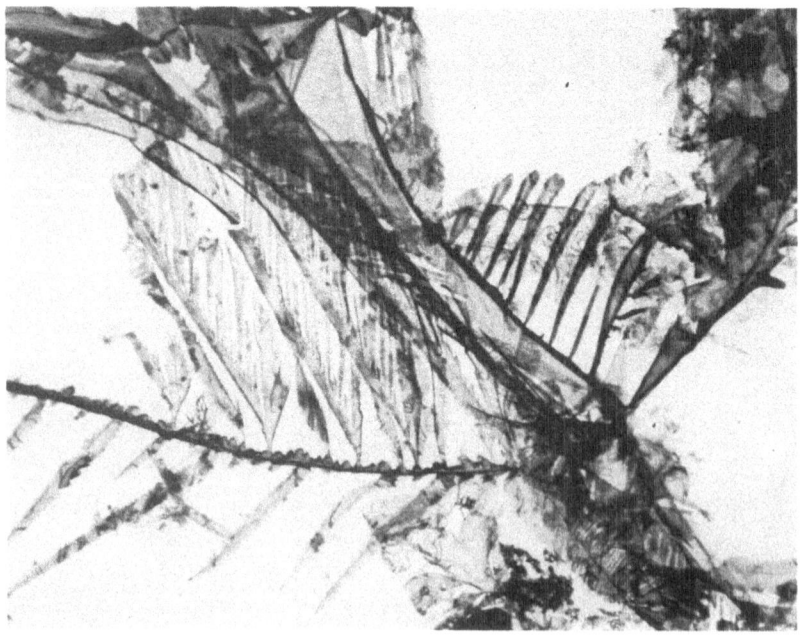

Figure 4. Isolated grain boundary films of V-57 forgings—5000×.

left in place. It can be seen that the amount and size of films being removed by the replica is increasing.

Since the initial grinding and polishing steps in the surface preparation cause a great deal of distortion in precipitation which is filmlike in nature, the sample was chemically digested in order that the true form of the grain boundary film be retained. In addition to large massive particles which settled quickly, it was found that copious amounts of two-dimensional films were floating in the isolation solution. This solution was allowed to stand for 48 hr to allow the films to settle. The residue of the films was examined by electron microscopy, electron diffraction, X-ray diffraction, and X-ray fluorescence analysis.

Figures 3 and 4 show the true size and shape of the grain boundary films. These films are so large that it is believed that they encompass the grains. Electron diffraction and X-ray diffraction studies show them to be TiC, while the chemical analysis gives high Ti with some Cr in solution. The more massive residue of the particles usually found in the matrix also shows TiC structure by X-ray diffraction, but has in solution copious amounts of Cr, Fe, Ni, and Mo. Similar results on the massive particles have been found in other studies of high-temperature alloys.[5]

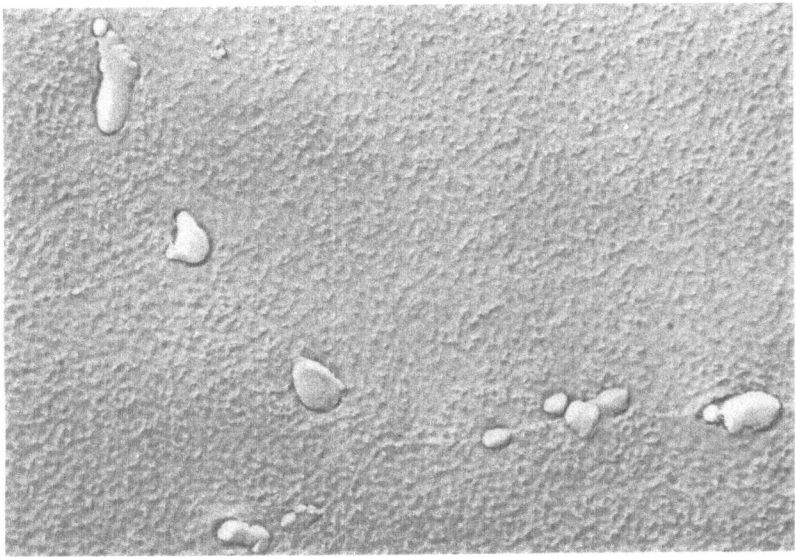

Figure 5. Grain boundary precipitation in V-57 forging after 2 hr at 1800°F— 8500×.

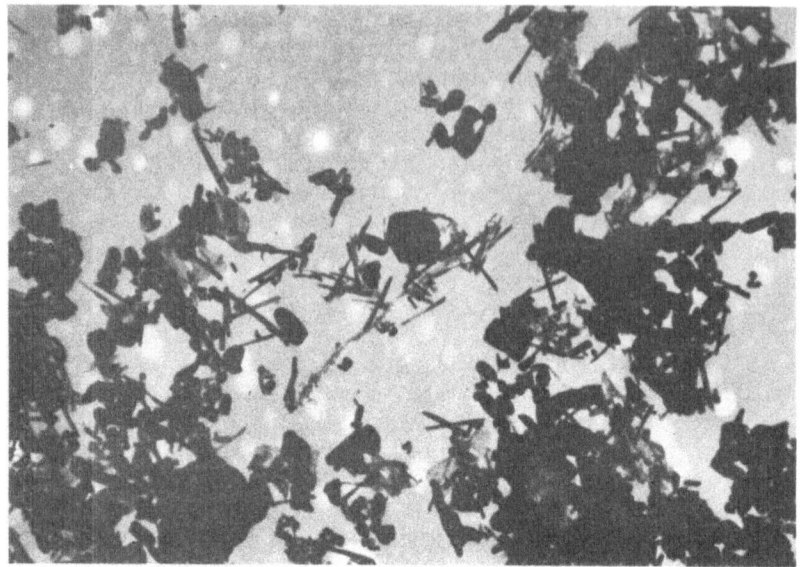

Figure 6. Isolated residue of V-57 forging after 2 hr at 1800° F — 5000×.

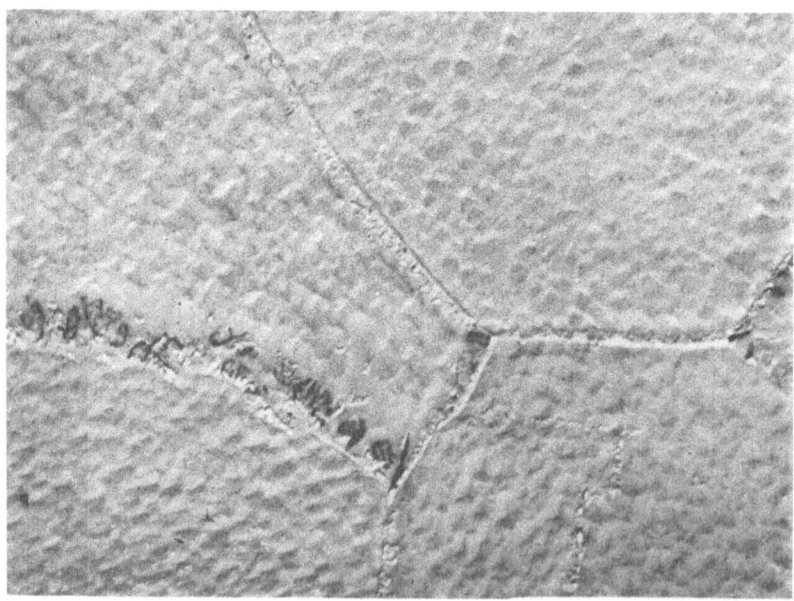

Figure 7. Grain boundary precipitation in heat-affected zone of cracked A-286 weld — 8500×.

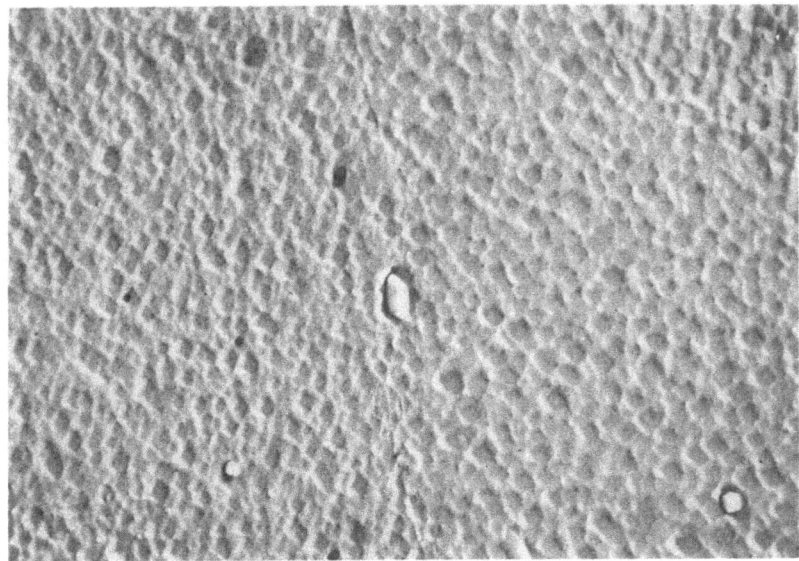

Figure 8. Grain boundary of A-286 matrix—8500×.

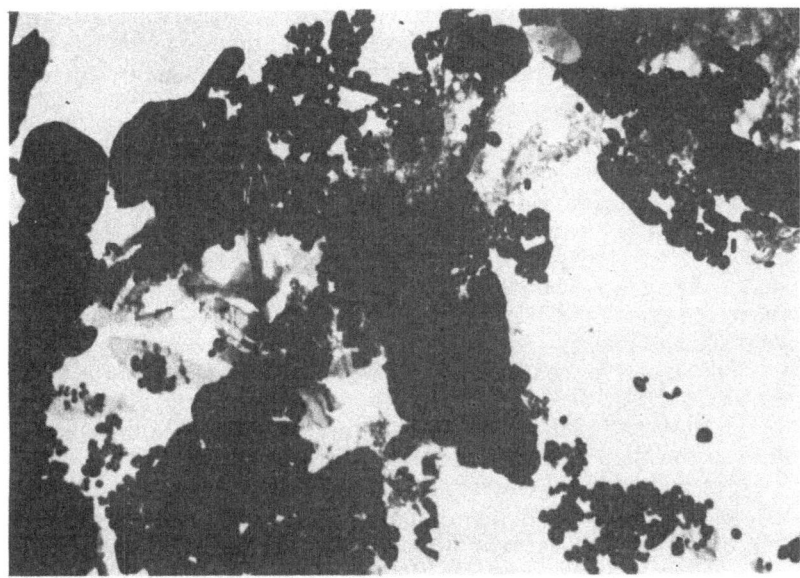

Figure 9. Isolated residue of A-286 weld showing evidence of thin TiC film—5000×.

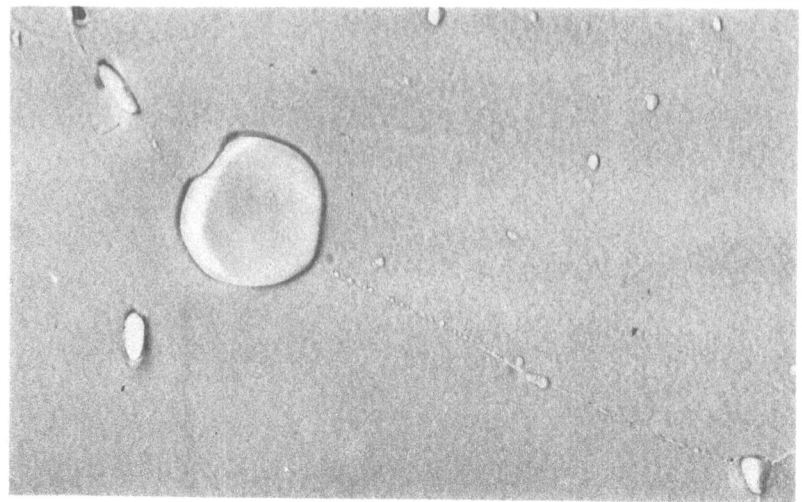

Figure 10. Grain boundaries of heat-affected zone of A-286 weld after heat treatment—8500×.

The samples showing the film precipitation were given a 2-hr 1800°F heat treatment, and then oil-quenched. Reexamination of the grain boundaries showed agglomeration of the precipitation, as seen in Figure 5. Examination of the heat-treated isolated residue shows more opaque particles, as seen in Figure 6.

2. Cracked Welds. In view of the results of the study on the forging materials at the high temperatures, the same examination techniques were tried on the heat-affected zone and the matrix of cracked welds of A-286 material. Electron microscope traverses across the heat-affected zone showed that some sort of film also exists in the grain boundaries. The region in which the grain boundaries contain the films is the same as where the cracks initiate. Figures 7 and 8 show the heat-affected zone and the matrix, respectively, of the welded material. Figure 9 is the isolated residue of the weld showing some thin films similar to those already found in the forging material. Upon suitable heat treatment of the welded sample, the grain boundaries show again the fine agglomeration as represented by Figure 10.

It is apparent that during the course of welding, the temperatures and times have approached those which initiate the formation of the thin film grain boundary constituent identified as TiC in the V-57 forging. Due to the large amount of other particles in the isolated residue of the welds, it is more difficult to identify the thin films.

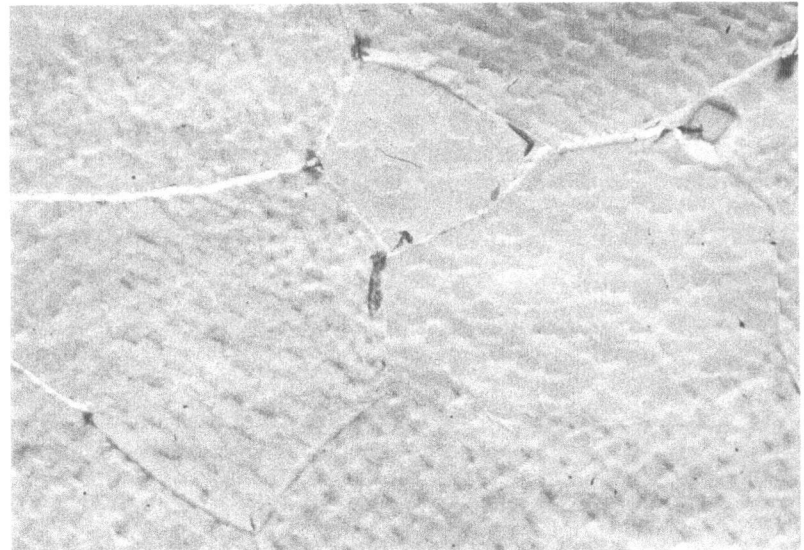

Figure 11. Grain boundary films in as-rolled V-57—8500×.

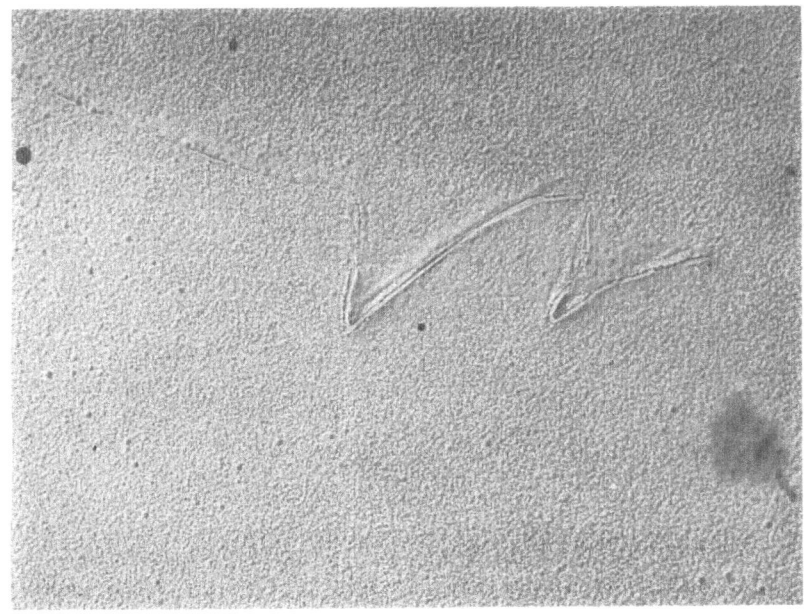

Figure 12. Grain boundaries of V-57 after 2 hr at 2100°F—8500×.

3. Furnace Heat Treatments. Careful examination of the as-rolled condition shows evidence of a grain boundary film constituent similar to those in the forgings and welds. Figure 11 shows small pieces of the film material at the grain boundaries. Samples of the as-rolled material were heat-treated for 2 hr at 1950 to 2150°F and air-cooled. The studies on the grain boundaries showed that as the temperature was increased, the amount of grain boundary constituent increased, such that at 2150°F the grain boundary showed nearly 100% precipitation, as seen in Figure 12. The thin film in the grain boundary at the lower temperatures, about 2000°F, grows much thicker at 2150°F, so that the grain boundaries must be in a highly stressed condition. The isolated residue of the 2150°F heat treatment shows very opaque material.

From the comparison of the electron microscope pictures of the embrittled forging and the furnace heat treatments, it seems that the forging temperature was near 2050°F since the grain boundary precipitations were very similar.

CONCLUSIONS

The appearance of a grain boundary TiC film introduces an embrittling nature to forgings or welds of V-57 and A-286 superalloys. The temperatures at which these films grow are in the range of 2000°F and the amount of film increases with increased temperature. Although nothing is known about the physical properties of very thin TiC, it must decrease the cohesion between the grains by acting as a barrier between the grains.

The occurrence of a grain boundary precipitation which agglomerates upon a lower heat treatment would influence the rate and amount of phase precipitation at the lower aging times. The breakdown of the thin grain boundary TiC and the more opaque matrix TiC with large amounts of alloying elements should be much different.

The appearance of the thin films in the cracked heat-affected zones of welds indicates that with improper welding procedures excessive grain boundary precipitation may occur, which may nucleate cracks upon application of stresses.

The forging temperatures used in the forging operations are critical, and should be kept as low as possible, preferably about 1950°F. Also, the temperatures used in the rolling and final finishing passes must be closely controlled as well. Overheating of the stock material would introduce the grain boundary condition quickly; subsequent heat treatments or forming operations could act in an abnormal manner from the physical properties point of view.

ACKNOWLEDGMENTS

This work has been carried out with the cooperation of the Jet Engine Department of the General Electric Company, and the Physics Department of Purdue University.

REFERENCES

[1] C. B. Craver, G. N. Aggen, and W. W. Dyrkacz, A Study of Precipitation Hardening of A-286 Alloy, Report H-0111-No. 8, Research Laboratories, Allegheny Ludlum Steel Corp., Watervliet, New York.

[2] H. J. Beattie, Jr., and W.C. Hagel, "Intermetallic Compounds in Titanium Hardened Alloys," Transactions AIME, J. of Metals, July, 1957, p. 911.

[3] M. Kaufman and C. E. Smeltzer, Jr., A Study of M308 and Super A-286—V-57, G.E. Report No. R58SE72.

[4] K. Metcalfe, "Solve Lamellar Phase Problem in A-286," Iron Age, July 3, 1958.

[5] J. F. Radavich and W. J. Pennington, The Occurrence of Carbides in Vacuum Melted Waspaloy (to be published).

APPENDIX

The papers listed in this appendix were presented as a part of the conference but were not available for inclusion in these Proceedings.

X-Ray Fluorescent Spectrometric Determination of Scandium in
  Ores and Related Materials
    Robert H. Heidel and Velmer A. Fassell
    Iowa State University of Science and Technology
    Ames, Iowa
Industrial Applications of an Automatic X-Ray Spectrograph
    William R. Kiley and John Croke
    Philips Electronics, Incorporated
    Mount Vernon, New York
Review and Evaluation of High-Temperature Furnaces for X-Ray
  Diffractometers
    W. J. Campbell, S. Stecura, and C. Grain
    U. S. Bureau of Mines, Region V
    College Park, Maryland
An X-Ray Diffraction Study of Radiation Damage in Molybdenum
    W. V. Cummings, D. L. Gray, and J. D. Barnett
    General Electric Company
    Richland, Washington
Recrystallization Habits in High-Purity Aluminum
    R.A. McCune, H.P. Leighly, Jr., and George P. Rauscher, Jr.
    Denver Research Institute
    University of Denver
    Denver, Colorado
Thermal Transformations and Properties of Cryptomelane
    G. M. Faulring, W. K. Zwicker, and W. D. Forgeng
    Union Carbide Metals Company
    Niagara Falls, New York
Analysis of Al–Zn Concentration Gradients by Back-Scattered
  Electrons
    R. E. Ogilvie* and R. K. Lewis**
    *Massachusetts Institute of Technology
    Cambridge, Massachusetts
    **Advanced Metals Research Corporation
    Somerville, Massachusetts
Internal Standard Method and Background Effect in X-Ray Spec-
  trography
    Maurice C. Lambert
    General Electric Company
    Richland, Washington